The Genetical Analysis of Quantitative Traits

The Genetical Analysis of Quantitative Traits

Michael J. Kearsey

and

Harpal S. Pooni
Plant Genetics Group
School of Biological Sciences
The University of Birmingham
Birmingham, UK

CHAPMAN & HALL

London · Glasgow · Weinheim · New York · Tokyo · Melbourne · Madras

Published by Chapman & Hall, 2–6 Boundary Row, London SE1 8HN, UK

Chapman & Hall, 2–6 Boundary Row, London SE1 8HN, UK

Blackie Academic & Professional, Wester Cleddens Road, Bishopbriggs, Glasgow G64 2NZ, UK

Chapman & Hall GmbH, Pappelallee 3, 69469 Weinheim, Germany

Chapman & Hall USA, 115 Fifth Avenue, Fourth Floor, New York, NY 10003, USA

Chapman & Hall Japan, ITP-Japan, Kyowa Building, 3F, 2-2-1 Hirakawacho, Chiyoda-ku, Tokyo 102, Japan

Chapman & Hall Australia, 102 Dodds Street, South Melbourne, Victoria 3205, Australia

Chapman & Hall India, R. Seshadri, 32 Second Main Road, CIT East, Madras 600 035, India

First edition 1996

© 1996 Michael J. Kearsey and Harpal S. Pooni

Typeset in 11/13pt Times by Academic & Technical Typesetting, Bristol
Printed in Great Britain at the Alden Press, Oxford

ISBN 0 412 60980 0

A catalogue record for this book is available from the British Library

Library of Congress Catalog Card Number: 95-72200

∞ Printed on permanent acid-free text paper, manufactured in accordance with ANSI/NISO Z39.48-1992 and ANSI/NISO Z39.48-1948 (Permanence of Paper)

Contents

Preface

The problem of analysing quantitative, continuously varying characters is encountered by people working in most areas of biology from evolution to plant or animal breeding. Such characters are often controlled by many genes together with the environment. This book sets out to explain why such characters are important, how they can be studied and what use can be made of the information so gained. It is aimed at anyone coming to the subject for the first time, either through University courses or through new research developments, and it has developed from our many years of teaching this subject to undergraduate and postgraduate students.

Most biologists prefer to avoid mathematics and statistics and thus quantitative genetics, with its emphasis on model building and hypothesis testing has a reputation for being a conceptually difficult subject. We have endeavoured to minimize the maths and statistics as far as possible and concentrate our explanation on examples and verbal descriptions. In so doing, we have focused on essential approaches which are simple to apply and lead to robust conclusions. Some knowledge of simple algebra and basic statistics, particularly analysis of variance, regression and correlation, are necessary, however.

We have attempted to make the book accessible as a self-instructional text, and for this reason have included sections on relevant supporting areas of genetics such as linkage and gene mapping. Where necessary we have explained additional statistical procedures and provided supporting statistical and other tables in the appendices. In so far as it is possible with this subject, we have tried to make the text chatty and readable while being reasonably rigorous. A number of problems are included to allow interested readers to test their understanding.

Notation is a subject that has bedevilled and divided the subject of quantitative genetics for generations. In this book we have decided, after considerable deliberation, largely to adopt the notation used by D.S. Falconer in his *Introduction to Quantitative Genetics*, with appropriate cross-reference in the appendices to the previous notation of the Birmingham School, as described by K. Mather and J.L. Jinks in *Biometrical Genetics*. We fully realize that this will make it more difficult for newcomers to understand the earlier literature and for previous students of the Birmingham School to follow this book. However, the changes involved are minimal and long overdue, and we do not anticipate our readers will find it more difficult than we did! Because of the very large number of symbols used in the text, we have had to use some characters in different contexts. In order to minimize confusion when we have had to use the same character more than once, we have distinguished them by italics or bold. A full list of symbols is given. Significance levels are indicated by *s and, in order to avoid breaking the text, references are kept to a minimum and indicated by a number in the text.

We were invited to write this book following the deaths of Sir Kenneth Mather and Professor John Jinks, to provide a substitute for their classic text, *Biometrical Genetics*, the third edition of which was no longer in print. We owe them both a debt of gratitude for developing our interest in the subject, but we knew that there was no way in which we could attempt to emulate their style and scholarship. So we opted for a simpler and, hopefully, less formal approach and if we have failed, the faults are entirely our own.

We are grateful to several colleagues for advice and assistance in producing the text. In particular our thanks go to Gina Hyne for her thorough and imaginative proof reading, and to Parminder Virk for help with the analyses and providing additional material. To generations of students, too numerous to mention, we express our gratitude for bringing their problems in understanding this subject to our attention.

The twins shown in Figure 1.3(ii) are James and Christopher Simcock, and William and Jonathan.

Dedication

This book is dedicated to the memory of Professor Sir Kenneth Mather, FRS, and Professor John L. Jinks, FRS, our teachers and mentors over many years.

Professor Sir Kenneth Mather, FRS, KB, CBE (1911–90)

Professor John L. Jinks, FRS, CBE (1929–87)

Introduction 1

Everything a living organism does throughout its life is ultimately determined by its genes. No matter whether the organism is a virus or a vegetable, a bacterium or a bat, a house fly or a human, its structure, physiology and lifestyle are ordained by the proteins produced by its genetic material. The genes do not act in total isolation, of course, but in response to triggers in the organism's internal and external environment. However, these environmental stimuli can only alter events to a limited degree, and may simply act to initiate or to stop an activity or to change its direction.

Consider a bird like the house martin, for example. In response to its genes it leaves a pleasant climate in central Africa to fly north over hostile mountains and seas. The genes instruct it to find and mate with an individual of the opposite sex of the same species. They tell it to build a nest, not of twigs or leaves like other birds, but of mud on the side of a house, preferably a white house. They go on to guide it to lay eggs in this nest, to incubate them and to work all the hours of daylight feeding the offspring that emerge, as well as to give instructions as to what to feed the young. Having guided the parents through all this they then tell them to turn around and fly south again. All this is achieved by proteins determined by genes; the house martin has no other instruction manual to guide it through this unimaginably complex set of behaviours, none of which can make any sense to the bird itself.

Not surprisingly, it requires many thousands of genes to produce and manage such a complex organism as a house martin, as indeed it does to produce a vegetable or a human being. It would be impossible to study all or even a small proportion of these genes at once, and so geneticists normally concentrate on just one characteristic or trait at a time. In order to study how these genes act, geneticists have to look at genetical differences; variation, in other words. It would be impossible to study how a house martin's genes instruct it to migrate without having access to house martins which have genes resulting in a different behaviour. Typically, the geneticist's approach to studying how a particular characteristic is controlled is to look for, or induce by mutation, genetical variants which behave abnormally or at least behave differently; variation is the essential prerequisite of genetical analysis.

Enormous advances have been made in our knowledge of genetics this century and the pace is accelerating all the time. It is a subject which has been fortunate in attracting the interests of particularly able people in all branches of science, including not only biology but also chemistry, physics and mathematics, and this has resulted in genetics holding a central role in late 20th century biology.

1.1 Qualitative, single gene differences

Since the earliest days the analytical approach used by geneticists has been a reductionist one. From Mendel's work to that of the molecular geneticists of today, insight has come from examining the inheritance and action of individual genes and more recently their molecular structure. Such approaches rely on the existence among individuals within a species of readily recognizable, distinct phenotypes arising from different forms of a gene, called alleles, one of which is either non-functional or at least has the normal function seriously impaired. Until quite recently such allelic differences between individuals had to be recognized at a distance from the gene, that is they had to cause the structure, physiology or behaviour of the organism to alter in such a way as to catch the eye of the experimenter. More minor changes would probably go unnoticed. Allelic differences of this sort produce phenotypic differences which are clear-cut and not greatly affected by environmental influences; differences such as horned *versus* polled cattle or tall *versus* dwarf rice plants (Figure 1.1). Such pronounced differences are called **qualitative** differences and arise from **major** allelic differences at one or two genes. With recent developments in molecular biology, however, all allelic differences are potentially identifiable with complete accuracy irrespective of the effect they may ultimately have on the phenotype. Unfortunately, the molecular procedures are relatively slow at present and therefore do not lend themselves to the study of many genes in large populations. Furthermore, the phenotypic effects of such allelic differences are generally difficult to identify or predict.

Figure 1.1 Qualitative differences in phenotype due to a single gene difference. (i) Normal *versus* polled cattle (reproduced with permission from The Rare Breeds Survival Trust photolibrary and archive). (ii) Normal *versus* dwarf rice.

(i)

(ii)

Our knowledge of gene action and interaction has been developed largely by comparing the phenotype of individuals carrying either functional or non-functional alleles [1]. However, much variation in natural or artificial populations is due to allelic variants which are functional and not readily amenable to the same methods of investigation, while the number of loci that can be simultaneously studied by these methods is limited.

1.2 What are quantitative traits?

Throughout the development of genetics, whenever a new phenotype has aroused interest, the first step in genetic analysis has invariably been to look for major gene control. Recent examples include insecticide resistance, heavy metal tolerance and enzyme activity [2]. However, although varying degrees of success have been achieved in such searches, it sooner or later becomes apparent that the one or two genes postulated actually fail to account for all of the observed variations in the trait and further analysis reveals considerable background variation at other unidentifiable genes. Although the individual effects of these other genes may be small their combined effects can often exceed that of the allelic variation at the major genes [3]. Once it is clear that there are several genes involved, the problems of identifying and locating them become so difficult that it is more convenient, if not essential, to pursue a completely different approach, namely that of **biometrical genetics**.

Biometrical or quantitative genetics has come to mean that area of genetics concerned with the inheritance of characters for which the genotype is but poorly identified from the phenotype, characters such as height and intelligence in humans, or yield and quality in crop plants [4,5]. With these characters there is no clear discontinuity between the phenotypes; the range of appearance of one genotype often overlaps that of others so extensively as to give the impression of a continuous distribution. Traits for which the variation shows such properties are referred to as quantitative, metrical or continuously varying traits to distinguish them from the qualitative traits described earlier, as shown in Figure 1.2.

The continuous distribution of a quantitative trait does not exclude the possibility that only one gene is involved in any particular instance but simply implies that, if it is so controlled, then the phenotypic differences between the genotypes at that locus are small relative to variation caused by non-genetical, or environmental, influences. Indeed, much of the theoretical framework to the subject has been developed on one or two gene models, only to be extended to many loci as a gesture to generality. In practice, however, it is likely that several and possibly very many genes are involved. As a consequence, Kenneth Mather [6] coined the terms **polygenes** and **polygenic traits** for

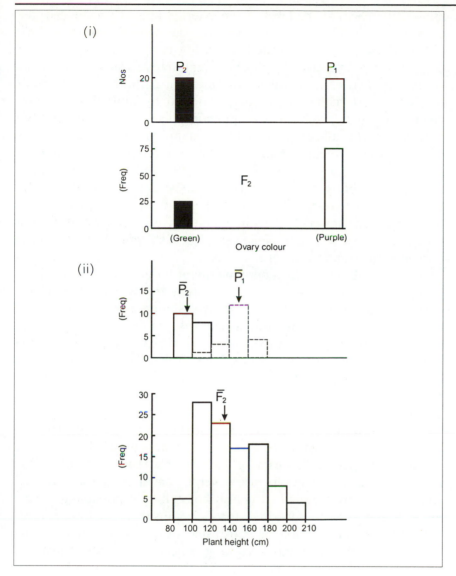

Figure 1.2 Comparison of F_2 and parental populations with (i) a qualitative, single gene difference and (ii) quantitative genetic variation.

the genes and traits they control. More recently, a new acronym has come to replace the term polygene; **QTL**, which stands for Quantitative Trait Locus [7]. However, as we will see later in Chapter 7, QTLs are not truly synonymous with polygenes.

There is no reason to believe that the genes involved in such traits are in any way different from the so-called major genes of classical genetics. Any one gene may have many allelic states. Some of the alleles of a given gene may have such marked effects as to be clearly recognizable as a classical, major allelic difference. Other alleles, though potentially separable at the DNA level, may well have only minor differences at the level of the external phenotype. Indeed, the extensive variability at protein and, particularly isozyme loci, identified by population geneticists using electrophoresis, has just the properties expected to produce quantitative traits. Thus, such genes have two or more alleles per locus in natural populations at frequencies far higher than those expected by mutation–selection balance, i.e. they exhibit genetic polymorphism, and the differential expression of the alleles in the biochemical properties of their proteins is small. The proteins produced may thus differ slightly in their preferred pH, temperature optimum or K_m, and confer upon their carriers the ability to cope better with certain environments [8]. Such loci also have non-functional alleles. We do not wish to give the impression, however, that the quantitative genetic variation in a trait is solely due to minor allelic variation in structural genes; regulatory genes no doubt also contribute.

Not only are the genes underlying quantitative characters likely to be representative of genes in general but they are also assumed to have the same properties. Thus, at meiosis, they should segregate and assort from other genes following the normal rules of linkage. They should also exhibit the features of gene action of other genes, including the property whereby the phenotype of the heterozygote lies somewhere between the ranges of the respective homozygotes in activity, i.e. we might expect partial dominance to complete dominance but not overdominance.

Thus, a **quantitative trait** is a trait for which the observed variation is due to the segregation of several to many naturally occurring polymorphic genes, for each of which the effects of the allelic differences on the phenotype are generally small compared with the effects of the environment. It is also normally not possible to determine the genotype of individuals for such traits from their phenotype. Quantitative traits should be contrasted to qualitative, major gene traits, for which the variation is due to allelic differences at just one or two genes and the effects of allelic differences on the phenotype are sufficiently large relative to differences due to the environment that the genotype of an individual can be easily determined from its phenotype (Figure 1.3).

1.3 Who studies quantitative traits?

Most biologists will find themselves having to work with quantitative characters at some time. There are, however, certain groups for

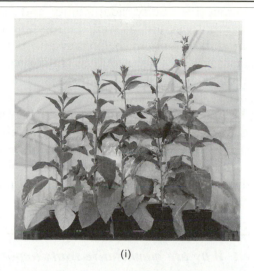

(i)

(ii)

whom a knowledge of such traits is essential. First among these are the breeders of plants and animals. Although there are still some advances to be made by manipulating single, major gene differences, particularly in the area of resistance to disease and herbicides, much of the present progress in breeding is achieved by improving quantitative characters such as quality, uniformity, growth rate, yield and response to fertilizers. Secondly, there are the population geneticists and students of evolution in natural populations. As with artificial, commercial selection, natural selection also acts largely on the pool of quantitative variation. Indeed, fitness, in that it is the cumulative result of the whole genome in a given environment, is the ultimate quantitative trait, while its component parts, such as

Figure 1.3 Examples of quantitative variation. Variation (i) in height in *Nicotiana* and (ii) facial features in twins. The differences between identical twins (left) are due to the environment while those between fraternal (two-egg) twins are due to genes as well as environment. (Reproduced with permission from the Multiple Births Foundation.)

competitive ability, adaptability and reproductive efficiency are all quantitative traits as well. Thirdly, we have the students of animal, particularly human, behaviour and personality – psychologists and sociologists. This group is interested in a variety of different, complex, quantitative characters such as intelligence, personality, social attitudes, etc. as well as abnormal behaviour, like schizophrenia. The fourth and final main group includes those interested in the genetics of human disease. Although many genetic diseases can be traced to identifiable major genes or chromosomal abnormalities, many others, such as propensity for heart disease, obesity and some cancers have a polygenic origin.

1.4 Why are quantitative traits important?

The groups of scientists outlined above are essentially distinguishable in terms of subject material and objectives, but there is considerable overlap between them and their separation is somewhat artificial. The nature of the information they require about the traits under investigation is influenced by their very different aims and it may be helpful at the outset to identify these disparate ambitions.

Animal and plant breeders are principally interested in improving the gene pool of the organism of interest in order to maximize productivity per unit of input. To this end breeders require information that will enable them to identify desirable genotypes efficiently, to select them and concentrate their genes in a strain, line or variety that is commercially acceptable. Thus they need to be able to decide on the type of end product they require – an inbred line, a hybrid or a population – and to choose the most efficient selection and breeding strategy to produce it. At the start of any new programme, however, it is necessary to enquire whether the extent of the genetical control is sufficient to warrant breeding at all; would it perhaps be more expedient simply to improve performance by modifying the environment? Further, given a particular type and intensity of selection, what will be the rate of response, how far will the trait respond (Figure 1.4) and what are the consequences on other traits?

Population and ecological geneticists, on the other hand, want to know how a trait(s) came to evolve in their population. They wish to demonstrate whether natural selection is acting, or has acted, on the character and, if so, what type of selection it is; directional,

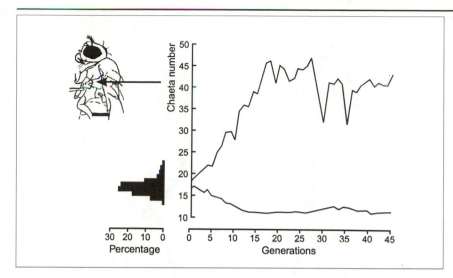

Figure 1.4 Results of long-term selection for a quantitative trait, sternopleural bristle number (arrowed) in the fruit fly, *Drosophila melanogaster* [9]. Lines can be selected (right) which considerably exceed the narrow range of observable, natural variation in the original population (left).

disruptive, stabilizing (Figure 1.5). In order to demonstrate selection it is necessary to show that the array of genotypes present when potential selective forces are removed (i.e. close to complete survival), is different from that found naturally. Given that selection is identi-fied and classified correctly, the genetical properties of the trait in terms of gene action, genotypic distribution, etc. should follow a pattern consistent with the selection mode. It may also be of interest to pursue the future outcome of continued natural selection of that type.

There are also many practical applications of population genetics which require a knowledge of biometrical genetics. The principal area here is that of the control of animal pests such as rodents and insects, although there is also a growing interest in plant sus-ceptibility to herbicides. In several instances resistance to pesticides can be traced to single genes but, in many others, multigenic resistance occurs and pest control procedures have produced genetical responses by the pest which can only be explained by a polygenic system.

Students of animal behaviour, too, are interested in answers to questions about past, present and future evolution of their traits, although for humans at least there are very real social issues. Thus a knowledge of the nature and extent of genetical control of traits such as intelligence or personality could have implications in the areas of education and attitudes to criminality, for instance. Many subsidiary problems also arise; for example, given that educational opportunities could, rightly, lead to people realizing their full genetic potential and attaining a particular social and economic position,

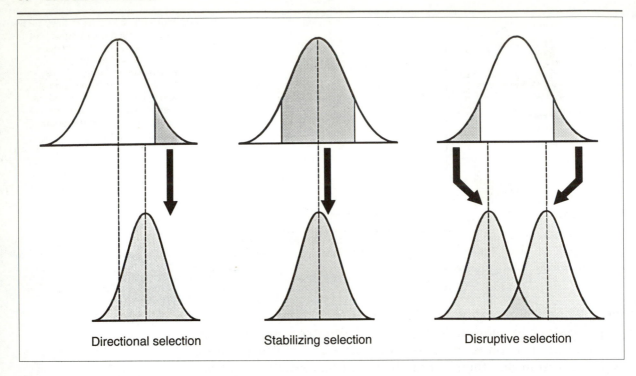

Directional selection   Stabilizing selection   Disruptive selection

Figure 1.5 Types of natural selection. The upper distributions are the parental populations, the lower ones the resulting progeny. Only those individuals in the shaded areas are selected and reproduce.

what, then, will be the long-term consequences of the inevitable assortative mating within these groups?

Finally, let us consider characters related to human health. In so far as the traits are quantitative it is important to discover whether the genetical control of the variation is large or small relative to that of the environment. If large, are there a few key genes which could be identified and used for genetic counselling? If the environment is a major factor in the variation, is it possible to identify important features such as diet or life style which can be modified to ameliorate the condition?

1.5 *What do we need to know about quantitative traits?*

All these groups are interested to various degrees, therefore, in understanding their traits in terms of the underlying factors controlling them. Various levels of understanding may be required and, in order to put later sections of this book into some general context, the following are some of the possible questions which could be posed, in increasing order of complexity.

1.5.1 Is the character inherited?

This apparently simple question can be quite difficult to answer because non-genetic variation can often be confused with genetical variation. Two plants from a natural habitat may look very different because they have been exposed to different environmental stresses. If each plant were to be propagated vegetatively and the cloned units grown in a uniform environment, differences between clones from each unit may still arise from a carry-over of the residual, environmentally induced, differences in their parents. Taking seed and looking at the sexual, rather than asexual, progeny would overcome this to some extent, but the quality of the seed parent could, in principle, still carry-over to the progeny, through its effect on the seed. This is called a **maternal effect**, and early seedling characters would be particularly susceptible to such effects. Common sense or intuition could well be a poor guide as to the importance of maternal transmission and defined experiments with controlled genotypes would be required to determine their magnitude and importance in any given instance.

An alternative approach might be simply to use pollen from each parent to produce progeny by crossing to a fixed tester, because the state of the wild parent is less likely to affect the pollen's contribution to the next generation. Yet, here again, the two wild plants may, as a result of their contrasting environments, flower at different times and hence the crosses and subsequent seed set on the testers would occur at different times. Such a time difference may result in detectable differences in the progeny, although the latter would probably not correlate with the original differences in the wild parents. Given the potential difficulties with an easily manipulated, experimental organism such as a plant, it is easy to imagine the problems of studying polygenic traits in man, where the effects of genes and environment are completely entwined.

1.5.2 How much variation is genetic?

Having detected genetical variation, it is of interest to estimate its magnitude relative to environmental variation and variation due to genotype by environment interactions. The proportion of the observed variation that is genetic is referred to as the **heritability** of the trait in that population.

1.5.3 What sort of genetic variation?

This is a much broader question with a variety of different levels of answer. The genetic variance can arise from nuclear or non-nuclear, cytoplasmic causes, but we will concentrate mainly on the nuclear causes. Of the variation controlled by nuclear genes, how much is potentially fixable by selection? This fixable variation is called the **additive genetic variation**, and the proportion of the variance in a trait in a particular population due to this source, is known as the **narrow heritability** and is central to predicting selection responses.

The remaining genetical variance, the non-additive part, involves dominance and the interactions between genes. A great deal of effort has been devoted to devising methods for studying these sources of variation, largely because of their important implications in terms of the ultimate goal of the breeder, that of choosing whether to produce an inbred line, a hybrid or an outcrossed variety. Important questions have centred on the magnitude and overall direction of dominance. Does the mean trait value of the heterozygotes at individual loci fall between the mean trait value of the alternate homozygotes, does it favour one homozygote, or is there true overdominance, i.e. is there inherent superiority of heterozygotes over homozygotes? A knowledge of the amount and direction of dominance also provides information on the history of natural selection and so is of interest to population and evolutionary geneticists. Gene interactions, too, where they exist, can either reduce or enhance selection limits, but in general distort predictions of genetic improvement. It is necessary, therefore, to be able to detect gene interactions, estimate their effects and predict their consequences in a breeding programme. As with dominance, the nature of the interactions is revealing about the selective forces which have moulded the character.

1.5.4 What is the breeding value of a particular genotype?
A plant or animal breeder is often faced with deciding which individual, line or hybrid to breed from and so needs to know the potential breeding value of that genotype. Information from the particular genotype may be limited. It may be unreliable because the phenotype is controlled by many genes plus the environment or, as in the case of sex-limited traits, like milk production in cattle or oil production in hops, one sex will not manifest the trait.

1.5.5 How is the genetic variation organized?
Under this heading we can include the number of genes controlling the trait, their allele frequencies, the location of genes on the chromosomes, the distribution of like alleles over chromosomes or gametes, the extent of homozygosity *versus* heterozygosity, and much more. Such information is useful in understanding the limits to selection response and interpreting hybrid superiority.

1.6 Historical development of quantitative genetics

It is not possible to cover the development of quantitative genetics thoroughly in the few pages available to us and so we will simply outline a few of the principal contributors and the factors that influenced them.

Many of the early attempts to understand the patterns of inheritance, particularly in the 18th and 19th centuries, centred around domesticated animals and crop plants on the one hand and man on the other. The commercial importance of understanding and exploiting the genetics of agricultural species was a major driving force in the former while a mixture of simple curiosity about our own species tinged with eugenic ambitions drove the latter. Unfortunately, many of the characters studied were quantitative and hence did not reveal their underlying causes very readily.

Charles Darwin was very much aware of the potential importance to his theory of evolution of the availability of small inherited differences that could be accumulated slowly over time. Indeed he carried out breeding experiments of his own. Frances Galton and Karl Pearson in the UK studied human variation in morphometric and behavioural characters in the late 19th and early 20th centuries. Not only were these characters polygenic but also the small size of human families and the uncertain pedigrees almost guaranteed that their **biometrical** approach, as it came to be called, would be unsuccessful in determining the mechanisms of inheritance. They did, however, develop a statistical methodology which was to prove invaluable for the subsequent development of the subject.

Although Gregor Mendel, too, worked with a crop plant, the garden pea, he was successful where others failed largely because he chose an inbreeding, and hence true breeding species, and he studied qualitative, major gene differences instead of quantitative ones. The use of true breeding lines provided repeatable results while the qualitative differences resulted in clear-cut categories of progeny which could be counted.

With the rediscovery of Mendel's work in 1900, there developed a conflict between the Biometricians and the Mendelianists as each saw the other's theory as being incompatible with their own. By 1909 a reconciliation became available following Johannsen's discovery that, in beans, a quantitative trait, seed weight, was influenced both by genes and environment and Nilsson-Ehle's demonstration that several different genes could act additively to control grain colour in wheat [5]. In 1918, Ronald Fisher published a classic paper laying down the basic theoretical framework for explaining, in genetical terms, the underlying relationship between the genes controlling quantitative traits and the phenotypic correlations among relatives observed by the biometricians [10].

Early demonstrations followed to show that the factors underlying quantitative traits were chromosomal. In 1923, Sax was able to show linkage of a factor(s) controlling seed weight, a quantitative trait, and seed colour which was controlled by a single, Mendelian gene. In the 1940s and 1950s, a series of elegant experiments with *Drosophila* by Mather and co-workers showed that genes controlling quantitative traits were distributed over many chromosomal sites. In 1949, Mather published the first edition of *Biometrical Genetics* which summarized much of what was then known about the inheritance of quantitative traits and he initiated a methodological approach for studying them.

Quantitative genetics can claim many diverse roots, but principal among its early pioneers were Ronald Fisher in the UK and Sewall Wright and Jay Lush in the USA. Most of the key developments since the 1940s have involved students and followers of these pioneers (Figure 1.6) while the demands of agriculture have largely influenced the development of the subject. Despite the exchange of individuals and ideas between the main groups, very different objectives and methodological approaches influenced their attitudes and theoretical constructs. In both the UK and USA, there were two principal centres which specialized in either animal or plant genetics. Thus Iowa State College at Ames (USA) and The University of Edinburgh (UK) were

Figure 1.6 Some of the key scientists in the early development of quantitative genetics pictured at the IXth International Congress of Genetics in Pavia, Italy, 1953. *Standing*: Mather, Da Cunha, Haldane, Dobzhansky, Waddington, Epling, Carlson, Robertson, Falconer. *Middle row*: Ford, Wallace, Lewis, Lerner, Scossiroli. *Sitting*: Mayr, Levine, Buzzati-Traverso, Fisher, Clausen, Prevosti.

particularly concerned with the improvement of hogs, poultry, cattle and sheep, while the North Carolina State University at Raleigh (USA) and The University of Birmingham (UK) concentrated on plants, particularly maize and tobacco in Raleigh and inbreeding species at Birmingham. The biology of the species affected the structure and complexity of the genetical models that needed to be used. Approaches were also shaped by whether a group subscribed to a purely descriptive, phenomenological view which avoided attempts to explore the underlying genes, or they believed that the underlying gene action and their interaction with the environment were all-important.

These differences in materials and objectives have led to differences in the theoretical models and assumptions made. These in turn have led to misunderstandings and, over the last 50 years, there have been lively, often acrimonious, disputes between the different groups. In particular, several different notations have developed, and these tend to divide rather than aid discourse [4,5]. Arguably the major difference in notation which still survives is that between the Birmingham School and the rest. This resulted principally from the use of crosses between inbred lines as the starting point for the theoretical development of the subject by Kenneth Mather and John Jinks in Birmingham. They were particularly interested in species which could be easily inbred such as tobacco, wheat and tomato which yield populations such as F_2s which are amenable to well-defined models and analyses. Thus there could only be two alleles segregating in the population and genotype frequencies were known. The animal breeders, on the other hand, were faced with much more complicated populations, with many alleles and unknown genotypes. Given the difficulties of defining what was actually happening at the level of individual genes, the purely descriptive, statistical approach advocated by Oscar Kempthorne had an obvious appeal [11]. Because of the major influence of scientists from Ames on the initiation and early development of quantitative genetics at Raleigh, it was inevitable that essentially the same basic methodology would be used to study hybrid vigour and selection response in plants there, also.

Although quantitative geneticists have been very successful in applying statistical approaches to solve agricultural and other problems, they have, until quite recently, been unable to study the individual genes underlying the variation. In the 1980s, however, the advent of molecular techniques has opened up new possibilities and horizons for quantitative genetical analysis [12]. We can now follow, with relative ease, the transmission of small, defined chromosomal regions and this facility enables us to locate and manipulate the polygenes themselves.

Summary

1. Quantitative traits are those controlled by naturally occurring allelic variation at several to many genes, polygenes, together with relatively major influences of the environment.
2. Such traits are of central importance in plant and animal breeding, evolutionary studies, behaviour and medicine.
3. A knowledge of such traits can enable breeders to plan improvement programmes and allow population geneticists to investigate the effects of natural selection as well as allowing us to assess the relative importance of nature to nurture in the control of behaviour and susceptibility to disease.
4. The development of the subject has been largely driven by the requirements of breeders but as a result of differing breeding materials and objectives, a variety of different approaches have been adopted.

References

1. Griffiths, A.J.F., Miller, J.H., Gebhart, W.M., Lewontin, R.C. and Suzuki, D.T. (1996) *An Introduction to Genetic Analysis*, 6th edn, W.H. Freeman & Co., New York.
2. Schat, H. and Ten Bookum, W.M. (1991) Genetic control of copper tolerance in *Silene vulgaris. Heredity*, **68**, 219–29.
3. Gibson, J. and Miklovich, R. (1971) Modes of variation in alcohol dehydrogenase in *Drosophila melanogaster. Experientia,* **27**, 99–100.
4. Falconer, D.S. (1989) *Introduction to Quantitative Genetics*, 3rd edn, Longman, UK.
5. Mather, K. and Jinks, J.L. (1982) *Biometrical Genetics*, 3rd edn, Chapman & Hall, London.
6. Mather, K. (1949) *Biometrical Genetics*, 1st edn, Methuen, London.
7. Geldermann, H. (1975) Investigation on inheritance of quantitative characters in animals by gene markers. I. Methods. *Theor. Appl. Genet.,* **46**, 300–19.
8. Place, A.R. and Powers, D.A. (1979) Genetic variation and relative catalytic efficiencies of lactate dehydrogenase allozymes in *Fundulus heteroclitus. Proc. Natl Acad. Sci. USA*, **76**, 2354–8.
9. Kearsey, M.J. and Barnes, B.W. (1970) Variation for metrical characters in *Drosophila* populations. II. Natural selection. *Heredity,* **25**, 11–21.
10. Fisher, R.A. (1918) The correlation between relatives on the supposition of Mendelian inheritance. *Trans. Roy. Soc. Edinburgh*, **52**, 399–433.

11. Kempthorne, O. (1957) *An Introduction to Genetic Statistics*, Wiley, New York.

12. Soller, M. and Beckmann, J.S. (1983) Genetic polymorphism in varietal identification and genetic improvement. *Theor. Appl. Genet.*, **67**, 25–33.

Further reading

Falconer, D.S. (1989) *Introduction to Quantitative Genetics*, 3rd edn, Longman, UK.

Mather, K. and Jinks, J.L. (1982) *Biometrical Genetics*, 3rd edn, Chapman & Hall, London.

Palladino, P. (1996) People, institutions and ideas: American and British Geneticists at the Cold Spring Harbor Symposium of Quantitative Biology, June 1955. History of Science (in press).

2 Basic generations – means

The genetical analysis of quantitative traits can not follow the standard procedures used to analyse major gene traits, such as looking for a one-gene $(3:1)$ or two-gene $(9:3:3:1)$ phenotypic ratio in an F_2, because it is not possible to follow the segregation of the separate, underlying polygenes. Instead, it is necessary to look at the degree of similarity or difference among related individuals and families using various statistics such as means, variances, covariances, regressions and correlations.

If we score a number of individuals from a particular family for a quantitative trait, the mean phenotype of those individuals and the variation among them will be due to the joint action and interaction of their genes and the environment. Different families will have different means and different variances because they contain different genotypes. The genetical contributions to these family means and variances can be investigated by searching for simple and plausible models which adequately explain the data.

In practice, there are a very large number of possible families that one might be working with but, for convenience and simplicity, we will start with those families that can be obtained from a pair of true breeding, homozygous lines. Table 2.1 gives an example of the means and within family variances of two such lines together with their F_1, F_2 and first backcross $(Bc_{1.1} = F_1 \times P_1$ and $Bc_{1.2} = F_1 \times P_2)$ families. These six families are often referred to as the **six basic generations**. Not only do such families provide a simple conceptual framework from which to illustrate the construction and testing of the genetical models but they also appear in such a wide variety of contexts that they are the linchpin of quantitative analysis in most situations. In this chapter we deal with models for analysing the family means while the variances will be considered in Chapter 3. Unless otherwise stated, we will always assume that the organism is either diploid or, at least, shows disomic inheritance.

Table 2.1 The means and variances of the six basic generations for a quantitative trait. Each mean is based on 100 individuals, and hence the variance has 99 df

Generation	Mean ($\bar{x}$)	Variance (s_x^2)
P_1	69.44	59.73
P_2	59.04	65.71
F_1	83.44	51.81
F_2	74.36	100.75
$Bc_{1.1}$	76.03	81.05
$Bc_{1.2}$	71.28	90.83

2.1 Single gene model with additive and dominance effects

Let us suppose initially that we have two homozygous, inbred lines which are to be used as parents, P_1 and P_2, which are identical except for one gene. This gene, which we shall call A, has two alleles, A^+ and A^-, which are responsible for the higher and lower phenotypic scores respectively of the character being studied. If we adopt the convention that P_1 is the parent with the higher phenotype for the character under study, then the genotypes of P_1 and P_2 for this gene will be A^+A^+ and A^-A^- while their F_1 hybrid (from a cross of $P_1♀ \times P_2♂$ or $P_2♀ \times P_1♂$) will be A^+A^-. Let us further suppose that these two parents and the F_1 are raised in a replicated experiment and that we calculate their average phenotypic values for the character, $\bar{P}_1, \bar{P}_2, \bar{F}_1$.

If we use the average phenotype of the two parents, P_1 and P_2 as a base line, m, we can define the expected mean of all three genotypes as linear deviations from this as follows [1],

$$\bar{P}_1(A^+A^+) \quad m + a_A$$
$$\bar{P}_2(A^-A^-) \quad m - a_A$$
$$\bar{F}_1(A^+A^-) \quad m + d_A.$$

The parameters a_A and d_A are deviations from the mean, m, due to the effects of homozygous or heterozygous genotypes respectively; a_A is called the **additive** genetic and d_A the **dominance** genetic component of means. This model implies that a_A will always be positive and P_1 will always be greater than or equal to P_2, while the sign of d_A will be determined by the direction of dominance. Thus, d_A will be positive when A^+ is dominant to A^-, negative when A^- is dominant to A^+ and zero when neither allele is dominant. It is further obvious that the absolute value of d_A ($|d_A|$) will be less than that of a_A when dominance is incomplete or partial, that $|d_A| = a_A$ for complete dominance and that $|d_A| > a_A$ for over-dominance or super-dominance.

These three parameters could be estimated using the following orthogonal comparisons among the three generation means,

$$m = \tfrac{1}{2}\bar{P}_1 + \tfrac{1}{2}\bar{P}_2$$

$$a_A = \tfrac{1}{2}\bar{P}_1 - \tfrac{1}{2}\bar{P}_2$$

$$d_A = \bar{F}_1 - \tfrac{1}{2}\bar{P}_1 - \tfrac{1}{2}\bar{P}_2.$$

The linear genetic model associated with each genotype $(m + a_A;\ m - a_A;\ m + d_A)$ will be referred to as the **genetic value** of that genotype because it defines the underlying genetic contribution to the observed phenotype.

Selfing or intercrossing the $F_1(A^+A^-)$ produces an F_2 family which will have the following constitution:

Genotype	A^+A^+	A^+A^-	A^-A^-
Frequency	$\tfrac{1}{4}$	$\tfrac{1}{2}$	$\tfrac{1}{4}$
Genetic value	$m + a_A$	$m + d_A$	$m - a_A$

The average phenotype of these genotypes obtained as,

$$\tfrac{1}{4}(m + a_A) + \tfrac{1}{2}(m + d_A) + \tfrac{1}{4}(m - a_A)$$

gives the expectation of the F_2 mean, which is equal to,

$$m + \tfrac{1}{2}d_A.$$

Backcrossing P_1 to F_1, on the other hand, produces the $Bc_{1.1}$ family which has two genotypes, A^+A^+ and A^+A^-, in equal proportions so that the average genetic value of these genotypes, the expected mean of $Bc_{1.1}$, is:

$$\tfrac{1}{2}(m + a_A) + \tfrac{1}{2}(m + d_A) = m + \tfrac{1}{2}a_A + \tfrac{1}{2}d_A.$$

Similarly, the two genotypes which will be present with equal frequency in the $Bc_{1.2}$ family, a cross between P_2 and F_1, are A^+A^- and A^-A^-, and their average genetic value, the $Bc_{1.2}$ family mean, is,

$$\tfrac{1}{2}(m + d_A) + \tfrac{1}{2}(m - a_A) = m - \tfrac{1}{2}a_A + \tfrac{1}{2}d_A.$$

In summary, when we consider just one gene, the means of the six basic generations are,

$$\bar{P}_1 = m + a_A$$

$$\bar{P}_2 = m - a_A$$

$$\bar{F}_1 = m + d_A$$

$$\bar{F}_2 = m + \tfrac{1}{2}d_A \qquad \text{[Eqns 2.1]}$$

$$\bar{B}c_{1.1} = m + \tfrac{1}{2}a_A + \tfrac{1}{2}d_A$$

$$\bar{B}c_{1.2} = m - \tfrac{1}{2}a_A + \tfrac{1}{2}d_A.$$

2.2 Two gene model with additive and dominance effects

Extension of the single gene model to generations derived from parents which differ at two genes, say A and B, is relatively simple because the expectations derived above for gene A, apply to gene B also. For example, the previous expectation of the F_2 mean,

$$m + \tfrac{1}{2}d_A$$

is now modified to

$$m + \tfrac{1}{2}d_A + \tfrac{1}{2}d_B$$

to accommodate the effects of segregation at the B gene. There is, however, one complication which affects both the parental and the backcross scores and this concerns the distribution of alleles in the parents. The $+$ and $-$ alleles could either be associated,

i.e. $P_1 = A^+A^+B^+B^+$, $P_2 = A^-A^-B^-B^-$

or dispersed,

i.e. $P_1 = A^+A^+B^-B^-$, $P_2 = A^-A^-B^+B^+$

and this situation affects the composition and size of the additive component which would be equal to $a_A + a_B$ for association or $a_A - a_B$ for dispersion. By definition, both a_A and a_B are positive, therefore the value of the additive component will obviously be larger if the genes are in association than if they are in dispersion.

The mean and dominance components, on the other hand, will remain independent of gene association and dispersion. The former will still be equal to m because all the parameters (a_A, a_B, d_A and d_B) are defined as deviations from the mid-parent, while the latter will become $d_A + d_B$. The dominance component $(d_A + d_B)$ will represent a net balance between the values of d_A and d_B because d may in some cases be positive and in others negative. Such a situation of positive and negative dominance at different genes is referred to as **ambi-directional dominance**, while dominance in one direction is called **unidirectional dominance**.

2.3 Multiple gene model with additive and dominance effects

One feature that is apparent from the one and two gene models is that whatever the number of genes, their additive and dominance effects will be measured by just two parameters. The effects of the individual genes will therefore be hidden in these parameters and it is difficult to visualize how the individual gene effects will be combined when there are several genes. This is particularly true of the additive genetic effects which will be affected by the distribution of the alleles and, by way of illustration, we will initially consider a special case of four genes controlling a trait. This will allow us to determine some general principles which can then be extended to accommodate any number of genes in our model. For simplicity, let us assume that the four genes are A, B, C and D, and that each gene has an additive genetic effect of $a = 2$ units.

Table 2.2 shows all possible combinations in which the alternative alleles can be present in P_1 and P_2 and gives the detailed composition of the additive component in the gene effects. However, we can present the expectations of the additive genetic effect in a more general and meaningful form which is based on just two parameters: (i) half the difference between the parental means $(a_A + a_B + a_C + a_D)$ when there is complete association; and (ii) half the additive effects of those genes that are dispersed in the parents. For example, we can write the expectation of the additive genetic effect for situation (b) in Table 2.2 as $(a_A + a_B + a_C + a_D) - 2(a_D)$, or situation (g) as $(a_A + a_B + a_C + a_D) - 2(a_B + a_D)$. In other words, we can write the expectations of $\bar{P}_1$ and $\bar{P}_2$ in terms of (i) the total effects of the k genes and (ii) those (k') genes that are dispersed between the parents, as:

$$\bar{P}_1 = m + \sum_{}^{k-k'} a_i - \sum_{}^{k'} a_i \quad \text{or} \quad m + \sum_{}^{k} a_i - 2\sum_{}^{k'} a_i;$$

$$\bar{P}_2 = m - \sum_{}^{k-k'} a_i + \sum_{}^{k'} a_i \quad \text{or} \quad m - \sum_{}^{k} a_i + 2\sum_{}^{k'} a_i.$$

Table 2.2 The effect of gene dispersion on the composition and size of the additive genetic component when there are four genes

	A	B	C	D	Score	Expected value of additive component
(i) Complete association						
(a) P$_1$	++	++	++	++	$2+2+2+2 = m+8$	$a_A + a_B + a_C + a_D = 8$
P$_2$	--	--	--	--	$-2-2-2-2 = m-8$	
(ii) Partial (one gene) dispersion						
(b) P$_1$	++	++	++	--	$2+2+2-2 = m+4$	$a_A + a_B + a_C - a_D = 4$
P$_2$	--	--	--	++	$-2-2-2+2 = m-4$	
or						
(c) P$_1$	++	++	--	++	$2+2-2+2 = m+4$	$a_A + a_B - a_C + a_D = 4$
P$_2$	--	--	++	--	$-2-2+2-2 = m-4$	
or						
(d) P$_1$	++	--	++	++	$2-2+2+2 = m+4$	$a_A - a_B + a_C + a_D = 4$
P$_2$	--	++	--	--	$-2+2-2-2 = m-4$	
or						
(e) P$_1$	--	++	++	++	$-2+2+2+2 = m+4$	$-a_A + a_B + a_C + a_D = 4$
P$_2$	++	--	--	--	$2-2-2-2 = m-4$	
(iii) Complete (two gene) dispersion						
(f) P$_1$	++	++	--	--	$2+2-2-2 = m+0$	$+a_A + a_B - a_C - a_D = 0$
P$_2$	--	--	++	++	$-2-2+2+2 = m-0$	
or						
(g) P$_1$	++	--	++	--	$2-2+2-2 = m+0$	$+a_A - a_B + a_C - a_D = 0$
P$_2$	--	++	--	++	$-2+2-2+2 = m-0$	
or						
(h) P$_1$	--	++	++	--	$-2+2+2-2 = m+0$	$-a_A + a_B + a_C - a_D = 0$
P$_2$	++	--	--	++	$+2-2-2+2 = m-0$	

Earlier we specified that P$_1$ should always take the larger score, which means that,

$$\sum_{}^{k} a_i \text{ must always be } \geq 2\sum_{}^{k'} a_i.$$

We now further simplify the equations above by defining a **coefficient of gene association/dispersion** [2], r$_a$, where:

$$r_a = \left(\sum_{}^{k} a_i - 2\sum_{}^{k'} a_i \right) \Big/ \sum_{}^{k} a_i.$$

Thus, r$_a$ = 1 when there is complete association (as in Table 2.2(a)), i.e.

$$k' = 0$$

$r_a = 0$ when there is complete dispersion (as in Table 2.2 (f to h)), i.e.

$$\sum_{i}^{k} a_i = 2 \sum_{i}^{k'} a_i$$

and $0 < r_a < 1$ when there is partial association (as in Table 2.2 (b to e)), i.e.

$$\sum_{i}^{k} a_i > 2 \sum_{i}^{k'} a_i.$$

We are now in a position to formulate a general definition of the additive component as,

$$[a] = r_a . \sum_{i}^{k} a_i.$$

The square brackets are used to denote that it is the net balance of additive genetic effects over all the genes which is being observed, after internal cancellations due to dispersion. The genetic values of the two parental lines thus become:

$$\bar{P}_1 = m + r_a . \sum_{i}^{k} a_i = m + [a]$$

$$\bar{P}_2 = m - r_a . \sum_{i}^{k} a_i = m - [a].$$

The genotype of the F_1 hybrid, on the other hand, will always be $A^+A^-B^+B^- \ldots K^+K^-$ irrespective of the degree of gene dispersion/ association in the parents. Thus, the F_1 score will be equal to $m + d_A + d_B + d_C + \ldots + d_K$, which can be summarized as:

$$\bar{F}_1 = m + \sum_{i}^{k} d_i = m + [d].$$

Parameter $[d]$ represents the net balance of the dominance effects and indicates the direction of dominance at the majority of the k genes, weighted by the magnitudes of their effects.

We can illustrate the application of this simple model by means of the data for P_1, P_2 and F_1 from Table 2.1. Thus,

$$m = \tfrac{1}{2}(\bar{P}_1) + \tfrac{1}{2}(\bar{P}_2) = \tfrac{1}{2}(69.44) + \tfrac{1}{2}(59.04) = 64.24$$

$$[a] = \tfrac{1}{2}(\bar{P}_1) - \tfrac{1}{2}(\bar{P}_2) = \tfrac{1}{2}(69.44) - \tfrac{1}{2}(59.04) = 5.20$$

$$[d] = \bar{F}_1 - \tfrac{1}{2}(\bar{P}_1) - \tfrac{1}{2}(\bar{P}_2) = 83.44 - \tfrac{1}{2}(69.44) - \tfrac{1}{2}(59.04) = 19.20.$$

The fact that $[a]$ is relatively small could imply either that the two parents hardly differ genetically for the trait under study or that there is gene dispersion, but the large positive value of $[d]$ clearly indicates that the parents are genetically different since $[d]$ is due solely to the heterozygosity of those genes for which the parents differ. A more detailed discussion of the interpretation and uses of these estimates will be deferred to a later section of this chapter.

2.4 Extension to other generations

The model we have just developed can easily be extended to other generations and the expectations of the six basic generations for a multiple gene situation are:

$$\bar{P}_1 = m + [a]$$

$$\bar{P}_2 = m - [a]$$

$$\bar{F}_1 = m + [d]$$

$$\bar{F}_2 = m + \tfrac{1}{2}[d] \qquad \text{[Eqns 2.2]}$$

$$\bar{Bc}_{1.1} = m + \tfrac{1}{2}[a] + \tfrac{1}{2}[d]$$

$$\bar{Bc}_{1.2} = m - \tfrac{1}{2}[a] + \tfrac{1}{2}[d].$$

These are clearly analogous to those given earlier for a single gene (Equations 2.1). There are, however, other generations frequently produced by breeders and geneticists as a result of selfing or crossing F_2 or backcross individuals. For example, if we consider a single gene case when we self a random sample of the F_2 individuals, we obtain the following types of F_3 families:

Family	Frequency	Progeny genotypes	Mean
A^+A^+ self	$\frac{1}{4}$	all A^+A^+	$m + a_A$
A^+A^- self	$\frac{1}{2}$	$\frac{1}{4}A^+A^+ : \frac{1}{2}A^+A^- : \frac{1}{4}A^-A^-$	$m + \frac{1}{2}d_A$
A^-A^- self	$\frac{1}{4}$	all A^-A^-	$m - a_A$

Thus, for a single gene difference, the F_3 mean has the expectation,

$$\tfrac{1}{4}(m + a_A) + \tfrac{1}{2}(m + \tfrac{1}{2}d_A) + \tfrac{1}{4}(m - a_A) = m + \tfrac{1}{4}d_A.$$

The mathematical relationship between F_1, F_2 and F_3 can be used to derive the expectation of any generation, F_n, which is obtained by applying $n - 1$ rounds of selfing to the above F_1 and the mean will be equal to:

$$\bar{F}_n = m + \left(\frac{1}{2}\right)^{n-1} d_A.$$

When a random sample of $Bc_{1.1}$ individuals are again crossed to P_1 giving the second generation ($Bc_{2.1}$) of recurrent backcrossing, this generation, like the F_3, has a family structure as follows:

Family	Frequency	Progeny genotype	Mean
$A^+A^+ \times A^+A^+$	$\frac{1}{2}$	all A^+A^+	$m + a_A$
$A^+A^- \times A^+A^+$	$\frac{1}{2}$	$\frac{1}{2}A^+A^+ : \frac{1}{2}A^+A^-$	$m + \frac{1}{2}a_A + \frac{1}{2}d_A$

Now, $\bar{Bc}_{2.1} = \frac{1}{2}(m + a_A) + \frac{1}{2}(m + \frac{1}{2}a_A + \frac{1}{2}d_A) = m + \frac{3}{4}a_A + \frac{1}{4}d_A$. In the case of repeated backcrossing to P_1 for n generations the generation mean has the expectation:

$$\bar{Bc}_{n.1} = m + \left(1 - \left(\frac{1}{2}\right)^n\right)a_A + \left(\frac{1}{2}\right)^n d_A.$$

By following the same approach we can derive the formulae for the first and the subsequent recurrent (n) backcrosses of P_2 (A^-A^-) to F_1 (represented by the symbol $Bc_{1.2}$ and $Bc_{n.2}$ respectively) and they are as follows:

$$\bar{Bc}_{1.2} = m - \frac{1}{2}a_A + \frac{1}{2}d_A$$

$$\bar{Bc}_{n.2} = m - \left(1 - \left(\frac{1}{2}\right)^n\right)a_A + \left(\frac{1}{2}\right)^n d_A$$

(where $n = 1$ for $Bc_{1.2}$).

Although we have derived these means for a single gene case, they translate simply into the multi-gene case and the general formulae for the expectations of other generations of the selfing and recurrent backcrossing series are:

$$\bar{F}_n = m + \left(\frac{1}{2}\right)^{n-1}[d];$$

$$\bar{Bc}_{n.1} = m + \left(1 - \left(\frac{1}{2}\right)^n\right)[a] + \left(\frac{1}{2}\right)^n[d] \quad \text{and} \qquad \text{[Eqns 2.3]}$$

$$\bar{Bc}_{n.2} = m - \left(1 - \left(\frac{1}{2}\right)^n\right)[a] + \left(\frac{1}{2}\right)^n[d].$$

Elaborations of these formulae for some of the frequently used generations in quantitative genetic studies are given in Table 2.3.

Table 2.3 Expectations of various generations on an additive-dominance model

Generation	m	$[a]$	$[d]$
P_1	1	1	0
P_2	1	-1	0
F_1	1	0	1
F_2	1	0	0.5
F_3	1	0	0.25
F_4	1	0	0.125
F_∞	1	0	0
$Bc_{1.1}$	1	0.5	0.5
$Bc_{2.1}$	1	0.75	0.25
$Bc_{1.2}$	1	-0.5	0.5
$Bc_{2.2}$	1	-0.75	0.25

It is clear from the above formulae that, with directional dominance, i.e. $[d] > 0$, the F_n mean will change with each generation of selfing, eventually reaching the mid-parental value, m. In the case of a cross exhibiting **hybrid vigour** or **heterosis**, that is where the F_1 mean exceeds the mean of the better parent, this decline in mean due to selfing is commonly referred to as **inbreeding depression**. This is well illustrated by the character final plant height in *Nicotiana rustica* (Figure 2.1).

2.5 Relationships between generation means

It is apparent from the theory so far that there are very simple relationships between the expected means of different generations. For example, the mean of $Bc_{1.1}$ has the expectation $m + \frac{1}{2}[a] + \frac{1}{2}[d]$ which is equal to the average of the means of the P_1 and F_1 generations. Similarly, the expectation of the $Bc_{1.2}$ mean is equal to the average of the means of P_2 and F_1. In fact, the expected mean of any generation that is derived from a cross between two pure breeding parents bears a simple relationship with the parental and F_1 means.

However, these expected relationships hold only if the generation means depend solely on the additive and dominance effects of genes. They would not be expected to hold if other factors such as differential viability, maternal effects or interactions between genes (often referred to as **epistasis** or **non-allelic interaction**) existed. Therefore, comparisons among these generation means can be used to provide tests

Figure 2.1 Heterosis and inbreeding depression for height in a cross between two inbred lines of *Nicotiana*. The solid line shows the expected decline in the mean with inbreeding.

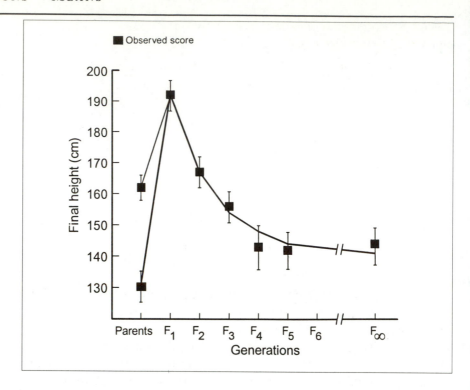

for the presence of such complex factors. For example, when the additive–dominance model is adequate we expect:

$$\bar{B}c_{1.1} = \tfrac{1}{2}\bar{P}_1 + \tfrac{1}{2}\bar{F}_1$$

or
$$2\bar{B}c_{1.1} = \bar{P}_1 + \bar{F}_1$$

or
$$2\bar{B}c_{1.1} - \bar{P}_1 - \bar{F}_1 = 0.$$

On the other hand, if the additive–dominance model is inadequate, then it is likely that,

$$2\bar{B}c_{1.1} - \bar{P}_1 - \bar{F}_1 = A.$$

Using the data in Table 2.1, for example, **A** is estimated as,

$$2 \times 76.03 - 69.44 - 83.44 = -0.82.$$

Clearly **A** is not zero, but in order to test if the additive–dominance model has failed in this case we need to know whether the estimate of **A** is significantly different from zero. In order to do this, we need the standard error of **A**.

Now, **A** is a linear sum of three means. It follows from basic statistical theory, that if some quantity, **Q**, is the sum of k_1y_1, k_2y_2, k_3y_3, etc. where the k_is are constants (i.e. $\mathbf{Q} = \sum k_iy_i$), then the variance (s^2) of $\mathbf{Q}$ is,

$$s_{\mathbf{Q}}^2 = \sum_i k_i^2 s_{yi}^2,$$

providing the y_is are independent.

For $\mathbf{A}$, the k_is are 2, -1 and -1 respectively, while the y_is are the generation means and hence the s_{yi}^2s are the variances of the relevant generation means. In Table 2.1 we have given the variance among the individuals of the various generations, s^2, and to obtain the variance of the generation means we need to divide these by the appropriate family size, n; e.g. for $Bc_{1.1}$, we obtain 0.8105 ($= 81.05/100$). The variance of $\mathbf{A}$ is therefore,

$$s_{\mathbf{A}}^2 = 4s_{Bc_{1.1}}^2 + s_{\bar{P}_1}^2 + s_{\bar{F}_1}^2$$

$$s_{\mathbf{A}}^2 = 4 \times 0.8105 + 0.5973 + 0.5181 = 4.3574.$$

We can now apply Student's t test where,

$$t_{(df)} = \mathbf{A}/\sqrt{s_{\mathbf{A}}^2}.$$

The **degrees of freedom** (df) of t are the sum of the df of the three variances, 297.

$$t_{(297)} = -0.82/\sqrt{4.3574}$$

$$= -0.39 \text{ n.s.}$$

This shows that $\mathbf{A}$ is not significantly different from zero and hence there is no reason to reject the null hypothesis (H_0). We therefore conclude that the additive–dominance model adequately explains the variation among these means.

Mather [3] derived the above relationship and called it the 'A Scaling Test'. He described two more scaling tests that are appropriately called $\mathbf{B}$ and $\mathbf{C}$ and they determine the conformity of the following relationships with the additive–dominance model.

$$\mathbf{B} = 2\bar{B}c_{1.2} - \bar{P}_2 - \bar{F}_1 \qquad s_{\mathbf{B}}^2 = 4s_{Bc_{1.2}}^2 + s_{\bar{P}_2}^2 + s_{\bar{F}_1}^2$$

$$\mathbf{C} = 4\bar{F}_2 - 2\bar{F}_1 - \bar{P}_1 - \bar{P}_2 \qquad s_{\mathbf{C}}^2 = 16s_{\bar{F}_2}^2 + 4s_{\bar{F}_1}^2 + s_{\bar{P}_1}^2 + s_{\bar{P}_2}^2.$$

Using the same procedure as for $\mathbf{A}$, we obtain:

$$\mathbf{B} = 0.08, s_{\mathbf{B}}^2 = 4.8084, t_{(297)} = 0.04, \text{n.s.}; \quad \text{and}$$

$$\mathbf{C} = 2.08, s_{\mathbf{C}}^2 = 19.4468, t_{(396)} = 0.47, \text{n.s.}$$

Since neither $\mathbf{A}$, nor $\mathbf{B}$ nor $\mathbf{C}$ is significant in the present example, we can accept that complicating effects, such as maternal effects or gene interaction, are not involved in the genetical control of the character in this cross. We can not say that such complicating factors do not exist, but simply that we have no evidence for their existence; the data are consistent with their absence and the additive–dominance model is adequate. It is possible that more complex factors are involved, but the experiment is not sufficiently large and powerful to detect them. This creates a dilemma. A very small experiment will invariably not detect any departures from a simple model while a very large experiment may be so sensitive that it may detect all sorts of trivial departures. We will return to this later. If any of the three scaling tests indicate that the simple additive–dominance model fails then it is necessary to consider the more complex situations described in Chapters 11, 12 and 13.

2.6 Estimating genetical parameters

If we have determined that the differences between generation means for a particular character are explained adequately by the additive and dominance effects of genes, then it is sensible to obtain reliable estimates of m, $[a]$ and $[d]$ and to test their significance before attempting their interpretation. Earlier we explained how these three parameters could be estimated when we had just P_1, P_2 and F_1 means, but we now have the problem of estimating three parameters from six generation means. Clearly, we would like to use all the genetical information that is available to us in order to obtain the most accurate estimates which have the smallest possible standard errors. In other words, we want maximum likelihood estimates of m, $[a]$ and $[d]$.

Such estimates can be obtained by regression procedures and so we need to translate our data into a form amenable to a regression analysis. Consider the data in Table 2.1 and the expectations given earlier in this section for the six basic generations as shown in Table 2.4. If, for simplicity, we consider parameters m and $[a]$ only, we can write the expectations of the various generation means (y_i) in the form:

$$y_i = c + bx_i$$

where the coefficients of $[a]$ are the x_is and the intercept, c, equals m and the regression coefficient, b, equals $[a]$ (i.e. $y_i = m + [a]x_i$). The x_is are, in fact, measures of the relative contribution of A^+ and A^- alleles to each generation. Thus, P_1 is all A^+ ($x_i = 1$), $Bc_{1.1}$ consists of $\frac{1}{2}A^+A^+$ and $\frac{1}{2}A^+A^-$ individuals and hence there are $\frac{3}{4}A^+$ to $\frac{1}{4}A^-$

Table 2.4 Means of six basic generations, together with m, $[a]$ and $[d]$ model to illustrate the use of regression for parameter estimation

Generation	Mean (y variable)	m	$[a]$ (x_1 variable)	$[d]$ (x_2 variable)
P_1	69.44	1	1	0
$Bc_{1.1}$	76.03	1	$\frac{1}{2}$	$\frac{1}{2}$
F_1	83.44	1	0	1
F_2	74.36	1	0	$\frac{1}{2}$
$Bc_{1.2}$	71.28	1	$-\frac{1}{2}$	$\frac{1}{2}$
P_2	59.04	1	-1	0

$(x_i = \frac{3}{4} - \frac{1}{4} = \frac{1}{2})$, F_1 is A^+A^- $(x_i = \frac{1}{2} - \frac{1}{2} = 0)$, etc. If there is no dominance, therefore, the generation means, y_i, will vary with the relative balance of A^+ to A^- (x_i), which is sometimes called the **gene dosage**.

Using linear regression analysis with the data in Table 2.4, using x_i, we obtain the estimates,

$$c = m = 72.27$$

$$b = [a] = 5.11.$$

Statistically, these estimates are those which minimize deviations between the observed and the predicted values and also they are independent of each other. We can test the significance of each estimate by a Student's t-test. We can also determine whether the parameters have accounted for all of the significant differences between the generation means by testing the significance of the remainder mean square (MS) in the analysis of variance (ANOVA). When the remainder MS is significant, it indicates that there are other factors, apart from m and $[a]$, which affect the means, and these need to be identified and included in the model. In these circumstances, we need to extend the analysis to include several regression variables and linear regression then converts into multiple regression. This takes the form:

$$y_i = c + b_1.x_{1i} + b_2.x_{2i} + \ldots,$$

where $b_1 = [a]$, $b_2 = [d]$ etc.

So far we have assumed that each of the y_i values (generation means) is known with equal precision. This implies that the variances of the generation means (s_i^2/n_i) are all the same, but this is not generally likely in practice because some generations, e.g. F_2, will have much larger variances among individuals (s^2) than will others due to genetic segregation (see Chapter 3), as the data from Table 2.1

show, while there may also be large differences in family size (n_i). This heterogeneity among variances of the generation means makes the accuracy of the means unequal, which must be adjusted in the regression analysis by **weighting** the means differently. These weights are the reciprocals of the variances of the generation means [4], i.e. for the ith generation mean,

$$\text{weight}(\text{wt}_i) = \text{family size}(n_i)/\text{variance}(s_i^2).$$

The regression equation now becomes:

$$\text{wt}_i \text{y}_i = \text{wt}_i(c + b_1.x_{1i} + b_2.x_{2i} \ldots \text{etc.}).$$

Most statistical software packages available for modern computers should be capable of handling such weighted regression problems, but for those who wish to understand the methodology, a short theoretical explanation is given in Appendix F. Briefly, because there are N generation means, it is possible to estimate a total of N parameters, of which one is the mean, m. This leaves a maximum of $N - 1$ genetical or other parameters that can be estimated, which is equivalent to the $N - 1$ df between the N means. The object is to explain the variation between the observed generation means with as simple a model as possible. If only m is fitted, then there will be $N - 1$ df remaining to test the adequacy of the model using a χ^2 or F test. Every time a further parameter is included in the model, there is one less df for testing the adequacy of the model.

Faced with a new set of data, how should the model fitting proceed? It always makes sense to try a model with just m, first. If this adequately explains the variation in the trait, then there is no need to fit any genetical parameters. If not, then further parameters should be introduced. For example, the next simplest model would involve the parameters m and $[a]$. If both parameters are significant and the model fits, then there is no need to proceed further. If both parameters are significant, but the χ^2 indicates that the model is still inadequate, then try a model comprising m, $[a]$ and $[d]$. At each step, only significant parameters are retained in the model, while only sufficient parameters are added in total to provide an adequate fit to the data. It is quite possible for two different models to fit the data, in which case it is customary to accept the one with the fewest and biologically most plausible parameters. The process is illustrated in Table 2.5 with the data taken from Table 2.1.

With multiple regression situations, the parameters are seldom completely independent, i.e. they are correlated, and this means that the magnitude of a particular parameter will depend on those other

Table 2.5 The generation means, variances, weights and additive–dominance model for the six basic generations of a cross between two homozygous lines. (i) Basic data and model; (ii) parameters from sequential model fitting; (iii) best model

(i)

Generation	Mean	s_x^2	df	$s_{\bar{x}}^2$	Weight	m	[a]	[d]
P_1	69.44	59.73	99	0.5973	1.6742	1	1	0
P_2	59.04	65.71	99	0.6571	1.5218	1	−1	0
F_1	83.44	51.81	99	0.5181	1.9301	1	0	1
F_2	74.36	100.75	99	1.0075	0.9926	1	0	0.5
$Bc_{1.1}$	76.02	81.05	99	0.8105	1.2338	1	0.5	0.5
$Bc_{1.2}$	71.28	90.83	99	0.9083	1.1010	1	−0.5	0.5

(ii)

(a) $m = 72.5420 \pm 0.3439^{***}$ $\chi^2_{(5)} = 542.73^{***}$

(b) $m = 72.4172 \pm 0.3442^{***}$

  $[a] = 4.8209 \pm 0.5148^{***}$ $\chi^2_{(4)} = 455.02^{***}$

(iii)

  $m = 64.2474 \pm 0.5151^{***}$

  $[a] = 5.1252 \pm 0.5149^{***}$

  $[d] = 19.1989 \pm 0.9005^{***}$ $\chi^2_{(3)} = 0.47\,\text{n.s.}$

parameters which are being estimated simultaneously. As additional parameters are added, it is to be expected, therefore, that the values of the parameters initially estimated will change. If the simple additive–dominance model fits, we can discount the presence of complicating factors such as maternal effects and interactions between genes. This approach thus provides a comprehensive test of the simple additive model which replaces the **A**, **B** and **C** scaling tests and has been termed the **Joint Scaling Test** [4].

2.7 Interpretation: heterosis and potence ratio

We have seen that it is possible to construct simple additive–dominance models for the means of generations derived from a pair of true breeding lines. We have also seen how it is possible to estimate the genetical parameters and test whether they alone are sufficient to explain the differences observed in the means of different generations. Where more complex models are required, such as with gene interaction, maternal effects, etc., the reader is referred to Chapters 11,

12 and 13. What insight, then, do these models and the parameters they generate give us into the genetical control of a trait?

Does $[a] = 0$ mean that there is no additive variation in the cross? No, not unless the parents differ for just a single gene (which is very unlikely), i.e. $[a] = a_A$. Similarly, we can conclude that there is no dominance only when $[d] = d_A = 0$. Further, for a single gene only, we can interpret $[d]/[a] = d_A/a_A$ as the **dominance ratio** and conclude the following:

if $d_A/a_A = +1$, allele A^+ is completely dominant to allele A^-;

if $d_A/a_A = -1$, allele A^- is completely dominant to allele A^+;

if $0 < d_A/a_A < +1$, allele A^+ is partially dominant to allele A^-;

if $-1 < d_A/a_A < 0$, allele A^- is partially dominant to allele A^+;

if $d_A/a_A = 0$, there is no dominance;

if $d_A/a_A > +1$, allele A^+ is over-dominant to allele A^-;

if $d_A/a_A < -1$, allele A^- is over-dominant to allele A^+.

With two or more genes, on the other hand, the situation becomes more complicated because

$$[d]/[a] = \left(\sum_{i=1}^{k} d_i \right) \bigg/ \left(r_a \sum_{i=1}^{k} a_i \right).$$

The numerator could be zero as a result of ambi-directional dominance while the denominator could be zero as a result of gene dispersion. Thus, this ratio could take almost any value irrespective of the true degree of dominance. It is sometimes referred to as the **potence ratio** because it indicates which parent has the most dominant alleles and is therefore the more potent in the cross. Therefore, we can rarely equate the ratio $[d]/[a]$ with the true dominance ratio which is obtained only when dominance is uni-directional and there is complete association of alleles in the parents ($r_a = 1$). In other words, the true magnitude of the dominance ratio,

$$\left(\sum_{i=1}^{k} |d_i| \right) \bigg/ \left(\sum_{i=1}^{k} a_i \right)$$

can only be estimated from the components of means when dominance is uni-directional at all the k genes for which the parents differ (i.e. $d_A, d_B \ldots d_K$, etc. have the same sign and $|[d]| = |d_A| + |d_B| \ldots + |d_K|$) and all the alleles with increasing effect are

Figure 2.2 Heterosis for height in *Nicotiana rustica*.

present in P_1 (i.e. $\bar{P}_1 = m + a_A + a_B + \ldots + a_K$) and their counterparts are in P_2 ($\bar{P}_2 = m - a_A - a_B + \ldots - a_K$).

These conclusions concerning dominance are also relevant to interpreting the causes of **heterosis** or **hybrid vigour**. This phenomenon is of central importance in breeding and in evolutionary genetics, and concerns the superiority of the F_1 over the better parent, P_1 (see Figures 2.1 and 2.2). In the absence of gene interaction and other complicating factors, for which the tests have been described earlier, heterosis implies $\bar{F}_1 > \bar{P}_1$, i.e.

$$m + [d] > m + [a]$$

$$\text{or} \qquad [d] > [a]$$

$$\text{or} \qquad \sum d_i > r_a \sum a_i.$$

This relationship tells us that, as long as the average dominance $(\sum d_i / \sum a_i)$ is greater than the degree of gene dispersion (r_a), then there will be heterosis. It follows, therefore, that it requires very little dominance at individual genes to produce quite considerable heterosis if the genes are dispersed in the two parents. For example, if the better parent has 70% of the increasing alleles, then r_a will be $0.4(= 1 - 2 \times 0.3)$ and hence the average amount of dominance needs to be just greater than 0.4 to cause the F_1 to out-perform the better parent. If the better parent has 55% of the increasing alleles, the average dominance needs to be little more than 0.15. Clearly, characters such as fitness or yield, that are controlled by many genes, would require only very small amounts of dominance to produce very major hybrid superiority without the need to invoke overdominance. This also implies that it should be possible to produce lines that out-yield the F_1 hybrid and this possibility will be discussed in Chapter 15.

Summary

1. Six basic generations derived from a cross between two inbred lines, P_1, P_2, F_1, F_2, $Bc_{1.1}$ and $Bc_{1.2}$, appear in a whole range of experimental designs and their genetic models are central to quantitative genetics.
2. The means of these six basic generations can be described in terms of just three parameters which measure the mean, the additive and the dominance effects, m, $[a]$ and $[d]$, respectively.
3. It is possible to use these generations to provide powerful tests of the adequacy of a simple genetical model and, in particular, to test for complex effects such as epistasis, maternal effects, etc.
4. The parameterization of the additive and dominance effects shows that considerable heterosis can result from small amounts of dominance.

References

1. Fisher, R.A., Immer, F.R. and Tedin, O. (1932) The genetical interpretation of statistics of the third degree in the study of quantitative inheritance. *Genetics*, **17**, 107–24.

2. Mather, K. and Jinks, J.L. (1982) *Biometrical Genetics*, 3rd edn, Chapman & Hall, London.
3. Mather, K. (1949) *Biometrical Genetics*, 1st edn, Methuen, London.
4. Cavalli, L.L. (1952) An analysis of linkage in quantitative inheritance, in *Quantitative Inheritance* (eds E.C.R. Reeve and C.H. Waddington), HMSO, London, pp. 135–44.

3 Basic generations – variances

Variation is central to the study of quantitative traits because little can be deduced from populations or generations that are monomorphic. From studies of variation among genotypes, various questions concerning the genetic control of a trait can be asked, such as how much of the variation is heritable, which factors contribute to the genetic variability and what are their relative magnitudes and importance. Variation may be measured either among individuals when the population of individuals has no discernible structure, or among family means where there are naturally occurring or artificially produced families or clones.

In this chapter, we will restrict ourselves to defining, analysing and interpreting variation in the six basic generations while the variation among generations produced by selfing and sib-mating will be considered in Chapters 4 and 5.

3.1 Variation in the non-segregating generations

We can divide the six basic generations into two distinct groups with respect to their variances. The first group includes those generations that consist of genetically identical individuals, such as the P_1, P_2 and F_1 families. Because individuals within these generations do not show any genetic differences, they are referred to as the **non-segregating generations**. The second group includes the F_2, $Bc_{1.1}$ and $Bc_{1.2}$ generations which contain a mixture of genotypes resulting from segregation, random assortment and recombination of alleles at those loci for which P_1 and P_2 differ and the F_1 is heterozygous. Hence they are referred to as the **segregating generations**.

3.2 Environmental variation

Because the individuals within the P_1, P_2 and F_1 families are genetically identical, any variation between them cannot be genetical and
it is conventional to refer to such variation as the **environmental variation within families**, V_E. This type of variation exists even between
individuals which may be exposed to virtually identical environments
and it is a major component of the total variation for quantitative
traits, which frequently exceeds the genetical variation and the
macro-environmental variation caused by such factors as blocks,
locations or plots. Because V_E is such a large component it is worth
exploring its underlying causes, of which there are several and not
all are environmental in the strict sense.

3.2.1 Pre-replication factors

These include a variety of factors which influence the phenotype of
the individual but which are determined directly by the mother or
environmental effects on the mother. Consider, for example, plant
height. This trait is affected as much by variation in, say, the quality
of the seed as it is by soil fertility; poor quality seed produces
weaker seedlings which, in turn, may lead to shorter plants. Seed
quality is affected by various factors, such as the age of the mother,
the age and position of the flower, age of the plant at the time of
pollination, harvest time, conditions at the time of maturity and
health of the maternal plant. For these reasons, even seeds harvested
from the same plant often show extensive variation in quality including size, maturity and vitality. No two seeds are equally endowed, even
those located next to each other on a maize cob, a wheat spike or in a
tobacco capsule. A large proportion of such variation in seed quality,
however, may be eliminated by taking precautions during seed production and grading the seed, but it is virtually impossible to obtain
completely uniform seed. Consequently, some effects of seed quality
will always be apparent on the plant height and they will appear
confounded with V_E. Similar factors operate in animals. For example,
monozygotic twins in humans are genetically identical; clones in fact.
Yet, such twins are never completely identical at birth and all pairs
show at least some differences in weight, height, health and personality
which must have arisen in the uterus (Figure 1.3).

3.2.2 Developmental factors

Another important factor which affects V_E is developmental variation. This is an inherent property of the individuals which is difficult
to measure, control or manipulate, but one example of where

developmental variation can be measured is the difference between duplicate measurements from the same individual. In animals, the two halves of the body are symmetrical and therefore provide duplicate measures on limbs, wings and bone structure, etc. In plants, differences between duplicate structures on opposite sides of a plant or repeat structures on different branches or tillers can be measured. Because these duplicate structures have received the same environment, the differences between them provide a measure of the developmental variation. In the case of traits such as height or weight, these developmental factors cannot be separated from the genuine within family variation.

3.2.3 Measurement errors

A third major factor which contributes significantly to V_E is measurement error which can be subdivided into that due to (a) measurement and (b) rounding off. The first group depends on the accuracy and consistency with which the data are scored and recorded, the magnitude of scoring error usually being largest when several people participate in scoring the experiment. Rounding off errors occur when a continuous variable is scored to a fixed interval of the measurement unit such as to the nearest centimetre or gram.

3.2.4 Statistical sampling error

Many quantitative traits are measured as a proportion, such as seed germination, disease attack or the answers to a personality test. These traits will additionally show statistical sampling variance because the proportion, p, will have the binomial variance, $p(1-p)/n$, where n is the number scored. Such variation is clearly related to p but is not otherwise determined by the environment.

3.2.5 True micro-environmental variation

The final factor which contributes to V_E is genuine micro-environmental variability. Because every individual occupies a unique part of the environment in the experimental area, it may be exposed to different levels of nutrition, temperature, sunlight, moisture, tillage, soil depth, etc.

It should be apparent that only the last of these five factors indeed represents differences due to the environment in which the genotypes are reared, and thus true micro-environmental variation often represents just a small part of the observed value of V_E (Table 3.1). However, there is generally no easy method of separating the contribution made to V_E by each component. The non-environmental components are difficult to manipulate while the true micro-environmental factors themselves may be very difficult to identify. It therefore follows

Table 3.1 Relative magnitudes of developmental variance as a proportion of V_E in *Drosophila* and *Nicotiana*

Organisms and traits	Developmental variance	V_E	% Developmental
Drosophila			
Sternopleural chaetae			
(a) Inbred lines	1.36	1.63	83
(b) F_1s	1.39	1.41	97
Nicotiana rustica			
(i) Height (12 weeks)	55.93	69.35	81
(ii) Leaf length	4.01	4.31	93

that V_E will always be present and will frequently be a major source of variation which will not be reduced easily by elaborate equipment for controlling the environment. In practice, therefore, one should obtain an accurate estimate of V_E, and reduce its effect on the variance of generation means by increasing the number of individuals. It is invariably easier and cheaper to increase the precision with which each mean is estimated by increased replication than by attempting to control the environment. The effect of greater replication on the variance of the means can be predicted and costed precisely while the results and cost of controlling the environment are unpredictable.

The components of V_E just discussed affect the phenotype of individuals within a family. Environmental differences can also exist between families, and such effects would be common to all progeny in that family. Thus, even though two mothers, plant or animal, may be genetically identical, differences in their environment such as nutrition or health, etc., may affect the phenotype of their progeny. Such environmental effects are represented by the symbol V_{EC}, which stands for the **common environment**. It is therefore very important in all studies of quantitative traits to take every precaution to ensure that the progeny are produced from parents that have been raised under similar conditions. For example, one should never raise F_1 plants from seed produced last year in the same experiment with P_1 plants from seed produced in the previous year.

3.3 Estimating environmental variance

Providing that the experiment is adequately randomized, the individuals of P_1, P_2 and F_1 are deemed to be exposed to the same range of environmental conditions. Their variances should thus provide independent estimates of V_E which are not expected to differ from each other. For this reason V_E is also called the **additive environmental variance**. For example, from the variances given in Table 3.2, we can estimate V_E by averaging the values of s^2 of the three non-segregating generations. Because each generation is based on the same number of observations in the present case,

$$V_E = \tfrac{1}{3}(59.73 + 65.71 + 51.81) = 59.08.$$

Had the number of observations in each generation been different, then the variances would be pooled by summing their SS and dividing by their combined df.

However, on many occasions inbred lines (P_1 and P_2) and hybrids (F_1) respond differently to the same micro-environment, and this estimate of V_E will be meaningful only when the variances of the P_1, P_2 and F_1 generations are homogeneous, i.e. there is no interaction between the micro-environmental variation and the genotype. We can check whether $s^2P_1 = s^2P_2 = s^2F_1$ by means of a Bartlett's [1] or a Levene's [2] test. Alternatively, and more simply, an F-test can be applied using the ratio of the largest (65.71) to the smallest (51.81) of the three variances.

$$F = 65.71/51.81 = 1.27; \quad P \approx 0.10.$$

Because we have chosen the two extreme variances out of n ($= 3$) and divided the largest by the smallest, the value of P has to be corrected by multiplying by n(n $-$ 1) ($= 6$). Thus, the correct probability that we obtain for the F test is P ≈ 0.6, which shows that the parental and F_1 variances do not differ significantly from each other in this case and so we can accept the pooled estimate of $V_E = 59.08$.

3.4 Variation in the segregating generations

The individuals in the segregating generations will be subject not only to non-genetical, but also to genetical variation, as is suggested by the larger variances of the F_2 and Bc generations in Table 3.2. All the genes that were heterozygous in the F_1 should contribute to these variances and we can determine the nature and magnitude of this variation by deriving theoretical expectations of these variances from the genetic values and their frequencies in the segregating generations.

Let the expected genetic variance of an F_2 population at a single segregating locus be V_G. From statistical theory we find that

Table 3.2 Within family variances for the six basic generations, each based on 100 individuals (data taken from Table 2.1)

Generation	Within-family variance, s_x^2 (99df)
P_1	59.73
P_2	65.71
F_1	51.81
F_2	100.75
$Bc_{1.1}$	81.05
$Bc_{1.2}$	90.83

$V_G = \sum f_i(g_i - \text{mean})^2 / (\sum f_i - 1)$ where f_i is the frequency of the ith genotype, and g_i is its genetic value. While this formula applies to small samples, it can be simplified to $V_G = \sum f_i g_i^2 - (\sum f_i g_i)^2$ when we are dealing with the F_2 population as a whole. The f_i now represents the proportion of individuals having the ith genotype and not the frequency, i.e. $\sum f_i = 1$ (not n). Applying this formula to gene A, for which the F_2 will consist of A^+A^+, A^+A^- and A^-A^- genotypes which will be present with the proportions of $\frac{1}{4} : \frac{1}{2} : \frac{1}{4}$ respectively, we get:

$$F_2 \text{ Mean} = \sum f_i g_i$$

$$= \tfrac{1}{4}(A^+A^+) + \tfrac{1}{2}(A^+A^-) + \tfrac{1}{4}(A^-A^-)$$

$$= \tfrac{1}{4}(m + a_A) + \tfrac{1}{2}(m + d_A) + \tfrac{1}{4}(m - a_A)$$

$$= m + \tfrac{1}{2}d_A \text{ (as in Chapter 2).}$$

$$F_2 \text{ Variance} = \sum f_i g_i^2 - \left(\sum f_i g_i\right)^2$$

$$= \tfrac{1}{4}(A^+A^+)^2 + \tfrac{1}{2}(A^+A^-)^2 + \tfrac{1}{4}(A^-A^-)^2 - (\text{mean})^2$$

$$= \{\tfrac{1}{4}(m + a_A)^2 + \tfrac{1}{2}(m + d_A)^2 + \tfrac{1}{4}(m - a_A)^2\}$$

$$- (m + \tfrac{1}{2}d_A)^2$$

$$= \tfrac{1}{2}a_A^2 + \tfrac{1}{4}d_A^2.$$

The expected genetical variance of an F_2, V_G, for a single gene, thus consists of two parts, and we can write this expectation in terms of V_A^* and V_D^*, where $V_A^* = \tfrac{1}{2}a_A^2$, the **additive genetic** component of variance, and $V_D^* = \tfrac{1}{4}d_A^2$, the **dominance** or **non-additive** genetic component of variance.

For many genes (k), each of which shows independent gene action (i.e. no epistasis) and independent gene transmission at gamete formation (i.e. no linkage), V_A^* and V_D^* will represent the sums of the additive and dominance variances of the individual genes, that is:

$$V_A^* = \tfrac{1}{2}(a_A^2 + a_B^2 + \ldots + a_K^2) = \tfrac{1}{2}\sum_{i=1}^{k} a_i^2$$

$$V_D^* = \tfrac{1}{4}(d_A^2 + d_B^2 + \ldots + d_K^2) = \tfrac{1}{4}\sum_{i=1}^{k} d_i^2.$$

(N.B. The * notation in V_A^*, V_D^* is to distinguish the special case of equal allele frequencies. See Chapter 9 for the general case.)

Thus, the expected variation among F_2 individuals is $V_A^* + V_D^* + V_E$, compared with just V_E for the non-segregating generations, P_1, P_2 and F_1. Consequently, the F_2 variance should be larger than V_E and significantly so where the F_2 is segregating at a large number of genes with small effect or a few genes with large effect.

In the case of $Bc_{1.1}$, a cross between F_1 and P_1, the genetic constitution of the family for gene A is $\tfrac{1}{2}A^+A^+ : \tfrac{1}{2}A^+A^-$, the scores of the genotypes are as given earlier, the mean genetic value is $m + \tfrac{1}{2}a_A + \tfrac{1}{2}d_A$ and the expected genetic variance is $\tfrac{1}{4}a_A^2 + \tfrac{1}{4}d_A^2 - \tfrac{1}{2}a_A d_A$. For k genes, this expectation becomes $\tfrac{1}{2}V_A^* + V_D^* - V_{AD}$ where V_A^* and V_D^* are as defined earlier and $V_{AD} = \tfrac{1}{2}(\delta_a a_A d_A + \delta_b a_B d_B + \ldots + \delta_k a_K d_K)$ [where $\delta_i = +1$ if the allele is in P_1 and $\delta_i = -1$ otherwise]; V_{AD} is thus a cross product of the additive and the dominance effects of the genes that are segregating in the cross. Thus, the phenotypic variance of $Bc_{1.1}$ will be $\tfrac{1}{2}V_A^* + V_D^* - V_{AD} + V_E$. Similarly, the phenotypic variance of $Bc_{1.2}$ ($= F_1 \times P_2$) will be $\tfrac{1}{2}V_A^* + V_D^* + V_{AD} + V_E$ while the average of the $Bc_{1.1}$ and $Bc_{1.2}$ variances will be equal to $\tfrac{1}{2}V_A^* + V_D^* + V_E$. This average phenotypic variance should be smaller than that of the F_2 providing the additive genetic, dominance genetic and additive environmental model is adequate.

A universal characteristic of V_A^*, V_D^* and V_E is that, being components of variances, they cannot be negative. Sometimes the estimates may be negative due to sampling errors but they are unlikely to be significant. V_{AD}, on the other hand, is a covariance and its sign will depend on the direction of dominance. When alleles which decrease the phenotype are dominant, V_{AD} will be negative, and it will be positive for the opposite situation. Further, the magnitude of V_{AD} will be maximum when the alleles are completely associated between the parental lines and dominance is unidirectional, otherwise it will reflect the net balance of the cross products after internal cancellations. Therefore, the backcross to the parent with the largest number of dominant alleles will have the smaller phenotypic variance and, in the extreme case of a parent which contains all the dominant alleles and where dominance is complete ($d = a$), then the genetical variance will be zero because $\tfrac{1}{2}V_A^* + V_D^* = V_{AD}$. The data in Table 3.2 show

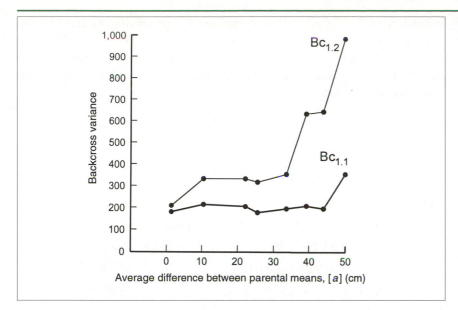

Figure 3.1 The observed relationship between backcross variance and the parental difference [a] for final height in *Nicotiana rustica* crosses.

that the backcross to P_1 ($Bc_{1.1}$) has the smaller variance, indicating that dominance is for the higher value of the trait and confirming what was shown by [d] in Chapter 2. With unidirectional dominance, therefore, V_{AD} will generally show a linear relationship with half the parental difference, [a], as is shown in Figure 3.1, for data collected on the basic generations derived from the recombinant inbred lines extracted from a cross between two pure breeding varieties of *Nicotiana rustica*.

3.5 Estimation of genetical components

Initially the breeder would like to know whether the cross shows significant genetic variation and, if so, how much of the variation is heritable and what types of gene effects are significant. The first and second questions are answered by comparing the variances of the segregating and the non-segregating generations. The simplest method is to use the pooled estimate of V_E derived earlier, 59.08, in an F-test to determine whether or not the F_2 and the backcross variances are significantly larger than V_E. Now,

for the F_2, $F_{(99,297)} = s^2_{F2}/V_E = 100.75/59.08 = 1.71^{**}$;

for the $Bc_{1.1}$, $F_{(99,297)} = s^2_{Bc1.1}/V_E = 81.05/59.08 = 1.37^{*}$;

for the $Bc_{1.2}$, $F_{(99,297)} = s^2_{Bc1.2}/V_E = 90.83/59.08 = 1.54^{**}$.

indicating that genetic variation exists in the generations derived from the cross. Having shown that the genetic variation exists, the breeder can then address the third question: what types of genetic variation are involved? This is achieved by estimating V_A^*, V_D^* and V_{AD} using the following equations.

$$V_A^* = (2s_{F_2}^2 - s_{Bc_{1.1}}^2 - s_{Bc_{1.2}}^2)$$

$$V_D^* = (s_{Bc_{1.1}}^2 + s_{Bc_{1.2}}^2 - s_{F_2}^2 - V_E) \qquad \text{[Eqn 3.1]}$$

$$V_{AD} = \tfrac{1}{2}(s_{Bc_{1.2}}^2 - s_{Bc_{1.1}}^2).$$

Therefore:

$$V_A^* = (2 \times 100.75 - 81.05 - 90.83) \qquad = 29.62$$
$$V_D^* = (81.05 + 90.83 - 100.75 - 59.08) = 12.05$$
$$V_{AD} = \tfrac{1}{2}(90.83 - 81.05) \qquad\qquad\quad = \;\; 4.89$$
$$\text{while } V_E \qquad\qquad\qquad\qquad\quad = 59.08.$$

The relative magnitudes of these estimates are typical for the components of variances. Two of the components (V_D^* and V_{AD}) are relatively small and unlikely to be significant. Estimates of variances are more variable than those of means and the parameters are highly correlated, so it is very probable that we will find some of the genetic components non-significant, even when the overall genetic variation is highly significant, as in the present case. Clearly, as with generation means (Chapter 2), we need to select the best model which provides an adequate fit to the data, has all the estimated parameters significant and makes biological sense. This can be achieved by applying a modified form of the weighted least squares procedure described for fitting models to means. This modification is necessary in order to determine the appropriate weights for each variance which are not directly available as they are for means. With generation means, there are replicate observations of individual scores which provide an empirical estimate of the variance of the mean and hence an empirical weight. With variances, on the other hand, no such replicate variances are available, and hence, we do not have empirical weights. The theoretical variance of an observed variance (s^2) is equal to $2(\epsilon s^2)^2/df$, but the expected variance, ϵs^2, is not known. Hayman [3] solved this problem by using $df/2(s^2)^2$ as the initial weight in an iterative process. After the first iteration the parameter estimates are used to calculate expected variances which will approximate to the true values. These are then used to calculate new weights for a second iteration. This process is repeated through successive iterations until the test statistic, χ^2, reaches a minimum.

Table 3.3 Expectations of the within-family variances in terms of the additive dominance genetic and the additive environmental components of variation

Generation	V_E	V_A^*	V_D^*	V_{AD}
		Parameters		
P_1	1	0	0	0
P_2	1	0	0	0
F_1	1	0	0	0
F_2	1	1	1	0
$Bc_{1.1}$	1	$\frac{1}{2}$	1	-1
$Bc_{1.2}$	1	$\frac{1}{2}$	1	1

Model fitting in fact provides a comprehensive solution to all the problems that we have outlined earlier. Firstly, fitting a V_A^*, V_D^*, V_{AD} and V_E model (Table 3.3) to six variances yields a test of goodness of fit in the form of χ^2 for two degrees of freedom which also determines if a single V_E fits the variances of P_1, P_2 and F_1 generations. In other words, it establishes whether the variances of the non-segregating generations differ significantly or not. If they do not differ, fitting a single parameter V_E, tests if the six variances differ significantly, which becomes an unambiguous test of the genetic variation. We then proceed with the model fitting to obtain the best statistical *cum* biological model. This is illustrated in Table 3.4 for the data of Table 3.2 using model fitting with different combinations of parameters. The most appropriate model requires just two parameters, V_E and V_A^*, both of which are significant and the $\chi^2_{(4)}$ test of goodness of fit is non-significant. Alternatively, when the variances of P_1, P_2 and F_1 are shown to be heterogeneous, we need to replace V_E with three separate parameters, V_{E1}, V_{E2} and V_{E3}, to account for differences between them. Now, there are six parameters in the model and their contributions to the variances will be:

$$s_{P_1}^2 = V_{E1}$$

$$s_{P_2}^2 = V_{E2}$$

$$s_{F_1}^2 = V_{E3}$$

$$s_{F_2}^2 = V_A^* + V_D^* + \tfrac{1}{4}V_{E1} + \tfrac{1}{4}V_{E2} + \tfrac{1}{2}V_{E3}$$

$$s_{Bc_{1.1}}^2 = \tfrac{1}{2}V_A^* + V_D^* - V_{AD} + \tfrac{1}{2}V_{E1} + \tfrac{1}{2}V_{E3}$$

$$s_{Bc_{1.2}}^2 = \tfrac{1}{2}V_A^* + V_D^* + V_{AD} + \tfrac{1}{2}V_{E2} + \tfrac{1}{2}V_{E3}.$$

On many occasions we may find that only one of the P_1, P_2 or F_1 variances is significantly different from the other two. In this case,

Table 3.4 Results of model fitting to the variances of the six basic generations

Model	Parameters				Test of fit	
	V_E	V_A^*	V_D^*	V_{AD}	χ^2 (df)	P
V_E	75.0***	–	–	–	15.9 (5)	**
$V_E, V_A^*, V_D^*, V_{AD}$	59.1***	29.6 n.s.	12.1 n.s.	4.9 n.s.	1.4 (2)	n.s.
V_E, V_A^*, V_D^*	59.1***	29.6 n.s.	12.6 n.s.	–	1.7 (3)	n.s.
V_E, V_A^*	59.6***	46.3***	–	–	2.0 (4)	n.s.

we need to use only two V_E parameters. For example, if $s_{P_1}^2 = s_{P_2}^2 \neq s_{F_1}^2$, we can replace V_{E1} and V_{E2} with the pooled estimate V_{E12}, which will then take the coefficients of 1, 1, 0, $\frac{1}{2}$, $\frac{1}{2}$ and $\frac{1}{2}$ for the six generations in the order shown.

3.6 Heritability, h^2

Arguably one of the most useful statistics that can be derived from the variance components is the **heritability**. Generally, it is defined as the proportion of the total phenotypic variance of a population, in the present case an F_2 population, that is attributable to the effects of genes (i.e. heritable) and is represented by the symbol h^2.

Two types of heritability are used. The first is based on the ratio of total genetic variation to the total phenotypic variation and is called the **broad sense heritability**, h_b^2. For the F_2 generation, it is estimated as:

$$h_b^2(F_2) = V_G/(V_G + V_E) \qquad \text{[Eqn 3.2]}$$

or, when estimates of the various components are available,

$$h_b^2(F_2) = (V_A^* + V_D^*)/(V_A^* + V_D^* + V_E). \qquad \text{[Eqn 3.3]}$$

The second type of heritability is more important because it provides a measure of the breeding value of a population. It is called the **narrow sense heritability** (h_n^2) and measures the proportion of the variation which is due to the additive effects of genes in a specific population. For an F_2 population, it is estimated as:

$$h_n^2 = (V_A^*)/(V_A^* + V_D^* + V_E). \qquad \text{[Eqn 3.4]}$$

Using our example data, we can calculate both types of heritability using the above equations as well as some other methods. From the perfect fit values of V_A^*, V_D^* and V_E, $h_b^2 = 0.41$ and $h_n^2 = 0.29$. When calculated directly from the variances of the parental, F_1 and

F_2 generations $h_b^2 = (F_2 \text{ variance} - V_E)/F_2 \text{ variance} = 0.41$, which should be and is identical to the first estimate. Finally, from the WLS estimates, $h_n^2 = V_A^*/(V_A^* + V_E) = 0.44$. Clearly, all estimates are close and show that around 40% of the total phenotypic variation is heritable and the remaining 60% is environmental or non-heritable. However, of all these estimates which is the most reliable? The answer is the one that we have obtained from the WLS estimates because it involves only those components that are shown to be significant statistically.

3.7 Relationships between [a] and V_A^*, and [d] and V_D^*

It is evident from the definition of V_A^* that its magnitude is not affected by gene dispersion. Further, being a component of variance, it is orthogonal to [a] and therefore should not be correlated with [a]. Consequently, we can expect no systematic relationship between [a] and V_A^*. For instance, [a] can be small (or zero) due to gene dispersion while V_A^* is large, as is the case in our example. Similarly, [a] can be large while V_A^* is non-significant (statistically zero), particularly when the effects of individual genes are very small. The latter case is demonstrated by a simple example of an F_2 derived from a cross between two parents that show complete association ($r_a = 1$) and differ at $k = 100$ genes. Assuming for simplicity that $a_A = a_B = \ldots = a_K = 0.25$, the least squares estimates of [a] and V_A^* from this cross will be approximately 25 and 3.125 respectively, and it will be difficult to detect V_A^* as significant unless V_E is very small and the heritability is almost absolute. Had the same [a] ($= 25$) been due to $k = 25$ genes, with effect $a = 1$, then V_A^* would have been much larger (12.5).

While the above unpredictable relationship between the components [a] and V_A^* will hold for individual crosses, these components will generally be positively correlated when the results of several crosses are interpreted simultaneously. This happens primarily because the coefficient of gene association/dispersion r_a, gene number k and gene effects $a_A, \ldots, a_K$, etc. are rarely the same between crosses, particularly when the parents are very diverse. Thus, V_A^* can take a wider range of values when r_a is small than when r_a is large, and this produces a positive correlation over crosses.

The relationship between [d] and V_D^* will also be similar except that the significance and magnitude of [d] is affected by ambi-directional dominance and not gene dispersion. Detection of V_D^* is difficult because basic generations are inefficient for determining dominance variance. Nevertheless, one of the following interpretations will apply:

[d]	V_D^*	Interpretation
significant, +ve	significant	directional dominance for increasing alleles
significant, −ve	significant	directional dominance for decreasing alleles
non-significant	significant	ambi-directional dominance
non-significant	non-significant	no dominance
significant, small	non-significant	low dominance

3.8 Dominance ratio

From a breeder's point of view, the relative values of the additive and the dominance effects are important because the information can be used to decide whether to choose inbred lines or F_1 hybrids as commercial varieties. An important statistic that is derived from the additive genetic and the dominance components of variation is called the **dominance ratio** and it is estimated as $\sqrt{\{4V_D^*/2V_A^*\}}$, (i.e. $\sqrt{\{\sum d^2 / \sum a^2\}}$). This ratio takes a minimum value of zero when dominance is zero and its maximum value can be larger than one but only in exceptional cases. When, as must often be the case, the degree of dominance is not constant across genes, $\sqrt{\{4V_D^*/2V_A^*\}}$ provides an **average dominance ratio** because it represents the weighted average of the dominance ratios of the various loci. For example, for a two gene case, assuming that $a_A = 2$, $a_B = 4$, $d_A = 1$ and $d_B = 3$, the two dominance ratios of 0.5 and 0.75 respectively are obtained for genes A and B while, overall, $\sqrt{\{4V_D^*/2V_A^*\}}$ takes a value of $\sqrt{\{(1 + 9)/(4 + 16)\}} = 0.71$ which is much closer to the dominance ratio of the gene with larger effect, B. In general, $\sqrt{\{4V_D^*/2V_A^*\}}$ is used to determine the importance of dominance effects in relation to the additive deviations of genes and is interpreted as partial, complete or over-dominance when the ratio is <1, 1 or >1 respectively.

Finally, in our example we have been unable to determine the dominance ratio because the dominance component is not significant. This situation is not unusual because most biometrical designs are rather inefficient at detecting dominance variance, unless the dominance component takes a large value. However, using the perfect fit estimates calculated earlier, we obtain

$$\sqrt{\{4V_D^*/2V_A^*\}} = \sqrt{\{48.2/59.24\}} = 0.9.$$

3.9 Variances and means

The information which can be obtained from the means of the six basic generations complements that provided by the variances. Generation means provide very sensitive tests for the presence of complicating factors such as epistasis and maternal effects, while the variances provide tests for genotype by environment interaction. Means tell us little about the amount of dominance present and it is necessary to turn to variances to estimate the dominance ratio. On the other hand, the parameter estimates obtained from means indicate the direction of gene action, such as dominance, which can be obtained from variances only by a comparison of the backcross variances. It may sometimes appear that the means and variances are indicating quite different types of genetic control, but this is illusory. Because of the problems of gene dispersion [a] may be small, while the additive genetic variance, V_A^*, which is not affected by dispersion, may be large and significant. Conversely, [d], which is independent of gene dispersion may appear large, particularly with directional dominance, while V_D^* may fail to reach significance.

3.10 Conclusions from the analysis of basic generations

Analysis of the present data set has shown the following. There is no evidence for epistasis or genotype by environment interaction. The additive effect of means [a] is small but that of variance $[V_A^*]$ large, suggesting gene dispersion. Dominance is for high score because [d] is positive and the variance of $Bc_{1.2}$ is greater than the variance of $Bc_{1.1}$. Although the dominance variance was not significant the dominance ratio was actually large, 0.9. Some 40% of the F_2 variance was under genetic control.

Summary

1. Basic generations provide estimates of all the genetic and environmental components of variances which can be used to determine the extent of genetic variability prevailing in a cross and to interpret the relative importance of the additive *versus* the non-additive variation.

continued

2. Comparisons between the components of means and of variances show that the two provide complementary information concerning the genetical control. Variances reveal the true nature of the genetic variation and together with means provide indications of the breeding potential of the source material.
3. The basic generations do, however, provide an inefficient design for obtaining reliable estimates of the components of variances, particularly of V_D^*, and more efficient designs are discussed in later chapters.

References

1. Bartlett, M.S. (1937) Some examples of statistical methods of research in agriculture and applied biology. *J. R. Statist. Soc. Suppl.*, **4**, 137–83.
2. Levene, H. (1960) Robust tests for equality of variance, in *Contributions to Probability and Statistics* (eds I. Olkin, S.G. Ghurye, W.G. Madow and H.B. Mann), Stanford University Press, Stanford, pp. 278–92.
3. Hayman, B.I. (1960) Maximum likelihood estimation of genetic components of variation. *Biometrics*, **16**, 369–81.

Further reading

Falconer, D.S. (1989) *Introduction to Quantitative Genetics*, 3rd edn, Longman, UK.

Selfing and full-sib mating

4

So far we have introduced the ideas behind the derivation of models to explain the means and variances of inbred lines, their F_1s, F_2s and first backcrosses, and shown how the data from these generations can be used to understand the genetical control of the trait. In this chapter, we look at generations which can be obtained by inbreeding the F_2, either by selfing or sib-mating.

These generations could occur within the natural breeding system of many species but, more importantly, they are generations which are normal components of plant and animal breeding programmes. Inbred lines are used as cultivars in their own right in crops such as wheat and rice and as the parents of F_1 hybrids, as in maize. In animals, highly inbred lines of mice and rats are used for medical purposes while animal breeders use controlled inbreeding to produce uniform strains of poultry and pigs.

Inbreeding can be achieved in various ways depending on the material available and the biology of the species, and involves matings between close relatives. Self-fertilization is the most extreme and rapid form of natural inbreeding, which requires self-compatible, hermaphrodite species. It is common in many plants and may be enforced in a wide range of crops. Mating of full-sibs, i.e. brother × sister, is the next most efficient system, taking approximately three times as long as self-fertilization to attain the same degree of homozygosity. However, it is the quickest method in animals and in fully self-incompatible plants, otherwise the breeder is forced to consider half-sib matings. It is also possible in many plant species to produce a collection of completely homozygous genotypes in a single generation by *in vitro* methods or by wide species crosses. They are generated from plants derived from male or female gametes and are known as **doubled haploid** (DH) lines. Their uses, together with their advantages and disadvantages over lines produced by conventional inbreeding, are considered in this chapter. The methods used to produce them are, however, beyond the scope of this book [1].

Because these inbreeding methods are so commonly used, we will now consider their consequences on the means and variances of a quantitative trait. This will provide both a method of obtaining useful information about the genetic control of the trait concerned and also a formal structure upon which to base improvement programmes. We will concentrate on selfing and full-sib mating because they are the most common types of inbreeding practised.

4.1 Selfing: F₃ families

If we take a random sample of n F_2 individuals and self-pollinate each we will obtain seed of n F_3 families. Suppose that r plants from each F_3 family are grown in a completely randomized trial and scored for a quantitative trait, then we have,

Phenotype of F_2 parent	Phenotype of F_3 progeny	F_3 Family mean	SS (r − 1) df
$X_1 \Rightarrow$	$Y_{11}, Y_{12}, \ldots, Y_{1r}$	Y_1	SS_1
$X_2 \Rightarrow$	$Y_{21}, Y_{22}, \ldots, Y_{2r}$	Y_2	SS_2
.	.	.	.
$X_i \Rightarrow$	$Y_{i1}, Y_{i2}, \ldots, Y_{ir}$	Y_i	SS_i
.	.	.	.
$X_n \Rightarrow$	$Y_{n1}, Y_{n2}, \ldots, Y_{nr}$	Y_n	SS_n

Such data can be analysed by a one-way ANOVA to test for differences between F_3 families. The skeleton ANOVA of such a data set is given in Table 4.1(i) and an actual analysis, based on n = 50 families of size r = 5, is presented in Table 4.1(ii). It shows that there are significant differences between families and, since the experiment was completely randomized, we can assume that these differences are genetic and not environmental. In order to clarify the nature of the genetic cause, we need to derive the expectations of the σ^2 in the ANOVA table.

Starting as before with a single gene model, the derivation of the σ^2s is illustrated in Table 4.2. The F_2 parent can be one of three genotypes, A^+A^+, A^+A^-, A^-A^- with frequencies $\frac{1}{4}$, $\frac{1}{2}$, and $\frac{1}{4}$. Selfing these produces P_1, F_2 and P_2 families respectively, the means and variances of which were derived in Chapters 2 and 3 and are given in Table 4.2.

Table 4.1 Analysis of F_3 families. (i) Skeleton ANOVA; (ii) ANOVA of n = 50 families of size r = 5; (iii) parameter estimates

(i)

Source	df	MS	ems
Between families	n − 1	MS_B	$\sigma_W^2 + r\sigma_B^2$
Within families	n(r − 1)	MS_W	σ_W^2

(ii)

Source	df	MS	F	ems
Between families	49	204.3790	2.87**	$\sigma_W^2 + 5\sigma_B^2$
Within families	200	71.2574		σ_W^2

Mean of F_2s = 73.49
Mean of F_3s = 69.80

(iii)

$$\sigma_B^2 = (MS_B - MS_W)/r$$
$$\sigma_W^2 = MS_W$$
$$V_A^* = \sigma_B^2; \quad V_E = \sigma_W^2 - \tfrac{1}{2}V_A^*$$
$$V_A^* = 26.6243 \pm 8.3802; \quad c = 3.18^{***}$$
$$V_E = 57.9452 \pm 8.8594; \quad c = 6.54^{***}$$
$$h^2 = V_A^*/(V_A^* + V_E) = 0.31$$

The parameter σ_W^2 is the expected, average variance within F_3 families and is, therefore,

$$\sigma_W^2 = (\tfrac{1}{4})V_E + (\tfrac{1}{2})(\tfrac{1}{2}a^2 + \tfrac{1}{4}d^2 + V_E) + (\tfrac{1}{4})V_E$$
$$= \tfrac{1}{4}a^2 + \tfrac{1}{8}d^2 + V_E.$$

Similarly, σ_B^2 is the expected variance of the F_3 family means, of which there are only three types for a single gene, P_1 with a mean $m + a$, F_2 with a mean $m + \tfrac{1}{2}d$ and P_2 with a mean $m - a$. Ignoring the constant term m, the variance of these three means, given their frequencies is,

$$\sigma_B^2 = \tfrac{1}{4}a^2 + \tfrac{1}{2}(\tfrac{1}{2}d)^2 + \tfrac{1}{4}(-a)^2 - (\tfrac{1}{4}d)^2$$
$$= \tfrac{1}{2}a^2 + (\tfrac{1}{16})d^2.$$

So far we have only derived the mean and variances of the F_3 for a single gene, A^+/A^-, but, providing the other genes are independent of each other in action (i.e. no epistasis) and in distribution (i.e. no linkage) the total genetic variance due to all genes is simply the sum

Table 4.2 Derivation of mean and σ^2s for an F_3

Genotype of F_2 parent	Freq. (f_i)	Type of family	Genotypes in F_3 family	Mean of F_3 family (y_i)	Variance of F_3 family (σ^2_{yi})
A^+A^+	$\frac{1}{4}$	P_1	A^+A^+	$m + a$	V_E
A^+A^-	$\frac{1}{2}$	F_2	$\frac{1}{4}(A^+A^+) : \frac{1}{2}(A^+A^-) : \frac{1}{4}(A^-A^-)$	$m + \frac{1}{2}d$	$\frac{1}{2}a^2 + \frac{1}{4}d^2 + V_E$
A^-A^-	$\frac{1}{4}$	P_2	A^-A^-	$m - a$	V_E

$$\bar{F}_3 = \sum f_i y_i$$
$$= \tfrac{1}{4}(m + a) + \tfrac{1}{2}(m + \tfrac{1}{2}d) + \tfrac{1}{4}(m - a)$$
$$= m + \tfrac{1}{4}d$$
$$= m + \tfrac{1}{4}[d] \text{ for many genes.}$$

$$\sigma^2_B = \sum f_i y_i' - \left(\sum f_i y_i'\right)^2 \text{ where } y_i' = y_i - m$$
$$= \tfrac{1}{4}(a)^2 + \tfrac{1}{2}(\tfrac{1}{2}d)^2 + \tfrac{1}{4}(-a)^2 - (\tfrac{1}{4}d)^2$$
$$= \tfrac{1}{2}a^2 + \tfrac{1}{16}d^2$$
$$= V_A^* + \tfrac{1}{4}V_D^* \text{ for many genes.}$$

$$\sigma^2_W = \sum f_i \sigma^2_{yi}$$
$$= \tfrac{1}{4}(V_E) + \tfrac{1}{2}(\tfrac{1}{2}a^2 + \tfrac{1}{4}d^2 + V_E) + \tfrac{1}{4}(V_E)$$
$$= \tfrac{1}{4}a^2 + \tfrac{1}{8}d^2 + V_E$$
$$= \tfrac{1}{2}V_A^* + \tfrac{1}{2}V_D^* + V_E \text{ for many genes.}$$

of their individual variances. Thus,

$$\sigma^2_W = \tfrac{1}{4}\sum a^2 + \tfrac{1}{8}\sum d^2 + V_E$$
$$= \tfrac{1}{2}V_A^* + \tfrac{1}{2}V_D^* + V_E \qquad \text{[Eqn 4.1]}$$
$$\sigma^2_B = \tfrac{1}{2}\sum a^2 + \tfrac{1}{16}\sum d^2$$
$$= V_A^* + \tfrac{1}{4}V_D^* + V_{EC} \qquad \text{[Eqn 4.2]}$$

while the mean $= m + \tfrac{1}{4}\sum d = m + \tfrac{1}{4}[d]$.

It is not possible to estimate all the genetical and environmental components of variation as we have four parameters, V_A^*, V_D^*, V_{EC} and V_E but only two statistics, σ^2_B, σ^2_W. It is therefore necessary to ignore two parameters and, because there is bound to be environmental variation (V_E), V_D^* and V_{EC} are set to zero. This may be an unreasonable assumption for which we have no evidence, but it will not seriously mislead us since the contribution of dominance effects, in general, will be small. This can be shown by the following argument. If the dominance ratio, d/a at each locus is a constant, f, i.e. $d = fa$, then,

$$V_D^* = \tfrac{1}{4}\sum d^2 = \tfrac{1}{4}\sum f^2 a^2,$$

and, given that the coefficient of V_D^* in the σ^2s is small, $\frac{1}{4}$ and $\frac{1}{2}$ as in the F_3, the dominance contribution will also be small. Even if there is complete dominance, $f = 1$, there is an eight-fold greater contribution of V_A^* than V_D^* in σ_B^2, while with $f = 0.7$, this difference increases by a further factor of two. So, ignoring dominance will cause little bias to our estimates unless $\sum d^2$ is very much greater than $\sum a^2$. Moreover, ignoring dominance also considerably increases the precision of the estimate of V_A^* [2,3].

The contribution of V_{EC}, however, can only be determined by independent experiments (see Chapter 13). Experience shows that it is usually negligible in plant experiments and we shall ignore it for the purposes of simplicity.

Table 4.1(iii) shows the estimates of V_A^* and V_E together with the estimate of heritability which, since it includes just the additive genetic variance, is the narrow-sense heritability. These estimates are very close to those obtained from the basic generations in Chapter 3. The method of obtaining the estimates and their standard errors by weighted least squares will be explained at the end of this chapter.

4.2 Selfing: F_4 families and beyond

Before considering how the information from F_3s can be used, it is appropriate to develop the analysis of the F_4 generation because it reveals a pattern in the means and variances that will enable us to handle any generation of selfing from an F_2.

Suppose that we selfed an equal number of plants, n', from each of the n F_3 families and raised r progeny from each in a completely randomized experiment, i.e. there are $nn'r$ plants altogether. The analysis of such data results in a 'nested' or 'hierarchical' ANOVA as we have F_4s, within F_3s, within F_2s as shown in Figure 4.1, and the basic ANOVA for an F_4 derived from n F_2s and n' F_3s per F_2 is shown in Table 4.3. The computations for the ANOVA of a small data set of a total $r = 5$ F_4 individuals from each of $n' = 2$ F_3 plants from $n = 3$ F_3 families is illustrated in Table 4.4, while the ANOVA of a much larger data set with $r = 5$, $n' = 2$, $n = 100$ is shown in Table 4.5.

The ANOVA has three levels, as can be seen from the relationships in Figure 4.1. The first, σ_1^2, represents the variation between the means of all those F_4 individuals which trace back to single F_2s; because there were n F_2s there are n such means and the MS has $n - 1$ df. The second, σ_2^2, relates to the means of all those F_4s that come from a given F_3, and measures the average variance of

Figure 4.1 F_4 families and their hierarchical structure in an inbreeding series. The sources of the variances in the ANOVA in Table 4.5(i) are shown.

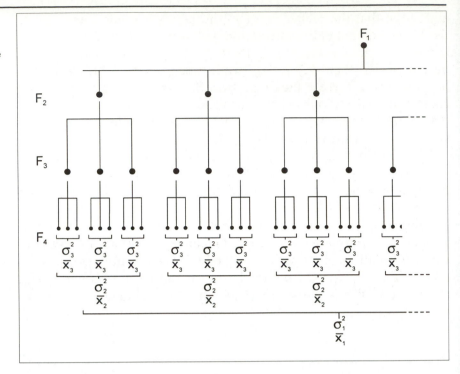

such means within each F_3 group; since there were $n'F_3$s selfed in every family, this item has $n(n'-1)$ df. Thirdly, σ_3^2, is the residual variation among F_4 individuals which have the same F_3 parent; this has $r-1$ df for every F_3, i.e. $nn'(r-1)$ df in total.

The expectations of these three σ^2s are:

$$\sigma_{1F4}^2 = V_A^* + \tfrac{1}{16} V_D^*$$

$$\sigma_{2F4}^2 = \tfrac{1}{2} V_A^* + \tfrac{1}{8} V_D^*$$

$$\sigma_{3F4}^2 = \tfrac{1}{4} V_A^* + \tfrac{1}{4} V_D^* + V_E.$$

Table 4.3 General ANOVA of a nested experiment involving F_4 families. It involves F_4 from n' F_3 individuals, from each of n F_2 with r individuals in each F_4 family

Source	df	MS	ems
Between F_2 groups	$n-1$	MS_1	$\sigma_3^2 + r\sigma_2^2 + n'r\sigma_1^2$
Between F_3 groups within F_2 groups	$n(n'-1)$	MS_2	$\sigma_3^2 + r\sigma_2^2$
Between F_4 individuals within F_3 groups	$nn'(r-1)$	MS_3	σ_3^2
Total	$nn'r-1$		

Table 4.4 Analysis of F_4 data based on three F_3 families. (i) Basic data; (ii) ANOVA; (iii) parameter estimates

(i)

F_2	1		2		3	
F_4\F_3	1	2	1	2	1	2
1	73.14	60.14	82.69	84.24	73.96	61.78
2	68.52	67.16	86.89	74.23	88.20	71.56
3	82.09	75.45	89.89	77.04	63.99	73.05
4	76.78	56.55	63.62	71.68	75.73	65.92
5	71.07	71.91	78.74	84.15	67.10	69.38
SS (4df)	112.0194	242.6634	421.7425	131.4779	352.2365	82.5241
Mean of F_4 family	74.320	66.250	80.366	78.268	73.796	68.338
Mean of all F_4 from each F_2	70.285		79.317		71.067	

(ii)

Source	df	SS	MS	F	P	ems
Between F_2 groups	2	500.8368	250.4184	3.03 / 4.48	n.s. / *	$\sigma_3^2 + 5\sigma_2^2 + 10\sigma_1^2$
Between F_3 groups within F_2 groups	3	248.2907	82.7636	1.48	n.s.	$\sigma_3^2 + 5\sigma_2^2$
Between F_4 individuals within F_3 groups	24	1342.6638	55.9443			σ_3^2
Total	29	2091.7913				

(iii)

	Full model				Reduced model		
Parameter	Estimate	SE	c	Parameter	Estimate	SE	c
V_A^*	18.78	38.00	0.49	V_A^*	14.25	16.06	0.89
V_D^*	−32.20	226.93	0.14	−			
V_E	59.30	53.46	1.11	V_E	52.07	17.03	3.06
				$\chi_{(1)}^2$		0.02 n.s.	

The first subscript after the σ^2 indicates the 'rank' of the statistic; rank 1 always relates back to an F_2 individual, rank 2 to an F_3 within an F_2, and rank 3 to an F_4 within an F_3, and so on. The expectations of these σ^2s are shown for each rank and generation in Table 4.6, from which it can be seen that as we move from generation n to $n+1$ within any one rank, the coefficient of V_A^* stays constant but that of V_D^* is reduced by a quarter. Also, the genetical component of the first value in any row is exactly half that of the first

Table 4.5 Analysis of large data set of F_4 families, based on 100 F_2s, two F_3s per F_2 and five F_4s per F_3. (i) ANOVA; (ii) parameter estimates

(i)

Source	df	MS	F	P	ems
Between F_2 groups	99	401.00	2.29	***	$\sigma_3^2 + 5\sigma_2^2 + 10\sigma_1^2$
Between F_3 groups within F_2 groups	100	175.00	2.72	***	$\sigma_3^2 + 5\sigma_2^2$
Between F_4 individuals within F_3 groups	800	64.28			σ_3^2
Total	999				

(ii)

	Full model				Reduced model		
Parameter	Estimate	SE	c	Parameter	Estimate	SE	c
V_A^*	15.37	10.07	1.5 n.s.	V_A^*	31.92	4.47	7.1***
V_D^*	115.67	72.98	1.6 n.s.	–			
V_E	31.52	16.62	1.9 n.s.	V_E	56.91	3.54	16.1***
				$\chi_{(1)}^2$	3.14 n.s.		

value in the previous row, i.e. as selfing proceeds, the genetical variation within families halves. Given this pattern, it is easy to derive the expectations for any further generations such as F_5, F_6, etc., as shown in Table 4.6.

Returning to the F_4, there are now three statistics and three unknown parameters, so it is possible, from the ANOVA in Table 4.4, to estimate the three σ^2s. Then, by equating the estimates of the σ^2s to their expected values, V_A^*, V_D^* and V_E can be estimated from the formulae above. These estimates are given in Table 4.4(iii), together with their standard errors. The data set used for illustration in Table 4.4 is, however, unrealistically small, therefore the corresponding ANOVA of F_4s derived from 100 F_2s is given in Table 4.5(i).

Table 4.6 Genetical and environmental components of σ_{iFn}^2 for the first four generations of the selfing series

Rank		Generation (n)		
(i)	F_2	F_3	F_4	F_5
1	$V_A^* + V_D^* + V_E$	$V_A^* + \frac{1}{4}V_D^*$	$V_A^* + \frac{1}{16}V_D^*$	$V_A^* + \frac{1}{64}V_D^*$
2		$\frac{1}{2}V_A^* + \frac{1}{2}V_D^* + V_E$	$\frac{1}{2}V_A^* + \frac{1}{8}V_D^*$	$\frac{1}{2}V_A^* + \frac{1}{32}V_D^*$
3			$\frac{1}{4}V_A^* + \frac{1}{4}V_D^* + V_E$	$\frac{1}{4}V_A^* + \frac{1}{16}V_D^*$
4				$\frac{1}{8}V_A^* + \frac{1}{8}V_D^* + V_E$

We see from Table 4.5(ii) that, although V_A^* and V_E are both significant when compared with their standard errors (c = 2.7 and 8.0), the estimate of V_D^* is not significant (c = 1.6). This is not unusual in an analysis of selfed generations because selfing decreases the frequency of heterozygotes and it is the heterozygotes which exhibit dominance. On the other hand, V_A^* can be estimated with good precision as shown by its small standard error. The selfing series does not provide a useful data set for studying dominance but it does provide reliable estimates of the additive genetical variation, V_A^*. Again, the values of V_A^* and V_E are consistent with those obtained from the basic generations.

Consecutive selfing as described above leads to a random sample of pure breeding lines called F_∞ lines. DH lines produced from the F_1 are equivalent to F_∞ lines in their mean and variance providing that the method of production has not directly affected the material and that there is no linkage of the genes governing the trait. The effect of linkage will be considered in Chapter 10, but the essential point to note here is that while F_∞ lines will display higher levels of recombination because there will be some chance of recombining during each round of inbreeding, albeit to a diminishing extent as the lines become more homozygous, DH lines result from just one round of recombination. Thus, F_∞ lines produced by selfing may produce more transgressive segregation than DH lines.

So far we have simply considered selfing from an F_2 but we could develop the same approach to inbreeding from a backcross. For example, inbreeding a random sample of $Bc_{1.1}$ individuals will yield a set of recombinant backcross lines with a mean $m + \frac{1}{2}[a]$ and variance $1\frac{1}{2}V_A^*$. Similarly, recombinant inbred lines derived from $Bc_{1.2}$ will have a mean and variance of $m - \frac{1}{2}[a]$ and $1\frac{1}{2}V_A^*$. Further elaborations of inbreeding from these and other sources are provided in Chapter 15.

4.3 Applications of selfing theory

How does a knowledge of the formulae we have derived for means and variances help us in genetical analysis or breeding? Most programmes in plant breeding are based on looking for transgressive segregants among the population of recombinant inbred lines derived by several generations, generally 6–10, of selfing from an F_1 hybrid. The way in which this is achieved varies. The breeder may collect seed separately from each F_2 to produce F_3 families called ear rows and continue to keep each F_2 pedigree distinct throughout successive generations. This is commonly known as pedigree inbreeding. The most extreme form of this is **single seed descent** (SSD) in which just one F_3 is raised from each F_2, one F_4 from each F_3 and so on. Thus by some later generation, e.g. F_6, there will be as many F_6 families as there were F_2 individuals at the start, and only one F_6 family has descended from each F_2 individual.

Pedigree and SSD methods require more effort and book-keeping, particularly in the early generations, thus a common alternative is to proceed from the F_2 to the F_4 by 'bulking' the seed so that it is not possible to tell which seeds are derived from which F_2 individual, and then to keep the lines distinct from the F_4 onwards. This may be technically simpler to handle as it does not require keeping the families distinct, but it can result in considerable loss of variability. For example, suppose seed is collected from 100 F_2 plants, bulked and 100 F_3 plants are grown the following season. On average, each F_2 will have produced one F_3 individual, but some will have no representative in the F_3, while others will have two or more. The distribution of family sizes will follow a Poisson distribution with mean, $\mu = 1.0$. The first term of this distribution, that is the proportion of F_2 which leave no F_3 representatives, is $e^{-\mu}$. But, $e^{-1} = 0.37$ and hence, of the original 100 F_2s, only 63, on average, will have F_3 progeny actually raised in the experiment. Were this procedure to continue to the F_4, there would, on average, be descendants from no more than 40 of the original F_2s, and therefore much of the variability will have been lost. Moreover, the plants that do contribute descendants to the sample are most likely to be those that produced seeds in large numbers.

Often, the breeder will attempt to select superior types from among the F_2 plants in the plot and continue selecting throughout subsequent generations. As we will show in Chapter 15, the effect of selection is directly related to the proportion of additive genetic variance in the population for the trait under selection, i.e. the narrow heritability, h_n^2. To be more precise, the response to selection (R) is a function of the additive genetic variation divided by the phenotypic standard deviation (σ_p) of the specific population under study; $R = ih_n^2\sigma_p = i\sigma_A^2/\sigma_p$, where i is the intensity of selection.

For many traits of agronomic importance, the narrow heritability, when measured on an individual basis in an F_2, is less than 0.4 and, for yield-related traits in particular, often less than 0.2. Thus most of the variation is not additive and hence selection will have very little effect; the breeder will be selecting largely for differences which are due to the micro-environment and non-additive genetical variation which will not produce a desired shift in the performance of the progeny. However, if selection is delayed until the F_3, and the breeder selects on family means, the selection is much more effective because the environmental variation is reduced by working with means and the non-additive variation is reduced by inbreeding.

This is best illustrated with an example. Consider a trait with a narrow heritability in the F_2 of 0.2 and a phenotypic variance (σ_P^2) of 1.0 (although the actual value of σ_P^2 does not affect the argument). For a given intensity of selection, i, response will be proportional to $h_n^2 \sigma_p$, or 0.2 in this case. Now, suppose we were to grow F_3 families of size 20 and base our selection on these family means. The observed variance of family means would be the 'Between families' MS in Table 4.3 divided by the family size, i.e. 20. The expected variance 'between family means' is, therefore, the *ems* from the ANOVA divided by 20, i.e.:

$$\{\sigma_W^2 + 20\sigma_B^2\}/20 = \{(\tfrac{1}{2}V_A^* + \tfrac{1}{2}V_D^* + V_E) + 20(V_A^* + \tfrac{1}{4}V_D^*)\}/20$$
$$= 1.025V_A^* + 0.275V_D^* + 0.05V_E$$

The narrow heritability of F_3 family means is therefore,

$$1.025V_A^*/(1.025V_A^* + 0.275V_D^* + 0.05V_E)$$

and its actual size would depend on V_D^* and V_E.

In the present illustration V_A^* is 0.2, so with complete dominance, $V_D^* = 0.1$. Because the phenotypic variance of the F_2 was set to unity,

$$\sigma_P^2 = V_A^* + V_D^* + V_E = 1.0$$

and hence, $V_E = 0.7$. Using these values, we find that $h_n^2 \sigma_p$ for the F_3 family means is 0.4, i.e. twice that of the F_2.

Similar calculations for traits with various heritabilities and dominance ratios are given in Table 4.7. The effect of selecting on family means clearly becomes more marked the lower the F_2 heritability, while the degree of dominance, on the other hand, hardly affects the result. This suggests that it might be worth delaying selection even further into the inbreeding programme, and this is indeed the case as is also shown in Table 4.7 for F_∞ lines, where the gain in response to selection varies from a factor of 2 at heritabilities of 0.5 to a factor of over 5 when the narrow heritability of the F_2 is only 0.05.

4.4 Variation between inbred lines derived from an F_2

The derivation of the F_∞ variance above does, perhaps, require some explanation. It can easily be obtained using the *ems* set out in Table 4.3 together with the expectations of the σ^2 set out in

Table 4.7 Response (R) to selection on F_3 or F_∞ family means relative to selecting within an F_2, based on family sizes in the F_3 and F_∞ of 20. $R = ih_n^2\sigma_P$

Heritability of F_2	Dominance ratio	Response relative to F_2 for:	
		F_3	F_∞
0.5	1.0	1.35	1.99
	0.0	1.41	1.98
0.3	1.0	1.69	2.52
	0.0	1.77	2.51
0.1	1.0	2.60	4.06
	0.0	2.71	4.04
0.05	1.0	3.24	5.23
	0.0	3.32	5.21

Table 4.6. Again, an illustration may help and we will use the F_4 ANOVA given earlier (Table 4.3) for this purpose.

Suppose we had produced only one F_3 from each F_2, so that $n' = 1$. The analysis would then lose the item 'Between F_3 groups, within F_2s', and the *ems* for the MS 'Between F_2 groups' would become,

$$\sigma_3^2 + r\sigma_2^2 + r\sigma_1^2.$$

Hence the expected variance of F_4 family means will be this MS divided by r, i.e.

$$(\sigma_1^2 + \sigma_2^2) + \sigma_3^2/r.$$

If we were to carry out SSD to the F_5 and raise F_6 families then, by the same argument, the expected variance of F_6 family means would be,

$$(\sigma_1^2 + \sigma_2^2 + \sigma_3^2 + \sigma_4^2) + \sigma_5^2/r.$$

The genetical expectations of the σ^2s are given in Table 4.6, from which it can be seen that the sum of the terms in brackets above approach the value $2V_A^*$ as the generations of inbreeding increase, and by F_8 it will have reached $1.96V_A^*$ which is effectively equivalent to F_∞. This explains the advantage of selecting at F_∞ because we are selecting on family means, the variance between these means being almost entirely additive genetic, irrespective of the variance of the F_2.

Had the seed been bulked at each generation until sown in a trial, the variance observed would be that between individual F_n plants, i.e.

$$\sigma_1^2 + \sigma_2^2 + \sigma_3^2 + \ldots + \sigma_{n-1}^2.$$

This would approximate to $2V_A^* + V_E$ as n approaches infinity and so, although the response to selection would increase, it would only do so to a small extent because of the large contribution of V_E to the phenotypic variance.

Using the information in Table 4.6 we can easily derive the expected variance of any type of selfed population starting from an F_1. Unless the heritability is high, selecting in the F_2 gives very little return on effort as indeed does inbreeding with bulking of seed. However, pedigree inbreeding, particularly SSD, followed by selection on family means can improve response by up to a factor of 5 or more [4].

The changes in mean and genetic variance with selfing are illustrated in Figure 4.2 for a trait exhibiting heterosis. This shows that although the mean declines, the variance among lines increases reaching a maximum at F_∞. Therefore, although there is inbreeding depression with respect to the mean, some of the inbred lines produced may still be as good as the F_1 hybrid, if not better. We will return to this point in Chapter 15. Inbreeding does, of course, reduce the genetical variation within lines and this will reach zero at F_∞.

There is another reason for a breeder carrying out selection at the F_3 or later generation and that is because it is possible to raise families in experimental plots. Selecting within a segregating population of plants such as an F_2 normally requires that the material is sown at lower than normal densities in order for the breeder to have access to the individual plants. This changes the environment from that in which the final pure stand will be grown commercially and, as a result, selection could well favour the wrong genotypes. Plots overcome this problem to some extent in that they attempt to simulate agronomic practice. This topic is discussed further in Chapter 16.

4.5 Sib-mating an F_2

The alternative to selfing the F_2 is to sib-mate the individuals. For the first generation this is equivalent to randomly mating the F_2 individuals but, in successive generations, it involves random mating within families. As before, we will start with the F_2.

Let us take a random sample of 2n individuals from an F_2 and cross them together in pairs to produce n families, and from each family r replicate individuals are raised in a completely randomized experiment of total size, nr. Such families are referred to as full-sib (FS) or bi-parental (BIP) families.

Figure 4.2 Inbreeding often results in a decline in the mean but also releases genetic variation. Thus, although the final mean of all the inbred lines will be lower than the F_1 there may well be lines which show transgressive segregation and exceed the F_1. (Experimental data from *Nicotiana*.)

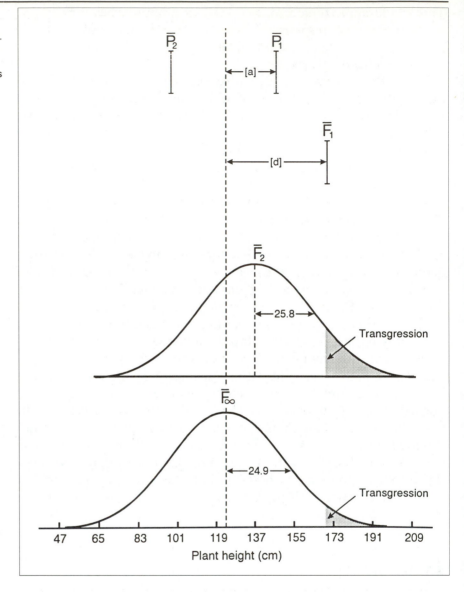

Figure 4.2 Inbreeding often results in a decline in the mean but also releases genetic variation. Thus, although the final mean of all the inbred lines will be lower than the F_1 there may well be lines which show transgressive segregation and exceed the F_1. (Experimental data from *Nicotiana*.)

Apart from the bi-parental origin of these families, the structure of the experiment is exactly the same as for the F_3s discussed earlier in this chapter and the ANOVA will therefore be as shown in Table 4.1(i). In order to determine the causal components of the σ^2 in this ANOVA, we need to consider the types of family that are produced for a single gene case, as shown in Table 4.8. Randomly mating an F_2 results in producing families of the six basic generations, P_1, P_2, F_1, F_2, $Bc_{1.1}$ and $Bc_{1.2}$, the means and variances of which were derived previously and whose frequencies are as shown in Table 4.8(i) and summarized in Table 4.8(ii) from which the genetical expectations of

Table 4.8 Sib-mating an F_2 population. (i) Structure of families; (ii) summary of means and variances; (iii) derivation of covariance of offspring onto mid-parental values

(i)

Female parent	Frequency		Male parent A^+A^+ $\frac{1}{4}$	A^+A^- $\frac{1}{2}$	A^-A^- $\frac{1}{4}$	Row means (= HS means)
A^+A^+	$\frac{1}{4}$	Family	P_1	$Bc_{1.1}$	F_1	
		(Freq.)	$(\frac{1}{16})$	$(\frac{1}{8})$	$(\frac{1}{16})$	
		Family mean	$m+a$	$m+\frac{1}{2}(a+d)$	$m+d$	$m+\frac{1}{4}d+\frac{1}{2}a$
		σ_W^2	V_E	$\frac{1}{4}(a-d)^2+V_E$	V_E	
A^+A^-	$\frac{1}{2}$	Family	$Bc_{1.1}$	F_2	$Bc_{1.2}$	
		(Freq.)	$(\frac{1}{8})$	$(\frac{1}{4})$	$(\frac{1}{8})$	
		Family mean	$m+\frac{1}{2}(a+d)$	$m+\frac{1}{2}d$	$m+\frac{1}{2}(-a+d)$	$m+\frac{1}{2}d$
		σ_W^2	$\frac{1}{4}(a-d)^2+V_E$	$\frac{1}{2}a^2+\frac{1}{4}d^2+V_E$	$\frac{1}{4}(a+d)^2+V_E$	
A^-A^-	$\frac{1}{4}$	Family	F_1	$Bc_{1.2}$	P_2	
		(Freq.)	$(\frac{1}{16})$	$(\frac{1}{8})$	$(\frac{1}{16})$	
		Family mean	$m+d$	$m+\frac{1}{2}(-a+d)$	$m-a$	$m+\frac{1}{4}d-\frac{1}{2}a$
		σ_W^2	V_E	$\frac{1}{4}(a+d)^2+V_E$	V_E	
Column means (=HS means)			$m+\frac{1}{4}d+\frac{1}{2}a$	$m+\frac{1}{2}d$	$m+\frac{1}{4}d-\frac{1}{2}a$	$m+\frac{1}{2}d$

(ii)

Type of family	P_1	$Bc_{1.1}$	F_1	F_2	$Bc_{1.2}$	P_2	Average
Frequency	$\frac{1}{16}$	$\frac{1}{4}$	$\frac{1}{8}$	$\frac{1}{4}$	$\frac{1}{4}$	$\frac{1}{16}$	1.0
Family mean (y_i)	$m+a$	$m+\frac{1}{2}(a+d)$	$m+d$	$m+\frac{1}{2}d$	$m+\frac{1}{2}(-a+d)$	$m-a$	$m+\frac{1}{2}d$
σ_W^2	V_E	$\frac{1}{4}(a-d)^2$ $+V_E$	V_E	$\frac{1}{2}a^2+\frac{1}{4}d^2$ $+V_E$	$\frac{1}{4}(a+d)^2$ $+V_E$	V_E	$\frac{1}{4}a^2+\frac{3}{16}d^2$ $+V_E$
Mid-parent (x_i)	$m+a$	$m+\frac{1}{2}(a+d)$	m	$m+d$	$m+\frac{1}{2}(-a+d)$	$m-a$	$m+\frac{1}{2}d$

(iii)

$$\text{Covariance } (y_i, x_i) = \tfrac{1}{16}\{a.a\} + \tfrac{1}{4}\{\tfrac{1}{2}(a+d).\tfrac{1}{2}(a+d)\} + \tfrac{1}{8}\{d.0\} + \tfrac{1}{4}\{\tfrac{1}{2}d.d\} + \tfrac{1}{4}\{\tfrac{1}{2}(-a+d).\tfrac{1}{2}(-a+d)\}$$
$$ + \tfrac{1}{16}\{(-a).(-a)\} - \{\tfrac{1}{2}d.\tfrac{1}{2}d\} = \tfrac{1}{4}a^2$$

σ_B^2 and σ_W^2 are easy to derive. Thus σ_W^2 is the expected average variance within FS families and by summing the σ_W^2 in Table 4.8(ii) according to the frequencies of the six types of family, we obtain,

$$\sigma_W^2 = \tfrac{1}{16}[0] + \tfrac{1}{4}[\tfrac{1}{4}(a-d)^2] + \tfrac{1}{8}[0] + \tfrac{1}{4}[\tfrac{1}{2}a^2 + \tfrac{1}{4}d^2]$$
$$+ \tfrac{1}{4}[\tfrac{1}{4}(a+d)^2] + \tfrac{1}{16}[0] + V_E$$
$$= \tfrac{1}{4}a^2 + \tfrac{3}{16}d^2 + V_E.$$

Similarly, σ_B^2 is the expected variance of the true family means,

$$\sigma_B^2 = \tfrac{1}{16}[a]^2 + \tfrac{1}{4}[\tfrac{1}{2}(a+d)^2] + \tfrac{1}{8}[d]^2 + \tfrac{1}{4}[\tfrac{1}{2}d]^2 + \tfrac{1}{4}[\tfrac{1}{2}(-a+d)]^2$$
$$+ \tfrac{1}{16}[-a]^2 - [\tfrac{1}{2}d]^2$$
$$= \tfrac{1}{4}a^2 + \tfrac{1}{16}d^2.$$

For many independent genes, i.e. unlinked and non-interacting, together with environmental variation, these variances become,

$$\sigma_W^2 = \tfrac{1}{2}V_A^* + \tfrac{3}{4}V_D^* + V_E \qquad \text{[Eqn 4.3]}$$

$$\sigma_B^2 = \tfrac{1}{2}V_A^* + \tfrac{1}{4}V_D^*(+V_{EC}) \qquad \text{[Eqn 4.4]}$$

giving a total variance,

$$\sigma_P^2 = V_A^* + V_D^* + V_E(+V_{EC}).$$

As with selfing, there is additive and dominance genetic variation within both σ^2s, but their relative proportions have changed, as we see by comparing Equations 4.1 and 4.2 with 4.3 and 4.4. In the case of the variation within families, σ_W^2, the change is very small indeed, being just $\tfrac{1}{4}V_D^*$ larger with sib-mating. We would thus not expect to find any discernible difference in practice between the variation within families from selfing or sib-mating. Conversely, the additive genetic variation in σ_B^2 is twice as large with selfing. This implies that the heritability of family means is greater with selfing than sib-mating and, hence, the response to selection on family means would also be greater. Given the choice, therefore, it would be preferable for the breeder to self rather than sib-mate as more variation would be available for selection.

The mean and total variance of the progeny of one generation of sib-mating an F_2 is the same as for the F_2 itself providing, of course, that there is no linkage or epistasis. This is because the F_2 population will be in Hardy Weinberg equilibrium and hence a further round of random mating does not change the relative frequencies of the three genotypes at a locus which determine the mean and variance of the population. As with F_3 families, it is not possible to estimate V_A^*, V_D^* and V_E because there are only two statistics. It is, therefore,

Table 4.9 Data from sib-mating an F_2. (i) Basic ANOVA; (ii) ANOVA including regression onto mid-parental scores

(i)

Source	df	MS	F	ems
Between FS families	49	148.9318	1.77 **	$\sigma_W^2 + 5\sigma_B^2$
Within FS families	200	84.1531		σ_W^2
Total	249			

Mean of F_2 parents = 72.3121

Mean of progeny = 73.3154

$$t_{FS} = \sigma_B^2/(\sigma_W^2 + \sigma_B^2)$$

$$= 12.9557/97.1088$$

$$= \underline{0.133}$$

$$h_n^2 = 2t_{FS}$$

$$= \underline{0.27}$$

$$V_A^* = 25.9115 \pm 12.4974; \quad c = 2.07^*$$

$$V_E = 71.1974 \pm 11.7554; \quad c = 6.06^{***}$$

(ii)

Source	df	MS	F
Regression onto F_2 parent	1	583.8769	4.17**
Residual	48	139.8704	1.66**
(Between FS families)	(49)	(148.9318)	(1.77)**
Within FS families	200	84.1531	

$b = h_n^2 = 0.26 \pm 0.13$

again necessary to assume that the dominance variation is negligible and so set V_D^* to zero. The analysis of a data set of identical size to that shown for the F_3s in Table 4.1(ii) is shown in Table 4.9(i). The major difference is the increased standard error of V_A^* which is due to the reduced coefficient of V_A^* in σ_B^2.

It is possible to obtain an estimate of the heritability directly from this ANOVA from the intra-class correlation t_{FS}, which is the ratio of σ_B^2 to σ_P^2 (Table 4.9(i)). Given the fact that we are looking at full-sib families, t_{FS} is the full-sib intra-class correlation and is,

$$t_{FS} = \sigma_B^2/\sigma_P^2 = \{\tfrac{1}{2}V_A^* + \tfrac{1}{4}V_D^*(+V_{EC})\}/\{V_A^* + V_D^* + V_E(+V_{EC})\}.$$

In fact t_{FS} is exactly half the mean of the broad and narrow heritabilities.

4.6 Sib-mating: regression of offspring onto the parents

One further useful statistic which can be calculated in order to increase the information available is the **covariance** between the offspring family means and the means of their parents. Intuition leads us to expect that the more highly heritable the trait, the higher the covariation between parents and their offspring; e.g. tall parents should have tall offspring while short begets short.

It is a simple matter to calculate the expected mean score of the parents that produced each of the six types of families shown in Table 4.8(ii). For example, the parents of a P_1 family are both A^+A^+ with mean $m + a$ whereas an F_1 has parents A^+A^+ and A^-A^- with mean m, etc. These mid-parent values and the means of their progeny are given in Table 4.8(ii) together with the derivation of the covariance. This table shows the close similarity between the offspring means and the mid-parental scores, any differences being due to dominance and, of course, sampling error which is not shown. The covariance, W, is purely additive variance, $\frac{1}{2}V_A^*$, and thus provides a third statistic which makes it possible to estimate all three components of variation, V_A^*, V_D^* and V_E.

This covariance can also be used to calculate the regression (b) of the full-sib family means (y_i) onto their mid-parental scores (x_i). This gives a direct measure of the narrow heritability as can be shown by the following argument. The regression slope is calculated as the SPxy/SSx and, because both the numerator and denominator have $n - 1$ df, this is equivalent to the covariance of x and y divided by the variance of x. The variance of the parents (x), which were F_2 individuals, is

$$V_P = V_A^* + V_D^* + V_E(+V_{EC}).$$

Now, recalling that the variance of means $= V_x/k$, where k = the number of items averaged, then the variance of the means of pairs of F_2 parents is $V_P/2$. The expectation of the covariance calculated above is $\frac{1}{2}V_A^*$ and therefore,

$$b = \frac{1}{2}V_A^* / \frac{1}{2}(V_A^* + V_D^* + V_E(+V_{EC}))$$
$$= V_A^* / V_A^* + V_D^* + V_E(+V_{EC})$$
$$= h_n^2.$$

This is a fundamental relationship in quantitative genetics and, as we will show later, it applies to a whole range of populations, not just an F_2.

It is possible to exploit the regression approach both to provide a test of significance for the heritability estimate and to test for the adequacy of the additive genetical model. The FS families provide a data set equivalent to that required for a regression with replicates. Fitting b allows for all the additive effects between FS family means, so what remains in the residual is sampling error plus non-additive effects. An F-test of the 'residual' MS against the 'within-families' MS is thus a test for non-additive effects.

The relationship between the full-sib family means and the mid-parental values for the data set analysed in Table 4.9(i) is illustrated in Figure 4.3, while the complete regression analysis of those data is shown in Table 4.9(ii). Thus, the data give an estimate of the narrow heritability of 0.26, while the significance of the residual demonstrates that non-additive variation (dominance and/or epistasis) must also be present. To put it another way, the regression of offspring onto parents has accounted for the $\frac{1}{2} V_A^*$ in σ_B^2, thus leaving the $\frac{1}{4} V_D^*$ unaccounted for in the residual.

If the trait under study is sex-limited or expressed differently in the two sexes then it is necessary to perform the regression of offspring onto parents on data from the sexes separately. To take an extreme example, with milk production in cattle one can do no more than regress the milk yield of the daughters onto that of their mothers. This regression estimates just half the narrow heritability because the daughters receive just half their genes from their mothers. Not only is the regression slope smaller but the residual mean square for the regression is larger because the mean performance of the progeny of any given maternal genotype will vary with the father. Such heritability estimates are therefore less reliable than those based on mid-parent offspring regression. Even where the trait is

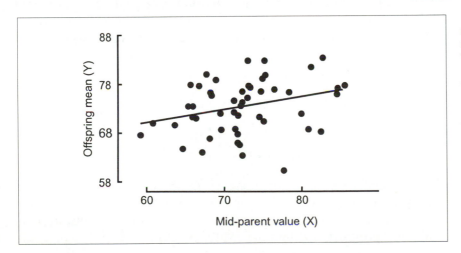

Figure 4.3 Offspring–parent regression. The observed relationship between offspring family means (Y) and their mid-parental values (X) obtained from an experiment involving FS, bi-parental families (BIPS). The regression line has a slope of $b = h_n^2 = 0.26$.

expressed in both sexes, it is good practice to check that the regression of daughter onto dam is not significantly different from that of sons onto sires before the mid-parent offspring regression is attempted.

Although such regression is common practice in animal genetics, it should be employed with caution. An underlying assumption is that the genetical and environmental variation in parents and offspring are the same. Because the offspring have to be raised at a later time than the parents, it is possible that the environments are different and that both the genetical and environmental variation are different, too. This will almost certainly be the case with plants where seasonal and annual variation in factors such as temperature, rainfall, solar radiation, etc. can vary considerably. With mammals and birds, which have reasonable environmental homeostasis and can be kept under constant feeding and ambient conditions, the problems may be smaller. Nonetheless, it is sensible to estimate heritabilities from contemporary relatives, raised together in the same trial whenever possible, otherwise extreme changes in the environment could actually lead to heritability estimates which are so variable as to be meaningless and some values could be in excess of one!

4.7 Further inbreeding by full-sib mating

Just as we can generate F_4 families by selfing we can generate S_4 families by sib-mating within the families obtained previously by sib-mating the F_2. As was shown in Table 4.6, there are six types of family, and to produce the S_4 generation, sib-mating is practised within each of these. The structure of the families is shown in Figure 4.4 and the ANOVA together with the expectations of the σ^2s is presented in Table 4.10.

These expectations are,

$$\sigma_1^2 = \tfrac{1}{2} V_A^* + \tfrac{3}{32} V_D^*$$

$$\sigma_2^2 = \tfrac{1}{4} V_A^* + \tfrac{5}{32} V_D^*$$

$$\sigma_3^2 = \tfrac{1}{2} V_A^* + \tfrac{11}{16} V_D^* + V_E.$$

Pedigree sib-mating as described here is far less common than pedigree selfing as a component of a breeding programme or a genetical analysis of a trait because its use tends to be confined to outbreeders and the progress to homozygosity is slow. Most outbreeding

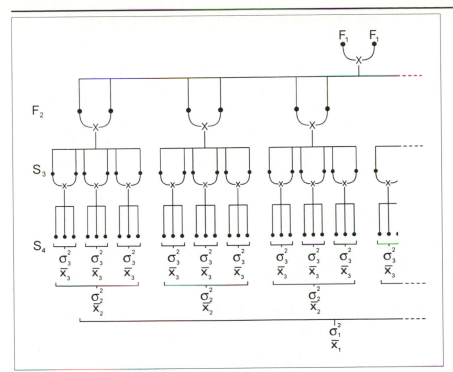

Figure 4.4 Inbreeding by full-sib mating. The hierarchical structure should be seen in conjunction with Table 4.10 and compared with Figure 4.1.

organisms carry a large number of recessive deleterious alleles in the heterozygous state and inbreeding results in these becoming homozygous, so reducing viability and reproductive success. This inevitably produces distorted segregations at linked loci for the quantitative traits under study. Because the means and variances derived above all assume normal diploid segregation, such distortion will result in the models becoming quite inappropriate.

4.8 Estimation of variance components by weighted least squares

Variance components can be estimated by weighted least squares (WLS) using the same approach as that used in Chapter 3. There is, however, an additional factor that needs to be considered here and that is the ANOVA structure. The MSs in the ANOVA are independent, i.e. orthogonal, but the σ^2s are not. Thus models are fitted to the MSs. For example, in Table 4.9 the 'Between' and 'Within FS' family mean squares (148.9318 and 84.1531) have expectations $\sigma_W^2 + 5\sigma_B^2$ and σ_W^2 respectively. Using the expectations of σ_W^2 and σ_B^2

Table 4.10 Derivation of σ^2 for the data obtained by sib-mating within FS families from an F_2

Type of F_2 FS family	Frequency	Mean	σ_2^2	σ_3^2
P_1	$\frac{1}{16}$	$m+a$	0	V_E
P_2	$\frac{1}{16}$	$m-a$	0	V_E
F_1	$\frac{1}{8}$	$m+\frac{1}{2}d$	0	$V_A^* + V_D^* + V_E$
F_2	$\frac{1}{4}$	$m+\frac{1}{2}d$	$\frac{1}{2}V_A^* + \frac{1}{4}V_D^*$	$\frac{1}{2}V_A^* + \frac{3}{4}V_D^* + V_E$
$Bc_{1.1}$	$\frac{1}{4}$	$m+\frac{1}{2}a+\frac{3}{8}d$	$\frac{1}{4}V_A^* + \frac{3}{16}V_D^* + \frac{1}{4}V_{AD}$	$\frac{1}{2}V_A^* + \frac{3}{4}V_D^* - \frac{1}{2}V_{AD} + V_E$
$Bc_{1.2}$	$\frac{1}{4}$	$m-\frac{1}{2}a+\frac{3}{8}d$	$\frac{1}{4}V_A^* + \frac{3}{16}V_D^* - \frac{1}{4}V_{AD}$	$\frac{1}{2}V_A^* + \frac{3}{4}V_D^* + \frac{1}{2}V_{AD} + V_E$
Combined	1.0	$m+\frac{3}{8}d$	$\frac{1}{4}V_A^* + \frac{5}{32}V_D^*$	$\frac{1}{2}V_A^* + \frac{22}{32}V_D^* + V_E$

Variation between S_4 family means, $\sigma_1^2 = \frac{1}{2}V_A^* + \frac{3}{32}V_D^*$

Total variation $= \sigma_1^2 + \sigma_2^2 + \sigma_3^2$

$$= 1\frac{1}{4}V_A^* + \frac{15}{16}V_D^* + V_E$$

given in Equations 4.3 and 4.4, then;

$$\epsilon MS_B = (\tfrac{1}{2}V_A^* + \tfrac{3}{4}V_D^* + V_E) + 5(\tfrac{1}{2}V_A^* + \tfrac{1}{4}V_D^*)$$

$$= 3V_A^* + 2V_D^* + V_E$$

$$\epsilon MS_W = \tfrac{1}{2}V_A^* + \tfrac{3}{4}V_D^* + V_E$$

As before, Chapter 3, estimation involves an iterative approach using the observed MSs to calculate the weights in the first iteration.

4.9 Unequal family sizes

It has been assumed so far that all family sizes are equal and that there are no missing values. Although such an assumption was useful in order to develop basic principles, it is not likely to be true in practice because individuals, if not whole families, may be lost through infertility, disease or random damage. The SS of the ANOVA is relatively straightforward to calculate in such circumstances, but the coefficients of the variance components vary for the same σ^2s in different levels of the table. As a result, exact tests of significance do not exist for any but the penultimate item in the ANOVA table, while approximate tests can be very imprecise. This is a problem with nested and factorial designs and partial solutions are discussed in most statistical texts [5]. Given the difficulties it is always worth designing the experiment to ensure that the number of families and the family sizes are equal even if it means discarding some data to restore the balance. Providing only a very small proportion of individual scores are missing, it is always possible to replace the missing values by the appropriate family mean. Computer-based statistical software, such as MINITAB [6] or SAS [7] can normally handle such unbalanced situations but do set restrictions and may not provide the estimates of the variance components.

Summary

1. Genetic and environmental models can be constructed for the means and variances of families produced by inbreeding.
2. Selfing is the quickest traditional method of inbreeding but it can be superseded by techniques which produce **doubled haploids** (DH) and can potentially achieve complete homozygosity in a single generation.
3. Recombinant inbred lines produced by SSD and DH techniques, however, can differ in their performance because the former allows marginally more recombination compared with the latter.
4. Sib-mating, on the other hand, is a far slower method of inbreeding and therefore is applied only in those cases where selfing cannot take place, e.g. animals, dioecious or monoecious but self-incompatible plants. However, sib-mating allows more recombination and provides more powerful tests of non-additive variance.
5. Heritability can be estimated directly from FS families by means of the intra-class correlation or by regression of the offspring onto the parents.
6. Selection is least effective in the F_2 unless the F_2 individuals can be replicated vegetatively; otherwise, maximum advance can be achieved by selecting among the F_∞ lines. Selection among the F_3 or F_4 families can also be moderately effective but only when family size is large and heritability is moderate to high.

References

1. Forough, B. and Wenzel, G. (1993) *Plant Breeding; Principles and Prospects* (eds M.D. Hayward, N.O. Bosemark and I. Romagoza), Chapman & Hall, London.

2. Jinks, J.L. and Pooni, H.S. (1980) Comparing predictions of mean performance and environmental sensitivity of recombinant inbred lines based upon F_3 and triple test cross families. *Heredity*, **45**, 305–12.
3. van Ooijen, J.W. (1989) Estimation of additive genotypic variance with the F_3 of autogamous crops. *Heredity*, **63**, 73–81.
4. Oeveren, A.J. van and Stam, P. (1993) Comparative simulation studies on the effects of selection for quantitative traits in autogamous crops: early selection versus single seed descent. *Heredity*, **69**, 342–51.
5. Sokal, R.R. and Rohlf, F.J. (1981) *Biometry*, 2nd edn, Freeman & Co., New York.
6. MINITAB, 1994. *Reference Manual: Release 10 for Windows*. Minitab Inc., State College, PA, USA.
7. SAS Institute (1988) *SAS Language Guide for Personal Computers*. Release 6.03 Edition, SAS Inst., Cary, NC.

Half-sib mating designs   5

We have considered various ways of studying the genetics of quantitative traits in generations derived from an F_1 either by backcrossing, selfing or full-sib mating. A further approach is to use information from half-sib (HS) families, i.e. sets of progeny which have either the same father but different mothers or *vice versa*.

It was shown in Chapter 3 that the total phenotypic variation in an F_2, σ_P^2, is,

$$V_A^* + V_D^* + V_E$$

and that the analysis of bi-parental progenies, BIPs, partitions this into variation between and within FS families, for which the expectations for the σ^2s are:

$$\sigma_B^2 = \tfrac{1}{2} V_A^* + \tfrac{1}{4} V_D^* (+V_{EC})$$

$$\sigma_W^2 = \tfrac{1}{2} V_A^* + \tfrac{3}{4} V_D^* + V_E.$$

We will now explain the use of five half-sib designs that can be applied to an F_2 population and which provide various partitions of the variation between FS families (σ_B^2 above). In so doing, they permit further genetical parameters to be tested and estimated. In all cases, we will assume that the parents are drawn from the F_2 population at random and are not a selected sample, and that the progeny are raised in a completely randomized experimental design. In fact all these HS experiments can be applied to populations other than an F_2, although their analysis and use in this wider context will be deferred to Chapter 9.

5.1 The North Carolina Experiment I: NCI

This is the first of three experimental designs whose analysis was first described by Comstock and Robinson [2] in Raleigh, North Carolina in 1952. Normally in this design, a number (n_1) of male parents are chosen and each is crossed to two or more different female parents, no female being crossed to more than one male. For example, if each male is crossed to a different set of n_2 females, then a total of $n_1 \times n_2$ females are used and hence $n_1 \times n_2$ full-sib (FS) families are produced.

The common parent would normally, though not invariably, be the male simply because it is easier, particularly in animals, to mate one male to several females and keep the progeny distinct. If each female animal was to be mated to several males, then the progeny from each mating would have to be raised separately in time, introducing birth order or parity effects, as well as additional environmental effects over time. In plants, this is less often a problem because separate flowers on the same plant can be artificially pollinated with pollen from different males. As we will see later, there are also sound genetical reasons for choosing males as the common parent.

Each set of families with the same father constitutes a HS family group. Thus we have n_1 HS groups, each made up of n_2 FS families, and each FS family contains r progeny. In total therefore the experiment consists of $n_1 \times n_2 \times r$ individual progeny to be scored and analysed. An illustrative set of data are presented in Table 5.1(i), based on eight males each crossed to four different females, with FS family sizes of 20. **It must be emphasized that such a small number of parents is far too low for most practical purposes and is used solely for illustrative purposes**.

The analysis of such data follows an hierarchical or nested ANOVA as shown for the general case in Table 5.2. The ANOVA of the data used for illustration is shown in Table 5.1(ii), from which it is apparent that there is significant variation both between and within male HS family groups, giving the estimates of the σ^2s in Table 5.1(iii). It is important to remember that, when F tests are performed in an hierarchical ANOVA, significant MSs become the error for items above them in the ANOVA table. In the present case, for instance, since the MS 'Between FS families within male HS family groups' is significant it must be used as the error to test the MS 'Between male HS family groups'.

What then are the expectations of these σ^2s? Again we need to refer to Table 4.8(i). The parameter σ^2_W is the average variance within FS families and was derived in Chapter 4 for BIPs:

$$\sigma^2_W = \tfrac{1}{2}V^*_A + \tfrac{3}{4}V^*_D + V_E.$$

However, the other component of variation for BIPs, σ^2_B is partitioned in the NCI into the components σ^2_M and $\sigma^2_{F/M}$.

The first, σ^2_M, is the variance of true HS family group means. We can see from Table 4.8(i) that there are just three types of HS family,

Table 5.1 Analysis of NCI experiment. (i) Family means from a NCI experiment involving eight males each crossed to four different females, with 20 progeny per family; (ii) ANOVA; (iii) parameter estimates

(i)

Female parents	Male parents							
	1	2	3	4	5	6	7	8
1	74.8100	78.2095	69.7960	70.8045	73.6725	75.5415	70.8565	79.4465
2	78.7560	79.4570	67.4205	69.7695	70.2030	77.1260	67.0955	72.2885
3	77.2325	66.9285	76.9345	71.3005	71.2740	77.9960	64.4375	81.6125
4	80.4400	75.3210	78.1480	75.4505	68.8610	74.4650	69.2285	76.9675
Total	311.2385	299.9160	292.2990	287.3250	284.0105	305.1285	271.6180	310.3150

(ii)

Source	df	MS	F	P	ems
Between male HS family groups	7	970.1798	3.81	**	$\sigma_W^2 + 20\sigma_{F/M}^2 + 80\sigma_M^2$
Between FS families within male HS family groups	24	254.8276	3.27	***	$\sigma_W^2 + 20\sigma_{F/M}^2$
Within FS families	608	77.9332			σ_W^2

(iii)

$$\sigma_M^2 = 8.9419 \qquad t_{HS} = 8.9419/95.7198 = 0.09; h_n^2 = 0.37$$

$$\sigma_{F/M}^2 = 8.8447 \qquad V_A^* = 35.7675$$

$$\sigma_W^2 = \underline{77.9332} \qquad V_D^* = 0.0$$

$$\sigma_P^2 = 95.7198 \qquad V_E = 60.0493$$

deriving from males which are, respectively, A^+A^+, A^+A^-, and A^-A^-, with frequencies $\frac{1}{4}$, $\frac{1}{2}$ and $\frac{1}{4}$. The expected means of all the progeny of these three types of male are derived in Table 4.8(i) and are,

$$A^+A^+ = m + \tfrac{1}{2}d + \tfrac{1}{2}a$$

$$A^+A^- = m + \tfrac{1}{2}d \qquad\qquad \text{[Eqn 5.1]}$$

$$A^-A^- = m + \tfrac{1}{2}d - \tfrac{1}{2}a.$$

Table 5.2 General ANOVA of a NCI experiment

Source	df	MS	ems
Between male HS family groups	$n_1 - 1$	MS_M	$\sigma_W^2 + r\sigma_{F/M}^2 + rn_2\sigma_M^2$
Between FS families within male HS family groups	$n_1(n_2 - 1)$	$MS_{F/M}$	$\sigma_W^2 + r\sigma_{F/M}^2$
Within FS families	$n_1 n_2(r - 1)$	MS_W	σ_W^2
Total	$n_1 n_2 r - 1$		

From these it is clear that the variation in the three HS means is entirely due to the additive effect a and, ignoring the constant term $(m + \frac{1}{2}d)$, the expected value of σ_M^2 can be derived as follows.

$$\sigma_M^2 = \frac{1}{4}(\frac{1}{2}a)^2 + \frac{1}{4}(-\frac{1}{2}a)^2 - 0$$
$$= \frac{1}{8}a^2$$
$$= \frac{1}{4}V_A^*, \text{ for many genes.}$$

Finally, $\sigma_{F/M}^2$ includes the residual variation between FS families within male HS groups and its expectation is easily derived by difference, i.e.

$$\sigma_{F/M}^2 = \sigma_B^2 - \sigma_M^2$$
$$= \frac{1}{2}V_A^* + \frac{1}{4}V_D^* + (V_{EC}) - \frac{1}{4}V_A^*$$
$$= \frac{1}{4}V_A^* + \frac{1}{4}V_D^* + (V_{EC}).$$

For the first time we have an item, σ_M^2, which is independent of environmental variation, providing the progeny are randomized, and thus represents purely additive genetical variation, $\frac{1}{4}V_A^*$. This is because the male parent only donates its pollen or sperm and cannot affect the environment of its offspring, unless, of course, the offspring of the same father are raised together and therefore are potentially in a different environment to other families.

The narrow heritability can be estimated from the HS intra-class correlation, where,

$$t_{HS} = \sigma_M^2/\sigma_P^2$$
$$= \frac{1}{4}V_A^*/(V_A^* + V_D^* + V_E(+V_{EC}))$$
$$h_n^2 = 4t_{HS}.$$

Thus, the narrow heritability can be estimated directly from the progeny alone. Using the data from Table 5.1(iii), for example, the heritability is estimated to be 0.37.

Provided that we have an independent estimate of V_{EC}, or $V_{EC} = 0$, then there are three statistics and hence we can estimate all three parameters (V_A^*, V_D^*, V_E) from the progeny alone (see Table 5.1(iii)). It is also possible to regress the HS family means on to the common (male) parent and this covariance can be derived as shown in Table 5.3. This covariance has the same expectation as for FS families (i.e. $\frac{1}{2}V_A^*$ as derived earlier) and suffers from the same problems with genotype by environment interaction. Assuming that this design is to be used specifically to determine the narrow heritability, then the optimum strategy is to maximize the number of half-sib groups,

Table 5.3 Derivation of covariance of HS family means onto common parent

Parent	A^+A^+	A^+A^-	A^-A^-	Mean
Frequency	$\frac{1}{4}$	$\frac{1}{2}$	$\frac{1}{4}$	
Parent score $- m$	a	d	$-a$	$\frac{1}{2}d$
Progeny mean $- m$	$\frac{1}{2}(a+d)$	$\frac{1}{2}d$	$\frac{1}{2}(-a+d)$	$\frac{1}{2}d$
Progeny mean $-(m+\frac{1}{2}d)$	$\frac{1}{2}a$	0	$-\frac{1}{2}a$	0

Covariance $\bar{o}/p = \frac{1}{4}(\frac{1}{2}a^2) + 0 + \frac{1}{4}(\frac{1}{2}(-a)^2) = \frac{1}{4}a^2 = \frac{1}{2}V_A^*$

although the number of females mated to each male should be at least 20 [1].

This design is appropriate for organisms with a wide range of breeding systems and is commonly used in animal breeding, for example with dairy cattle. It is important to note that, although we can estimate V_D^*, we have no direct test of its significance, nor do we have any tests of epistasis, G × E interaction or maternal effects.

If females are used instead of males as the common parent, then the variance component for female HS groups will contain an additional term due to maternal effects. Unlike the male, the female can, in principle, exert a considerable effect on her progeny over and above the additive effects of her genes as discussed in the section on environmental effects in Chapter 3. Thus the heritability estimate could be inflated. This is another reason why males are normally used as the common parent of HS families.

Female HS families can occur naturally in plants and animals; one can normally be sure who the female parent is, but not necessarily the male parent. Thus with wind- or insect-pollinated plant species, particularly if they are self-incompatible, the progeny of a single seed parent will be half-sibs. Unlike the situation described above, however, it is not possible to distinguish the progeny of different males. Thus if seed are collected from a number (n) of female parents and their progeny raised in a randomized experiment, a number of HS families could be recognized. The ANOVA of such data is as illustrated in Table 5.4. Now, as before, for female HS families,

$$\sigma_B^2 = \frac{1}{4}V_A^*(+V_{EM}).$$

Table 5.4 ANOVA of experiment with female HS families

Source	df	MS	ems
Between female HS family groups	$n-1$	MS_1	$\sigma^2 + r\sigma_B^2$
Within HS families	$n(r-1)$	MS_2	σ^2

But, since we cannot distinguish different FS families among the progeny of individual females,

$$\sigma^2 = \sigma_W^2 + \sigma_{F/M}^2 = \tfrac{3}{4}V_A^* + V_D^* + V_E + (V_{EC} - V_{EM}).$$

However

$$t_{HS} = (\tfrac{1}{4}V_A^* + V_{EM})/(V_A^* + V_D^* + V_{EC} + V_E)$$

$$= \tfrac{1}{4}h_n^2.$$

Hence, estimates of heritability can be obtained from such data but they will be inflated if maternal effects are a major factor.

5.2 The North Carolina Experiment II: NCII

This experimental design is similar to the NCI except that every male is now crossed to every female creating female HS groups as well as male HS groups [2,3]. Table 5.5 illustrates the general ANOVA of this cross-classified experiment when n_1 males are crossed to n_2 females and r progeny are raised from every FS family. An illustrative data set involving five males crossed to five females and ten FS progeny are presented in Table 5.6(i) and the corresponding ANOVA in Table 5.6(ii). As with the NCI, the use of such a small number of parents is solely for illustrative purposes, and much larger numbers should be used in practice.

The item 'Between male HS family groups' in this ANOVA is equivalent to that in the NCI given earlier and thus σ_M^2 has the same expectation in both, $\tfrac{1}{4}V_A^*$. In the NCII, however, the item which appeared in the NCI as $\sigma_{F/M}^2$ (variance 'between FS families within male HS family groups') has been further partitioned into σ_F^2 and σ_{MF}^2 because of the cross-classified nature of the experiment.

Table 5.5 General ANOVA of a NCII experiment

Source	df	MS	ems
Between male HS family groups (M)	$n_1 - 1$	MS_M	$\sigma_W^2 + r\sigma_{MF}^2 + rn_2\sigma_M^2$
Between female HS family groups (F)	$n_2 - 1$	MS_F	$\sigma_W^2 + r\sigma_{MF}^2 + rn_1\sigma_F^2$
M × F	$(n_1 - 1)(n_2 - 1)$	MS_{MF}	$\sigma_W^2 + r\sigma_{MF}^2$
Within FS families	$n_1 n_2(r - 1)$	MS_W	σ_W^2
Total	$n_1 n_2 r - 1$		

Table 5.6. Analysis of a NCII experiment. (i) Family means based on FS family size, r = 10; (ii) ANOVA; (iii) parameter estimates

(i)

Female parents	Male parents					Means	GCA
	1	2	3	4	5		
1	76.135	72.740	73.299	80.013	74.816	75.4006	1.8181
2	75.465	76.690	77.253	72.542	85.190	77.4280	3.8455
3	70.852	67.334	74.068	74.542	74.996	72.3584	−1.2241
4	72.114	63.554	74.889	66.603	79.415	71.3150	−2.2675
5	76.132	59.700	74.284	72.899	74.038	71.4106	−2.1719
Means	74.1396	68.0036	74.7586	73.3198	77.6910	73.5825	
GCA	0.5571	−5.5789	1.1761	−0.2627	4.1085		

(ii)

Source	df	MS	F	ems
Between male HS family groups (M)	4	622.0810	3.93*	$\sigma_W^2 + 10\sigma_{MF}^2 + 50\sigma_M^2$
Between female HS family groups (F)	4	368.1311	2.32 n.s.	$\sigma_W^2 + 10\sigma_{MF}^2 + 50\sigma_F^2$
M × F	16	158.6206	1.79*	$\sigma_W^2 + 10\sigma_{MF}^2$
Within FS families	225	88.5053		σ_W^2

(iii)

$$\sigma_W^2 = 88.5053 = \tfrac{1}{2}V_A^* + \tfrac{3}{4}V_D^* + V_E$$

$$\sigma_{MF}^2 = 7.0115 = \tfrac{1}{4}V_D^*$$

$$\sigma_M^2 = 9.2692 = \tfrac{1}{4}V_A^*$$

$$\sigma_F^2 = 4.1902 = \tfrac{1}{4}V_A^*$$

$$V_A^* = 26.92$$

$$V_D^* = 28.05$$

$$V_E = 54.01$$

By analogy with NCI, $\sigma_F^2 = \tfrac{1}{4}V_A^*$ as did σ_M^2, hence

$$\sigma_{MF}^2 = \{\tfrac{1}{4}V_A^* + \tfrac{1}{4}V_D^* + (V_{EC})\} - \{\tfrac{1}{4}V_A^*\}$$

$$= \tfrac{1}{4}V_D^* + (V_{EC}).$$

In the absence of epistasis and common environment effects, σ_{MF}^2 is a function of dominance variance (V_D^*) only.

However, if there is environmental variation between FS families, this could be due either to general or specific maternal effects. The general maternal effects (V_{EM}) will appear in σ_F^2 ($= \tfrac{1}{4}V_A^* + V_{EM}$)

while that specific to individual FS families $(V_{EC} - V_{EM})$ will be confounded with V_D^*, i.e.

$$\sigma_{MF}^2 = \tfrac{1}{4}V_D^* + (V_{EC} - V_{EM}).$$

With random sampling of parents, σ_M^2 should equal σ_F^2. So, providing the number of males and female parents are the same, i.e. $n_1 = n_2 = n$ we have a test of maternal effects by comparing MS_F/MS_M as a variance ratio.

$$F = MS_F/MS_M = (\sigma_W^2 + r\sigma_{MF}^2 + rn\sigma_F^2)/(\sigma_W^2 + r\sigma_{MF}^2 + rn\sigma_M^2).$$

So, this design allows tests of significance and estimates of V_A^* and V_D^*. Hence we can calculate heritabilities and dominance ratios although the test for dominance is not very powerful.

5.3 General and specific combining ability

It is normally not possible to deduce much about the genotype of an individual in an F_2 population from its own phenotype because of the combined effects of environment, gene distribution, dominance and epistasis. Indeed, for sex-limited traits such as milk yield or egg production, the male will have no relevant phenotype though he will contain genes for the trait. The mean performance of his HS progeny does, however, give a more precise measure of his genetic potential as Equations 5.1 show. Thus the male with the most increasing alleles will, on average, produce the best progeny when crossed to a random sample of females.

The difference between the mean performance of the progeny of a given male and the mean of the progeny from all the males is called his **General Combining Ability** (GCA). It reflects how well his genes combine, on average, with those of all females in the population. For this reason, the MS 'Between HS family groups' is often referred to as the GCA MS. This is illustrated in Table 5.6 using the small set of 25 NCII families. Subtracting the overall mean of the 25 families, 73.5825, from the means of the progeny of the five male and five female parents gives the GCA values shown. We see that male 5 has the highest breeding value with a GCA of 4.1085, while male 2 has the lowest GCA, -5.5789.

If there is no dominance or epistasis, it should be possible to predict the mean performance of the progeny of a cross between the ith male and the jth female as,

$$\epsilon Y_{ij} = m + g_i + g_j,$$

where g_i and g_j are their GCA effects. Again, using the data in Table 5.6(i), we can predict the expected mean of the cross between the male and female with the best GCAs, male 5 and female 2, as,

$$\epsilon Y_{52} = 73.5825 + 4.1085 + 3.8455 = 81.5365$$

whereas the observed mean was 85.190. Any significant deviation from the observed Y_{ij} must be due to dominance or epistatic effects. These deviations, specific to individual crosses, are referred to as Specific Combining Ability (SCA) and are measured by the 'Male × Female' MS in the ANOVA of the NCII.

5.4 Multiple NCIIs

With some organisms, particularly animals, it may be feasible to use the female in crosses to two males, but no more. In such cases it is possible to construct a number, k, of 2 × 2 NCIIs. Each individual 2 × 2 NCII would yield the standard ANOVA of Table 5.7(i). Although each would be inadequate on its own, the corresponding SS of the k individual ANOVAs can be combined into a single analysis as shown in Table 5.7(ii). Such an ANOVA has the added advantages that the df for the main effects MS_M and MS_F are equal to those of the interaction, MS_{MF}, while the number of parents that can be used increases geometrically.

Table 5.7 ANOVA for multiple NCIIs. (i) ANOVA for a single 2 × 2 NCII; (ii) general ANOVA of k sets of 2 × 2 NCII experiments

(i)

Source	df	MS	ems
Between male HS family groups (M)	1	MS_M	$\sigma_W^2 + r\sigma_{MF}^2 + 2r\sigma_M^2$
Between female HS family groups (F)	1	MS_F	$\sigma_W^2 + r\sigma_{MF}^2 + 2r\sigma_F^2$
M × F	1	MS_{MF}	$\sigma_W^2 + r\sigma_{MF}^2$
Within FS families	$4(r-1)$	MS_W	σ_W^2

(ii)

Source	df	MS	ems
Between sets	$k-1$	MS_S	—
Between male HS family groups (M)	k	MS_M	$\sigma_W^2 + r\sigma_{MF}^2 + 2r\sigma_M^2$
Between female HS family groups (F)	k	MS_F	$\sigma_W^2 + r\sigma_{MF}^2 + 2r\sigma_F^2$
M × F	k	MS_{MF}	$\sigma_W^2 + r\sigma_{MF}^2$
Within FS families	$4k(r-1)$	MS_W	σ_W^2

As with the other HS designs and BIPs, the NCII provides no tests of epistasis or $G \times E$ interaction. Moreover it is a more difficult crossing scheme to perform. Its only merits are that it provides tests of non-additive and maternal effects, but the power of both tests is low [4] while the effects themselves are more easily detected by other methods.

5.5 The North Carolina Experiment III: NCIII

As first proposed [2] the NCIII involved taking a number of F_2 individuals from a cross between two inbred lines (P_1 and P_2) and crossing each back to these same two lines. The two parental inbreds thus act as '**testers**' against which the F_2 are assessed. However, being the progenitors of the F_2 they are very special testers because the F_2 is segregating at all the loci for which the testers differ but for no other loci (see Table 5.8(i)).

Unless the F_2 individuals can be replicated clonally, as in the case of a perennial plant, they are all unique. The testers on the other hand, being inbred lines, can be replicated sexually and produced in large numbers. For these reasons it is common practice to use the F_2 as male and the testers as female parents in the experiment. Any one female need only be involved in one mating and other crosses to the same tester should use different females. There are sound genetical reasons, too, for setting the cross up this way as will be shown later although there are circumstances such as, for example, the presence of a self-incompatibility system that might preclude it. The experimental design is equally feasible with animals or plants.

The NCIII is clearly a very special case of the NCII, a $2 \times n$ NCII, in fact. Not surprisingly, therefore, the ANOVA of the data is similar to that of the NCII although it differs in one important feature; the two testers are not a random sample from any population but are two very particular lines – the grand parents of the F_2. Hence the item 'Testers' in the ANOVA must be treated as a statistically 'fixed' effect, which has major consequences for the analysis as we shall now see.

Let us turn to a specific experiment in which n F_2 individuals are all crossed to the two testers and that r full sib progeny are raised from the resulting 2n families in a completely randomized design. If these are then scored for some trait, we can represent the mean of the r sibs from the cross between the ith F_2 with P_1 and P_2 as $\bar{L}_{1i}$ and $\bar{L}_{2i}$ respectively (see Table 5.8(i)). The analysis of variance would take the form shown in Table 5.8(ii). A data set involving 20 F_2s is presented in Table 5.9(i).

Table 5.8 Analysis of an NCIII. (i) Data structure for family means; (ii) general ANOVA; (iii) general ANOVA with σ_{TS}^2 redefined

(i)

F_2	Testers	
	P_1	P_2
1	$\bar{L}_{11}$	$\bar{L}_{21}$
2	$\bar{L}_{12}$	$\bar{L}_{22}$
...	...	...
i	$\bar{L}_{1i}$	$\bar{L}_{2i}$
...	...	...
n	$\bar{L}_{1n}$	$\bar{L}_{2n}$

(ii)

Source	df	MS	ems
Testers (T)	1	MS_T	$\sigma_W^2 + r\sigma_{TS}^2 + nrk_T^2$
F_2 (S)	$n-1$	MS_S	$\sigma_W^2 + 2r\sigma_S^2$
T × S	$n-1$	MS_{TS}	$\sigma_W^2 + r\sigma_{TS}^2$
Within families	$2n(r-1)$	MS_W	σ_W^2
Total	$2nr-1$		

(iii)

Source	df	MS	ems
Testers (T)	1	MS_T	$\sigma_W^2 + 2r\sigma_D^2 + nrk_T^2$
F_2 (S)	$n-1$	MS_S	$\sigma_W^2 + 2r\sigma_S^2$
T × S	$n-1$	MS_{TS}	$\sigma_W^2 + 2r\sigma_D^2$
Within FS families	$2n(r-1)$	MS_W	σ_W^2
Total	$2nr-1$		

Because the 'testers' are fixed, the interaction item (T × S) and the main effect (S) are both tested against the basic error as the *ems* show. We can derive the genetical components of the σ^2s of this analysis simply by recalling that when we have a $2 \times n$ table, the sums of squares for the analysis of variance can be computed from the sums and differences of the paired observations. The same approach can also be followed in the derivation of the expected mean squares.

Using a single gene model, the two testers P_1 and P_2 are A^+A^+ and A^-A^- respectively and the means of the various types of family produced, $\bar{L}_{ij}$, are given in Table 5.10. The sums and differences of the two family means for each of the three possible F_2 parents shown in Table 5.10, indicate an important similarity. The sums $(\bar{L}_{1i} + \bar{L}_{2i})$ all

Table 5.9 NCIII data. (i) Family means for a NCIII with 20 F_2 and 10 sibs per FS family; (ii) ANOVA; (iii) parameter estimates

(i)

F_2 (i)	Testers		$\bar{L}_{1i} + \bar{L}_{2i}$	$\bar{L}_{2i} - \bar{L}_{1i}$
	P_1	P_2		
	$(\bar{L}_{1i})$	$(\bar{L}_{2i})$		
1	74.107	73.090	147.197	−1.017
2	76.117	65.518	141.635	−10.599
3	72.920	76.204	149.124	3.284
4	70.340	76.410	146.750	6.070
5	77.516	68.326	145.842	−9.190
6	79.287	74.602	153.889	−4.685
7	78.031	77.282	155.313	−0.749
8	77.564	79.147	156.711	1.583
9	70.905	71.220	142.125	0.315
10	76.243	73.359	149.602	−2.884
11	77.664	69.423	147.087	−8.241
12	70.339	71.342	141.681	1.003
13	71.292	72.012	143.304	0.720
14	72.699	75.357	148.056	2.658
15	78.140	72.107	150.247	−6.033
16	75.439	66.268	141.707	−9.171
17	83.148	76.132	159.280	−7.016
18	72.941	76.748	149.689	3.807
19	76.198	70.617	146.815	−5.581
20	64.098	69.501	133.599	5.403
Total	1494.9880	1454.6650	2949.6530	−40.3230
Mean	74.7494	72.7333	147.4827	−2.0162
Within FS family				
SS (180df)	14449.3542	12587.0442	–	–

(continued on next page)

have a common term $(2m + d)$ and only vary in a, while the differences $(\bar{L}_{2i} - \bar{L}_{1i})$ have a common term $(-a)$ and only vary in d. Thus any variation in the sums is solely due to additive (a) effects while variation in the differences is due solely to dominance (d) effects and we can ignore the constant terms since they cannot affect the variance.

The variance of the sums and differences are derived thus:

$$\text{variance of sums} = \tfrac{1}{4}(a)^2 + \tfrac{1}{2}(0) + \tfrac{1}{4}(-a)^2 - 0 = \tfrac{1}{2}a^2$$

$$\text{variance of differences} = \tfrac{1}{4}(d)^2 + \tfrac{1}{2}(0) + \tfrac{1}{4}(-d)^2 - 0 = \tfrac{1}{2}d^2.$$

However, when we compute the MSs for the analysis of variance on the actual data the variances of the sums and differences would be 'corrected' by multiplication by (r/2), since each mean ($\bar{L}_{ij}$) is obtained

Table 5.9 (*continued*)

(ii)

Source	df	MS	F	ems
Testers (T)	1	406.4861	2.98 n.s.	$\sigma_W^2 + 20\sigma_D^2 + 200k_T^2$
F_2 (S) (Additive)	19	181.7804	2.41**	$\sigma_W^2 + 20\sigma_S^2$
T×S (Dominance)	19	136.4894	1.82*	$\sigma_W^2 + 20\sigma_D^2$
Within FS families	360	75.1011		σ_W^2
Total	399			

(iii) Parameter estimates

$\sigma_W^2 = 75.1011 = \frac{1}{4}V_A^* + \frac{1}{2}V_D^* + V_E$

$\sigma_S^2 = 5.3340 = \frac{1}{4}V_A^*$

$\sigma_D^2 = 3.0694 = \frac{1}{2}V_D^*$

$V_A^* = 21.3360$

$V_D^* = 6.1388$

$V_E = 66.6977$

Dominance ratio $= 0.76$

$h_n^2 = 0.23$

by averaging r sibs, while every sum or difference is obtained from two such means.

By implementing these corrections to the genetical expectations, we obtain:

$$\text{corrected variance of sums} = \frac{1}{2}a^2 \times (r/2)$$

$$= r(\tfrac{1}{4}a^2)$$

$$\text{corrected variance of differences} = \frac{1}{2}d^2 \times (r/2)$$

$$= r(\tfrac{1}{4}d^2).$$

Table 5.10 Expectations of family means for a NCIII

		Testers			
		P_1	P_2	Sums	Differences
F_2 genotype	Frequency	A^+A^+	A^-A^-	$\bar{L}_{1i} + \bar{L}_{2i}$	$\bar{L}_{2i} - \bar{L}_{1i}$
A^+A^+	$\frac{1}{4}$	$m + a$	$m + d$	$2m + d + a$	$-a + d$
A^-A^+	$\frac{1}{2}$	$m + \frac{1}{2}a + \frac{1}{2}d$	$m - \frac{1}{2}a + \frac{1}{2}d$	$2m + d$	$-a$
A^-A^-	$\frac{1}{4}$	$m + d$	$m - a$	$2m + d - a$	$-a - d$
Mean		$m + \frac{1}{2}a + \frac{1}{2}d$	$m - \frac{1}{2}a + \frac{1}{2}d$	$2m + d$	$-a$

It follows from this that:

$$\epsilon MS_S = \sigma_W^2 + 2r\sigma_S^2 = \sigma_W^2 + r(\tfrac{1}{4}a^2)$$

and

$$\epsilon MS_{TS} = \sigma_W^2 + r\sigma_{TS}^2 = \sigma_W^2 + r(\tfrac{1}{4}d^2).$$

Hence

$$2r\sigma_S^2 = r(\tfrac{1}{4}a^2)$$

therefore

$$\sigma_S^2 = \tfrac{1}{8}a^2$$

and

$$r\sigma_{TS}^2 = r(\tfrac{1}{4}d^2)$$

therefore

$$\sigma_{TS}^2 = \tfrac{1}{4}d^2.$$

There are several features which are now apparent concerning MS_S and MS_{TS}:

(i)   They both have the same df, $(n-1)$.
(ii) They both have the same error, MS_W.
(iii) Their genetical expectations are identical apart from the one being a function of a^2 and the other d^2.

In order to emphasize this last similarity it is conventional to depart from the traditional formulation of the interaction and to define a term σ_D^2 which is just one half σ_{TS}^2, i.e.

$$\sigma_D^2 = \tfrac{1}{2}\sigma_{TS}^2 = \tfrac{1}{8}d^2$$

$$\text{just as } \sigma_S^2 = \tfrac{1}{8}a^2.$$

Thus, the *ems* can be rewritten from those shown in Table 5.8(ii) as given in Table 5.8(iii). The expectation of σ_W^2 is easily obtained from the variation within the various families shown in Table 5.11. We notice that only two of the families, both of which are backcrosses, have any genetical variance for a given gene and hence, $\sigma_W^2 = \tfrac{1}{4}V_A^* + \tfrac{1}{2}V_D^* + V_E$.

Finally, there is no great importance attached to k_T^2 in the ANOVA as this simply reflects the difference in overall mean of the progeny from P_1 and P_2. The more dissimilar the parents, the larger will be k_T^2 and it merely reflects [a].

Table 5.11 NCIII experiment: genetical variation within families

		Testers	
F_2 genotype	Frequency	A^+A^+	A^-A^-
A^+A^+	$\frac{1}{4}$	0	0
A^-A^+	$\frac{1}{2}$	$\frac{1}{4}(a-d)^2$	$\frac{1}{4}(a+d)^2$
A^-A^-	$\frac{1}{4}$	0	0
Average variance		$\frac{1}{8}(a-d)^2$	$\frac{1}{8}(a+d)^2$

$$\sigma_W^2 = \tfrac{1}{2}[\tfrac{1}{8}(a-d)^2] + \tfrac{1}{2}[\tfrac{1}{8}(a+d)^2] + V_E$$

$$= \tfrac{1}{8}a^2 + \tfrac{1}{8}d^2 + V_E$$

$$= \tfrac{1}{4}V_A^* + \tfrac{1}{2}V_D^* + V_E \text{ for many genes}$$

So far we have considered only a single gene segregating in the F_2 but the normal extension to many genes results in the following:

$$\sigma_S^2 = \tfrac{1}{8}\Sigma a^2 = \tfrac{1}{4}V_A^*$$

$$\sigma_D^2 = \tfrac{1}{8}\Sigma d^2 = \tfrac{1}{2}V_D^*$$

$$\sigma_W^2 = \tfrac{1}{8}\Sigma a^2 + \tfrac{1}{8}\Sigma d^2 + V_E = \tfrac{1}{4}V_A^* + \tfrac{1}{2}V_D^* + V_E.$$

A very important point to note is that it does not matter whether the original parents, i.e. the 'testers', were in association or dispersion. In the absence of linkage the F_2 would be the same, whatever the gene distribution in the parents and hence the expectations of the σ^2 are unaffected. The testers merely have to differ at all the loci segregating in the F_2 which, since they are the grandparents of the F_2, must be true.

This design, therefore, enables us to test for and estimate Σa^2 $(= 2V_A^*)$ and Σd^2 $(= 4V_D^*)$ independently with equal precision and even provides a variance ratio test for over-dominance, $F = MS_{TS}/MS_S$. This follows from the component σ^2s, MS_{TS} having the same expectation as MS_S apart from being a function of Σd^2 rather than Σa^2. In the data set used for illustration (Table 5.9(i)) the ANOVA (Table 5.9(ii)) indicates significant additive and dominance variation with the dominance being incomplete since the dominance ratio is estimated as 0.76 (see Table 5.9(iii)).

It may be useful to note that the crosses to P_1 and P_2 respectively correspond to backcrosses and hence the means of these half-sib groups are equal to $Bc_{1.1}$ and $Bc_{1.2}$ (see Table 5.10). They can thus be used as additional generation means to supplement an analysis of the basic generations.

5.6 The Triple Test Cross: TTC

The basic NCIII fails to provide a test for non-allelic interaction, although as was suggested above, the existence of two inbred lines makes possible very powerful tests for epistasis based on generation means. However, the addition of a third tester, the F_1 between the first two, provides an independent test for epistasis [5]. The genetic values of families derived from this third tester together with those families derived from P_1 and P_2 are shown in Table 5.12.

It is apparent from this table that for every F_2 parent the mean of their progeny from the cross to the F_1 is exactly one half the sum of the means of their crosses to P_1 and P_2, i.e.

$$\epsilon(\bar{L}_{3i}) = \epsilon(\bar{L}_{1i} + \bar{L}_{2i})/2$$

or

$$\epsilon(\bar{L}_{1i} + \bar{L}_{2i} - 2\bar{L}_{3i}) = \epsilon(E_i) = 0.$$

This is true for all i provided that there is no non-allelic interaction and thus we have an independent scaling test for every F_2, although it is more convenient to build these n scaling tests into a single ANOVA for epistasis. If we call E_i the observed value of $\bar{L}_{1i} + \bar{L}_{2i} - 2\bar{L}_{3i}$ for the ith F_2, then the epistasis SS is simply $\Sigma E_i^2/6$, where the divisor of 6 derives from the squared coefficients of the $\bar{L}_{ij}$s, i.e. $1^2 + 1^2 + (-2)^2$. Because we have not removed any correction term this SS has n and not n − 1 df. However, we could remove the correction term, $(\Sigma E_i)^2/6n$, which will have 1 df, leaving a remainder SS with n − 1 df as illustrated in Table 5.13(ii), using the data from Tables 5.9(i) and 5.13(i). When we come to consider genetical models with epistasis in Chapter 11, we will see that the correction factor detects additive × additive epistasis while the remainder MS with n − 1 df tests for additive × dominance and dominance × dominance epistasis.

Table 5.12 Expectations of family means for a Triple Test Cross

		Testers				
		P_1	P_2	F_1	$\frac{1}{2}$ Sums	Differences
F_2 genotype	Frequency	A^+A^+ $(\bar{L}_{1i})$	A^-A^- $(\bar{L}_{2i})$	A^+A^- $(\bar{L}_{3i})$	$\frac{1}{2}(\bar{L}_{1i} + \bar{L}_{2i})$	$(\bar{L}_{1i} + \bar{L}_{2i}) - 2\bar{L}_{3i}$
A^+A^+	$\frac{1}{4}$	$m + a$	$m + d$	$m + \frac{1}{2}a + \frac{1}{2}d$	$m + \frac{1}{2}a + \frac{1}{2}d$	0
A^+A^-	$\frac{1}{2}$	$m + \frac{1}{2}a + \frac{1}{2}d$	$m - \frac{1}{2}a + \frac{1}{2}d$	$m + \frac{1}{2}d$	$m + \frac{1}{2}d$	0
A^-A^-	$\frac{1}{4}$	$m + d$	$m - a$	$m - \frac{1}{2}a + \frac{1}{2}d$	$m - \frac{1}{2}a + \frac{1}{2}d$	0
Mean		$m + \frac{1}{2}a + \frac{1}{2}d$	$m - \frac{1}{2}a + \frac{1}{2}d$	$m + \frac{1}{2}d$	$m + \frac{1}{2}d$	0

Table 5.13 Triple Test Cross illustration, to be read in conjunction with Table 5.9. (i) Means of progeny from crosses to the F_1 tester together with additive and epistatic comparisons; (ii) ANOVA (see Chapter 11 for explanation of epistasis)

(i)

F_2	Tester $F_1 (= \bar{L}_{3i})$	Additive $\bar{L}_{1i} + \bar{L}_{2i}$	Epistasis $\bar{L}_{1i} + \bar{L}_{2i} - 2\bar{L}_{3i}$
1	77.930	147.197	−8.663
2	77.112	141.635	−12.589
3	74.791	149.124	−0.458
4	78.041	146.750	−9.332
5	68.393	145.842	9.056
6	79.332	153.889	−4.775
7	72.740	155.313	9.833
8	79.744	156.711	−2.777
9	66.044	142.125	10.037
10	70.320	149.602	8.962
11	69.839	147.087	7.409
12	71.767	141.681	−1.853
13	68.920	143.304	5.464
14	74.557	148.056	−1.058
15	77.866	150.247	−5.485
16	74.350	141.707	−6.993
17	75.916	159.28	7.448
18	70.954	149.689	7.781
19	76.688	146.815	−6.561
20	61.699	133.599	10.201
Total	1467.0030	2949.6530	15.647
Mean	73.3502	147.4827	0.7824
Within FS family SS (180df)	15445.4850	−	−

(ii)

Source	df	MS	F	ems
Additive by additive epistasis	1	20.4024	<1 n.s.	$\sigma^2 + 60\sigma_e^2 + 1200k_i^2$
Dominance by dominance and dominance by additive epistasis	19	99.1463	1.21 n.s.	$\sigma^2 + 60\sigma_e^2$
(All epistasis)	20	95.9091	1.15 n.s.	$\sigma^2 + 60\sigma_{e'}^2$
Within FS families	540	82.2392		σ^2

There is one further feature of this ANOVA for epistasis that requires explanation. Normally in an ANOVA we assume that all the errors are homogeneous, but in a TTC this is not necessarily true. Thus the families derived by crossing the F_2s to P_1 and P_2 will have less genetical variation than those families derived from crosses to the F_1. This means that when we calculate the error MS for the epistasis ANOVA we cannot simply average all the within family mean squares as we did for the NCIII ANOVA. Because $E_i = \bar{L}_{1i} + \bar{L}_{2i} - 2\bar{L}_{3i}$, the error variance of E_i is $V(\bar{L}_{1i}) + V(\bar{L}_{2i}) + 4V(\bar{L}_{3i})$. Had all these three variances been homogeneous and equal to $V(\bar{L})$, then the error would have been $6V(\bar{L})$ which is why we divided by 6 earlier in calculating the epistasis SS. Since they are likely to be heterogeneous, therefore, we should combine the three variances in the proportion $1:1:4$ and divide them also by 6. Using the data in Tables 5.9(i) and 5.13(i), we see that the pooled within family SS for L_1, L_2 and L_3 are 14449.3542, 12587.0442 and 15445.4850 respectively, each with 180 df. These yield the within family MS of 80.2742, 69.9280 and 85.8083. Combining these we obtain,

$$(80.2742 + 69.9280 + 4 \times 85.8083)/6 = 82.2392$$

as shown in Table 5.13(ii).

If there is no significant epistasis, then the L_1 and L_2 data can be analysed by the NCIII analysis, as we have in fact already done with the illustrative data set. Alternatively, we can use the L_3 data as well as the L_1 and L_2 to compute the additive MS, since the epistasis test has shown that $\bar{L}_3$ does not differ significantly from $(\bar{L}_1 + \bar{L}_2)/2$. Thus we can calculate the MS for $\bar{L}_{1i} + \bar{L}_{2i} + \bar{L}_{3i}$ instead of $\bar{L}_{1i} + \bar{L}_{2i}$, as we did for the NCIII, the MS for dominance staying as before. Such an approach will avoid wasting data and will provide a more reliable estimate of the additive variance, but it will lose the advantage of the NCIII of having equally reliable estimates of V_A^* and V_D^*.

It is unlikely that we could ever be lucky enough to have a trait for which there is no epistasis, though it may evade detection in any particular experiment. If epistasis is detected, however, how do we proceed? We clearly need to know whether it is sufficiently large to mislead us were we to ignore it and how we would be misled. This problem and its possible resolution will be discussed in Chapter 11, although it must be obvious, however, that estimates of V_A^*, V_D^* and V_E will be biased to some extent.

To optimize efficiency for detecting dominance, theoretical studies [4,6] suggest that at least 20 and up to 40 F_2s may be necessary. These studies also showed that the NCIII/TTC design was much more efficient than any other design for estimating additive and dominance effects.

5.7 The diallel cross

Although diallel crosses are not normally used to study F_2 populations, it is convenient to introduce them under the heading of HS designs. They have been very widely used, particularly by plant geneticists, and possibly more publications exist on the theory and analysis of diallels than any other design [7].

The complete diallel involves taking a set of n genotypes, i.e. individuals in the case of an F_2, and selfing and intercrossing them in all combinations including reciprocals to produce n^2 FS families. Clearly, this imposes severe restraints on the type of organism that can be used; in the case of an F_2, or indeed any other population, only self-compatible plants can be both selfed and crossed reciprocally. Also, the number of FS families increases geometrically with n, and for this practical reason, diallels seldom involve many more than ten parents, while the largest known to the authors involves 30. Various partial diallel crosses are, however, used to overcome some of these difficulties. They may lack selfs and/or reciprocals, or simply involve a balanced random sample of the crosses. We will concentrate on the full, n^2, diallel.

A set of data from a 5×5 diallel using F_2 parents is given in Table 5.14(i). This could be treated as a NCII analysis but it would ignore the fact that the males and female parents are the same individuals. Thus the families in the ith row have the same common parent as those in the ith column. The most complete analysis of such data was developed by Hayman [8] and so is referred to as the Hayman analysis. Essentially it involves treating the analysis in two parts, that on the reciprocal differences and that on the reciprocals summed, together with the selfs.

Since there are n^2 FS families there are $n^2 - 1$ df between families. Of these, $\frac{1}{2}n(n - 1)$ relate to differences between the pairs of reciprocal crosses, and reflect variation due to non-nuclear causes, while the remaining $\frac{1}{2}n(n + 1) - 1$ df reflect genetical variation among the selfs and crosses. Taking the reciprocal differences first, these can be partitioned into overall differences arising from using the ith individual as a mother rather than as a father, and this will have $n - 1$ df, because there are n parents (c in Table 5.14 (ii) and (iii)). The remaining variation, with $(\frac{1}{2}n - 1)(n - 1)$ df $[= \frac{1}{2}n(n - 1) - (n - 1)]$ reflects specific differences, unique to particular reciprocal crosses (d in Table 5.14). The genetical effects can be similarly partitioned into a GCA MS with $n - 1$ df and a SCA MS with $\frac{1}{2}n(n - 1)$ df (a and b in Table 5.14). Finally, it is possible to further partition the SCA MS into three components. The first, b_1 with 1 df, contrasts the mean of the

Table 5.14 Analysis of a 5×5 Diallel cross. (i) Family means based on ten individuals per FS family; (ii) Hayman's ANOVA; (iii) derivation of SS

(i)

Female parent (i)	Male parent (j)					Total (yi.)	yi. − y.j
	1	2	3	4	5		
1	59.092	69.797	65.231	68.384	71.439	333.943	−5.583
2	72.530	62.663	67.267	75.131	70.688	348.279	0.436
3	65.919	64.542	69.720	67.356	64.616	332.153	−18.557
4	71.965	75.689	77.153	65.225	70.562	360.594	13.611
5	70.020	75.152	71.339	70.887	76.893	364.291	10.093
Total (y.j)	339.526	347.843	350.710	346.983	354.198	1739.260 (Y..)	
(yi. + y.j)	673.469	696.122	682.863	707.577	718.489		
(yi. + y.j−5y.)	378.009	382.807	334.263	381.452	334.024		

Y. = sum of selfs = 333.593

(ii)

Source	df	MS	F
a	4	329.8965	4.17**
b	10	278.1616	3.52**
b_1	1	508.2980	6.42*
b_2	4	436.5890	5.52**
b_3	5	105.3922	1.33 n.s.
c	4	165.7125	2.09 n.s.
d	6	49.2997	<1.0 n.s.
Within FS families	225	79.1021	

(iii)

$$\text{Total SS (between family means)} = \sum Y_{ij}^2 - \{Y_{..}^2/n^2\}$$

$$(c + d)SS = \sum_{i>j} (Y_{ij} - Y_{ji})^2/2$$

$$aSS = \left\{\sum (Y_{i.} + Y_{.j})^2/2n\right\} - \{2Y_{..}^2/n^2\}$$

$$bSS = \text{Total SS} - aSS - (c + d)SS$$

$$cSS = \sum (Y_{i.} - Y_{.j})^2/2n$$

$$dSS = (c + d)SS - cSS$$

$$b_1 SS = (Y_{..} - nY_.)^2/n^2(n - 1)$$

$$b_3 SS = \sum_{i<j} (Y_{ij} + Y_{ji})^2/2n - \sum (Y_{i.} + Y_{.j} - 2y_i)^2/2(n - 2)$$

$$+ (Y_{..} - Y_.)^2/(n - 1)(n - 2)$$

n(n − 1) crosses with the mean of the n selfs and is a test for directional dominance or inbreeding depression. The second, b_2 with n − 1df, essentially tests the variation in the difference between the selfs and crosses among parents. Finally, b_3, with the remaining $\frac{1}{2}$n(n − 3) df, measures the residual dominance variation and is equivalent to the M × F MS in the NCII. The actual computation of these MS is rather non-standard and their derivation is beyond the scope of this text. However, the formulae used to calculate them are presented in Table 5.14(ii) and the ANOVA obtained is given in Table 5.14(iii). Other analyses have been proposed and for a full account one should read Griffing [9].

We see from the ANOVA in Table 5.14(ii) that there is significant additive and dominance variation but no evidence for non nuclear, maternal or cytoplasmic effects. The b_1 MS indicates directional dominance and, if we compare the mean of the crosses (70.2833) with the mean of the selfs (66.706) we see that there is dominance for high score. Furthermore, the significance of b_2 indicates that the extent of directional dominance varies between the five parents reflecting the fact that they carry different numbers of dominant alleles. There is no evidence for any residual dominance variation, b_3.

There has been much discussion in the literature about the interpretation of the σ^2s from this ANOVA; however, it is the view of the present authors that there is little point in trying to estimate and interpret the components of variation in a diallel. Their interpretation is very dependent on the model used but, more importantly, the restricted number of parents, and hence the low df, means that the standard errors of any estimates would be so large as to render them almost useless. We would recommend that interpretation be confined to the sort of qualitative statements which we have used above but we will return to expand this theme in Chapter 9.

5.8 HS designs using inbred lines from an F_2 as parents

We have used an F_2 population to explain the analyses and interpretations of the various HS designs mainly to keep the algebra simple, although their principal use has always been in the context of natural or artificial populations as will be explained in Chapter 9. It is, however, appropriate at this point to mention their use to study the genetics of populations derived from inbreeding an F_2; SSD or DH lines, for example. Such random, homozygous lines can be used as parents for all the HS experiments considered in this chapter and, indeed, the FS designs described in Chapter 4. Although the data are analysed in the same way the genetical and environmental expectations of the σ^2s change, albeit in a very simple and predictable fashion resulting in much more powerful tests of the model and more reliable estimates of the genetical parameters. Intercrossing a random collection of homozygous lines derived from an F_2 recreates the F_2, thus the total variation partitioned by the analyses, σ_P^2, is still $V_A^* + V_D^* + V_E$. But because we are crossing homozygous lines, the variation within FS families in every experimental design is solely environmental and therefore the genetical variation is entirely between families. The expectations of the various σ^2 which we developed in this chapter and Chapter 4 now become as follows:

$$\sigma_M^2 = \tfrac{1}{2}V_A^*$$

$$\sigma_F^2 = \tfrac{1}{2}V_A^*(+V_{EM})$$

$$\sigma_{MF}^2 = V_D^*(+V_{EC} - V_{EM})$$

$$\sigma_W^2 = V_E$$

$$\sigma_{F/M}^2 = \tfrac{1}{2}V_A^* + V_D^*(+V_{EC})$$

$$\sigma_B^2 = V_A^* + V_D^*(+V_{EC}).$$

Because the expected variation within FS families is reduced and that between FS families is increased, the power of the F tests in the relevant ANOVA will also be increased considerably. It is therefore far preferable to use inbred lines derived from the F_2 rather than the F_2 itself as parents for any of these designs.

We have illustrated the analyses for experiments involving complete individual randomization. The experiments could equally well involve randomized plots and/or blocks or indeed any other design which provides a measure of error against which to test the main effects of the ANOVAS.

Summary

1. Half-sib mating designs collectively provide ample choice for analysing any species, irrespective of the constraints of the breeding system.
2. The NCI is the most restriction-free design which allows the analysis of all kinds of populations particularly those that have imbalance between sexes, e.g. cattle and sheep. However,

continued

it is also the one which yields minimum genetic information and does not allow an independent test for the dominance effects.

3. The NCII is more restrictive in respect of the material and can not be applied to those species that either produce only one or two progeny per female or are difficult to cross, but it allows independent tests and estimates of the additive, dominance and maternal effects.

4. The NCIII and Triple Test Cross designs are more suited for studying the genetical control of variation in a cross of two pure lines. They provide tests and estimates of all the major sources of phenotypic variation and allow us to determine the relative importance of the additive, dominance and epistatic effects. These designs can also be applied to other populations provided suitable testers can be found. A major constraint to their application, however, is the rather extensive and costly crossing programme on which they are based and this further restricts their suitability to only a few species which are highly reproductive and easily crossable.

5. The Diallel is a design which is not very suitable for studying variation in a segregating population because the parental sample sizes are generally too small to be reliable. Its value comes in the analysis of crosses between inbred lines.

6. The analytical powers of all these designs are increased many fold when they are applied to the inbred families instead of individuals of a population.

References

1. Robertson, A. (1959) Experimental design in the evaluation of genetic parameters. *Biometrics*, **15**, 219–26.
2. Comstock, R.E. and Robinson, H.F. (1952) Estimation of average dominance of genes, in *Heterosis* (ed. J.W. Gowen), Iowa State College Press, Ames, Iowa, pp. 494–516.
3. Kearsey, M.J. (1965) Biometrical analysis of a random mating population: a comparison of five experimental designs. *Heredity*, **20**, 205–35.
4. Kearsey, M.J. (1970) Experimental sizes for detecting dominance. *Heredity*, **25**, 529–42.
5. Kearsey, M.J. and Jinks, J.L. (1968) A general method of detecting additive, dominance and epistatic variation for metrical traits: I Theory. *Heredity*, **23**, 403–9.

6. Kearsey, M.J. (1980) The efficiency of North Carolina Experiment III and the selfing, backcrossing series for estimating additive and dominance variation. *Heredity*, **45**, 73–82.
7. Pooni, H.S., Jinks, J.L. and Singh, R.K. (1984) Methods of analysis and the estimation of genetic parameters from a diallel set of crosses. *Heredity*, **52**, 243–53.
8. Hayman, B.I. (1954) The analysis of variance of diallel tables. *Biometrics*, **10**, 235–44.
9. Griffing, B. (1956) Concept of general and specific combining ability in relation to diallel crossing systems. *Aust. J. Biol. Sci.*, **9**, 463–93.

Genes, genetic markers and maps

6

The location of genes to positions on specific chromosomes is referred to as **gene mapping** and genes which are located on the same chromosome are said to be **linked** or exhibit **genetic linkage**. Because it is important to map the genes controlling quantitative traits, QTL, and to understand the effect of linkage between QTL on the means and variances of quantitative traits and their response to selection, an understanding of the principles of gene mapping is essential. In this chapter, therefore, we explain the basic theory and practice of gene mapping in eukaryotes. We will concentrate on the theory as it applies to single, major genes and in Chapter 7 discuss how the theory can be extended to QTL.

An individual chromosome contains a single molecule of double-stranded DNA. Its length is approximately constant for any given chromosome in a species but varies between chromosomes. Although it is dangerous to make general statements about the genetical organization of chromosomes, the following is probably not too unreasonable. Typically, they each consist of something of the order of 10^7 to 10^8 base pairs (bp), i.e. 10^4 to 10^5 kilo-base pairs (kbp). A typical structural gene, coding for a polypeptide chain, is between 1 and 2 kbp long, and only about 10% of the genome is actually coding, much of the rest being spacer DNA, although the amount of spacer DNA can vary considerably between species. Thus a chromosome probably contains something of the order of 1000 to 10 000 genes. Fortunately, the fact that the chromosome is a linear molecule means that the genes need to be mapped in only one dimension.

As we will see in Chapter 7, any attempt to locate and study the genes controlling quantitative traits requires the use of independent and easily recognizable genetic marker loci as flags to identify the regions of chromosome containing the QTL. Such markers have to exist in two or more allelic forms at the DNA level which can be distinguished so that precise genotypes at all marker loci can be identified.

6.1 Genetic markers

During the first half of this century, most genetical analysis of variation involved looking at fairly gross morphological, anatomical or behavioural differences; major mutants in other words. Subsequently, in the 1950s and 1960s, it was possible to look at more subtle variation in the structure of polypeptides and, more recently, since the 1980s it has become possible, fairly easily, to explore variation at the fundamental level of DNA itself. No matter what level one is studying, from a subtle difference in the DNA to an observable change in the phenotype, such differences can be used as genetic markers to identify particular chromosomal regions, providing their positions on the chromosomes can be mapped.

In order to fully understand the concepts of gene mapping and its uses, it is necessary to make a short digression into the nature of these various types of genetical variation. The coverage will be superficial and designed to indicate the general principles and methods employed. If the reader is not familiar with the basic ideas of DNA and RNA structure and the concepts of gene action through transcription and translation, they are advised to read any basic genetics text [1]. This section is also not intended to be a treatise for the molecular biologist and such readers are invited to show forbearance at the rather broad generalizations.

6.2 Structural and regulatory genes

Most genes with which we are familiar, such as those controlling eye colour in humans or dwarfing in plants, are concerned with the production of polypeptides, each gene defining the primary structure of one of these long molecules. They are called structural genes, for this reason, and the great majority of them are located on the chromosomes in the nucleus. Other genes, however, are to be found in cytoplasmic organelles such as mitochondria and chloroplasts, but we will not concern ourselves with these. Although the genes are in the nucleus, the polypeptides are synthesized in the cytoplasm by ribosomes and, once produced, they form proteins either alone or together with other polypeptides produced by other genes. In their turn, proteins have a variety of key roles acting as enzymes, hormones, transporter molecules, storage and structural elements as well as regulating the expression and action of other genes. Their roles are so fundamental to life that genetical variation that disrupts protein structure often produces a drastic change in the phenotype of the organism.

Each structural gene typically consists of a length of chromosome approximately 1000–2000 bp long. At one end it has a promoter, which is the binding site for an enzyme, a DNA-dependent RNA polymerase, which will transcribe the DNA code of the gene into RNA. At the far end of the gene is a terminator which tells the enzyme where to stop transcribing. The RNA produced by this process is termed messenger RNA (mRNA), and it carries a complementary copy of the DNA code of that particular gene from the nucleus to the cytoplasm. In the cytoplasm, this RNA message is read and translated by the ribosomes into a linear molecule of amino acids, a polypeptide chain. However, the RNA from some genes is used directly as RNA, either as building blocks for ribosomes or for the transfer RNAs (tRNAs) that carry the amino acids to the ribosomes. Some DNA sequences are never transcribed yet are essential for the control and regulation of other genes and act, for example, as binding sites used in the control of transcription or replication. Because their sequence fidelity is essential, they will be included under the present heading, although they are not genes in the strict sense because they are not transcribed.

Within many genes of eukaryotes, there may exist one or more intervening sequences of non-coding DNA known as **introns** in order to distinguish them from the coding parts or **exons**. The function and origins or these are not known precisely, although they appear to be tightly conserved in size and position over long periods of evolutionary time. These introns are transcribed along with the exons but are cut out before translation.

6.3 Mutations in structural genes

In order to illustrate the consequences of molecular variation in the DNA on the action of a typical structural gene, we will use the well-studied constituent of red blood cells, haemoglobin, as a model. This is the protein which is associated with an iron-binding haem group and transports oxygen. Normal adult haemoglobin consists of two molecules each of two different polypeptide chains, α and β, which are known to be encoded by genes on human chromosomes 16 and 11, respectively. The α chain consists of 141 amino acids and the β chain 146, the first seven of the β

chain being,

Normal β chain	NH₂-	Val-	His-	Leu-	Thr-	Pro-	Glu-	Glu-
Position		1	2	3	4	5	6	7

During translation, each of the amino acids is added in sequence as determined by the sequence of triplets of bases in the mRNA. For example, the mRNA sequence might read,

- GUA - CAU - CUA - ACC - CCU - GAG - GAG ...
  1   2   3   4   5   6   7 ...

where A, C, U, G stand for the nucleotides adenine, cytosine, uracil and guanine. Reference to the genetic code in Table 6.1 shows that the first mRNA triplet codes for the amino acid valine, because a transfer RNA with the anticodon for GUA will be the transfer RNA carrying valine. Similarly the tRNA carrying histidine will have the anti-codon for CAU, etc. Given this mRNA sequence, it follows from the rules of complementary base pairing that the corresponding DNA sequence of the sense strand must have been,

- CAT - GTA - GAT - TGG - GGA - CTC - CTC
  1   2   3   4   5   6   7

Although the processes of DNA replication are very accurate and constantly being checked for errors, it is inevitable given the vast number of bases being replicated, that over generations errors will occur, i.e. there is mutation at the base level. The most common will be base changes, additions or deletions. With 146 amino acids in the β chain there are 438 bases that could mutate, and each has four alternative forms. Therefore, the number of possible single base variants of the gene, called **alleles** or **allelomorphs**, that could exist is 4^{438}. This is an enormous number and exceeds the number of fundamental particles in the universe many times over. Fortunately, in practice, only a very small number of these alleles actually exist in a population. Let us now look at the possible consequences of such genetic mutation by base changes in the DNA and we will use changes in the 6th triplet of human β haemoglobin as an example as shown in Table 6.2

The normal allele, Table 6.2(a), results in the insertion of glutamic acid at position 6. We see in mutant (b) that although there has been a base change in the DNA, the same amino acid is encoded because of the redundancy in the genetic code. This is called a **same-sense** mutation because the same polypeptide is produced, and so such a mutation should have no effect on the individual that carries it. The

Table 6.1 The genetic code. The margins give the bases in the mRNA codons and the body of the table contains the amino acids for which they code

First base	Second base				Third base
	U	C	A	G	
U	Phe	Ser	Tyr	Cys	U
	Phe	Ser	Tyr	Cys	C
	Leu	Ser	STOP	STOP	A
	Leu	Ser	STOP	Trp	G
C	Leu	Pro	His	Arg	U
	Leu	Pro	His	Arg	C
	Leu	Pro	Gln	Arg	A
	Leu	Pro	Gln	Arg	G
A	Ile	Thr	Asn	Ser	U
	Ile	Thr	Asn	Ser	C
	Ile	Thr	Lys	Arg	A
	Met	Thr	Lys	Arg	G
G	Val	Ala	Asp	Gly	U
	Val	Ala	Asp	Gly	C
	Val	Ala	Glu	Gly	A
	Val	Ala	Glu	Gly	G

Amino acids:
Ala, alanine; Arg, arginine; Asn, asparagine; Asp, aspartic acid; Cys, cystine; Gln, glutamine; Glu, glutamic acid; Gly, glycine; His, histidine; Ile, isoleucine; Phe, phenylalanine; Pro, proline; Ser, serine; Thr, threonine; Trp, tryptophan; Tyr, tyrosine; Val, valine.
STOP, chain terminator.

Bases:
U, uracil; C, cytosine; A, adenine; G, guanine.

The corresponding DNA codons can be obtained from the mRNA codons as follows:

mRNA	DNA
U	A
C	G
A	T
G	C

Table 6.2 Consequences of mutations at the DNA base level using the 6th triplet of the Hb(A) gene in humans for illustration

Allele	DNA codon	mRNA codon	Amino acid	Type
Normal (a)	CTC	GAG	Glutamic acid	Wild-type
Mutant (b)	CTT*	GAA	Glutamic acid	Same-sense
Mutant (c)	CA*C	GUG	Valine	Mis-sense
Mutant (d)	A*TC	UAG	Nil-STOP	Non-sense

* indicates the site of change from the normal base.

only possible disadvantage might be in the preferred codon usage because not all alternative tRNAs are equally available and so it may be easier to use the GAG mRNA codon to insert glutamic acid than to use GAA.

In (c) the mutation results in a different amino acid, namely valine, replacing glutamic acid at position 6. This is called a **mis-sense** mutation and the consequences of such mutations are very variable. At one extreme, it could be negligible if the amino acid concerned had no more than a spacing role in the protein molecule while at the other extreme it could be lethal if it affected a critical structural site, the haem site for example. In the present case the effect is to produce the form of haemoglobin that results in **sickle-cell anaemia**, the consequences of which are well known. Had the mutation been at the first base, on the other hand, then lysine would have replaced glutamic acid and the consequences would still be anaemia although of a less severe form.

Mis-sense mutations, as the above example shows, can result in substituting amino acids of different charge, because glutamic acid, valine and lysine are acidic, neutral and basic amino acids respectively. Such differences in charge enable the resulting proteins to be separated in an electric field by a procedure known as electrophoresis, as shown in Figure 6.1. If a mixture of the two proteins is placed in a suitably buffered gel of starch or polyacrylamide, through which an electric current is passed, they will separate into two bands. This principle can be used to distinguish genotypes from blood samples alone. Thus blood from individuals who were homozygous for normal haemoglobin, Hb(A), would give a single slow band while that from individuals homozygous for the sickle-cell form, Hb(S), would give a fast band. Heterozygotes would produce both bands and hence all three genotypes are distinguishable.

This procedure has been used extensively to investigate genetical variation in a wide range of proteins in natural and artificial populations of plants and animals, including humans, and has revealed very

Figure 6.1 An illustration of gel electrophoresis equipment. Protein or DNA samples are placed in the wells in the buffered gel and migrate in response to an electric current. The distance travelled depends on molecular size and charge.

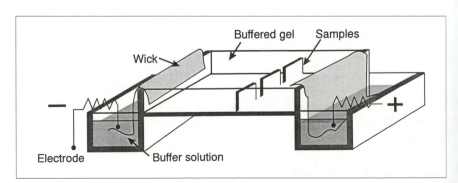

considerable genetical variation for mis-sense mutations in which two or more alleles are common in populations. The situation where two or more allelic forms of a gene exist at frequencies greater than expected by mutation pressure alone is called **genetic polymorphism**. It would appear that such mutations, while not having the debilitating effect of anaemia, may slightly modify the activity of the protein's response to environmental factors such as temperature or oxygen level. This in turn may give the two forms an advantage in different environments and hence maintain the two forms at moderate frequencies.

The final type of base substitution in Table 6.2, case (d) results in the introduction of a terminator codon, and is called a **non-sense mutation**. This codon is not recognized by any tRNA and so stops translation. Instead of the β chain being 146 amino acids long, it consists of the first seven amino-acids only. Clearly the consequence of this is to produce an inactive protein.

The effect of base addition or deletion at the DNA level is to put the triplet reading frame out of sequence and cause **frame-shift** mutations. They result in a nonsense sequence of amino acids being built up, followed sooner or later by a premature stop codon. They are therefore likely to behave like non-sense mutations.

Thus we can see that mutation at the base level of a gene can have a variety of effects from the almost inconsequential same-sense to the drastic non-sense mutation, depending on where it occurs and what the changes are. The so-called 'major mutants' involve base changes which result in non- or mis-functioning polypeptides, while the variation that underlies quantitative traits and other polymorphisms probably consists of the same-sense and innocuous mis-sense mutations plus regulatory mutants.

6.4 Molecular genetic markers

It is possible therefore to recognize mutations at different levels; the gross phenotype (e.g. sickle-cell anaemia), charge changes in the substituted amino acids (i.e. the rate of migration of proteins in response to electrophoresis), the primary structure of the protein (i.e. the amino acid sequence) and finally the DNA sequence itself. Until the 1980s it was difficult to look specifically at the DNA sequence of different individuals directly and for many reasons this is the most useful level at which to work. Only about 10% of DNA is coding DNA and hence looking at structural genes limits the variation available for analysis. Moreover, as we have seen, a great deal of DNA variation in structural genes may be cryptic at higher levels. However, new techniques have made it increasingly easy to look at variation in the DNA in both coding and non-coding regions of the chromosome.

Essentially these techniques involve breaking the DNA into small fragments, using enzymes extracted from bacteria called restriction enzymes [1]. The bacteria use these enzymes as part of their defence mechanism against viruses, bacteriophages, which attack them and the enzymes function by identifying viral DNA and cutting it. This defence is based on the enzyme recognizing particular 4- to 6-bp lengths of unmethylated DNA which are unique to the virus and do not occur in the bacterium. For example, a restriction enzyme called *Eco*RI comes from the bacterium *Escherichia coli* and searches out unmethylated double-stranded DNA with the sequence,

$$\Downarrow$$
$$5' - G - A - A - T - T - C - 3'$$
$$3' - C - T - T - A - A - G - 5'$$
$$\Uparrow$$

which it cuts at the point indicated by the arrows. A very wide range of restriction enzymes have now been isolated from different bacteria, each cleaving a different DNA sequence.

Given the very large number of base pairs in the genome, any one of these short sequences is likely to occur, by chance, very frequently. Thus, a given set of six bases such as that recognized by *Eco*RI will have an expected frequency of 1 in 4^6, i.e. 1 in 4096 randomly drawn sets of six bases. Since a single chromosome may contain 10^8 bases, such a sequence could occur some 24 000 times. DNA sequences are not entirely random over the whole of their length, but clearly the likely number of occurrences of a given site is large. Multiply this by the number of different enzymes, and the potential cleavage sites across any genome are enormous. Moreover these sites do not have to be within structural genes, indeed since most DNA is non-coding, most sites are likely to be between genes.

If DNA is digested with a specific restriction enzyme, it will be cut into a large number of fragments, called **restriction fragments**, of different length, depending on the distances between the sites along the DNA as shown in Figure 6.2. If these DNA fragments are subjected to electrophoresis, they will be separated according to their length into bands, the shortest fragments travelling farthest as shown diagrammatically in Figure 6.2. These fragments can be transferred to a special nylon membrane by a process known as Southern blotting, in order to stabilize them and fix their position.

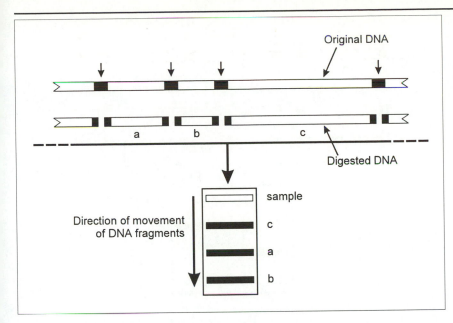

Figure 6.2 The use of restriction enzymes to cut DNA into fragments of different length that can be separated on a gel. The arrows above the DNA indicate restriction sites e.g. for *Eco*RI. The DNA fragments a, b and c are of various sizes, and they migrate in the gel according to their length, shorter fragments travelling farthest, as shown.

In practice, if this were done with total genomic DNA, there would be so many different-sized fragments, that, were the membrane to be treated with a DNA stain such as ethidium bromide, one would simply observe a smear without recognizing individual bands. Thus it is necessary to have a label that will specifically identify particular bands. Such specific labels can be produced by isolating individual small fragments of DNA and replicating them in a source of radio-active bases. If these labelled DNA fragments, called **probes**, are spread over the smear of DNA restriction fragments under defined conditions, they will attach by complementary base pairing to any fragment that contains the same sequence. If the probes are made from the same DNA as the restriction fragments then at least one fragment, i.e. that containing the sequence of chromosome from which the probe was originally made, will hybridize with it. However, there may be other fragments which contain the same or similar sequences, as well. The presence of the probe-labelled band(s) can be identified either by autoradiographic or fluorescence techniques and a typical plate is shown in Figure 6.4. By using different probes, different bands can be identified. Because there has been considerable conservation of DNA sequences during evolution, it is possible to use probes obtained from one species to identify restriction fragments in another and there are now a wide variety of useful probes available.

The value of this technique in genetical analysis lies in its ability to identify polymorphism at the DNA base level in a population. Consider a region of chromosome with three *Eco*RI restriction sites X, Y_1 and Z, as shown in Figure 6.3, and for which we have a diagnostic

Figure 6.3 Restriction fragment length polymorphism (RFLP). X, Y_1, Z are *Eco*RI restriction sites while Y_2 is a mutated restriction site, not recognized by the restriction enzyme. The down arrow (▽) indicates the probe site. Chromosomes containing Y_1 will produce two restriction fragments, rfl_a and rfl_b, while chromosomes containing Y_2 will produce only one, rfl_{ab}. These will result in different band patterns on electrophoresis, but only fragments rfl_a and rfl_{ab} will be recognized and labelled by the probe.

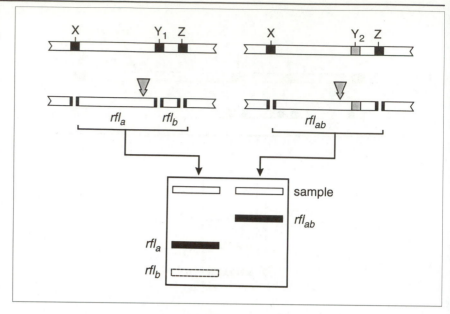

probe. This particular DNA sequence will be cut into two restriction fragments, rfl_a and rfl_b, with rfl_a being the largest and labelled by the probe to give the bands shown. Let us suppose that this population also contained chromosomes with a single base substitution in site Y such as,

$$-C-T-T-T-A-G-$$

$$-G-A-A-A-T-C-$$

This sequence, which we will call Y_2, will no longer be recognized and cut by the restriction enzyme and, as a result, a single long fragment rfl_{ab} made up of $rfl_a + rfl_b$ will be produced which will be labelled by the probe. Such DNA will give a single slow-moving band as shown in Figure 6.3. If we were to compare the DNA of individuals which are respectively Y_1Y_1, Y_2Y_2 and Y_1Y_2 we will observe the patterns shown in Figure 6.4(i), the heterozygote showing both bands found in the homozygotes. It is therefore possible to recognize all three genotypes simply and unambiguously and the single base change in what is possibly a non-coding region, can be handled as though it was a single, co-dominant gene. It is not a gene, because a gene codes for something and the restriction fragment site does not. So, it is conventional simply to call it a **locus** in order to distinguish it from a true gene. Genetic variation that can be recognized in this way is referred to as a **Restriction Fragment Length Polymorphism**, or RFLP. There are a growing number of alternative approaches and techniques based on identifying small differences in base sequence

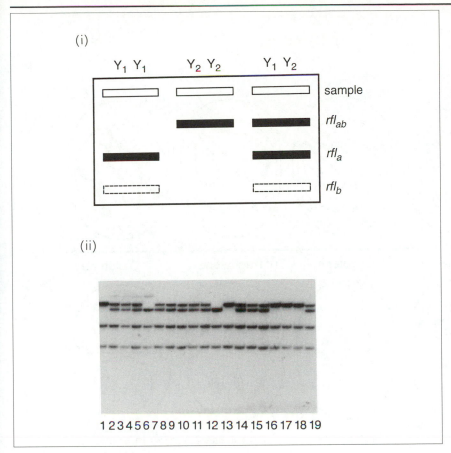

Figure 6.4 Restriction fragment length polymorphism (RFLP). (i) RFLP banding patterns for the three diploid genotypes, Y_1Y_1, Y_2Y_2 and Y_1Y_2, where Y_1 contains a restriction site not present in Y_2 (see Figure 6.3 for explanation of bands). (ii) A typical autoradiograph showing variation similar to that in (i).

becoming available such as **RAPD** (Random Amplified Polymorphic DNA), **AFLP** (Amplified Fragment Length Polymorphisms) and microsatellites [2,3].

6.5 Chiasmata, crossing over and genetic exchange

There are essentially two stages to gene mapping; locating genes to particular chromosomes and then identifying their positions along that chromosome. We will consider the second problem first because it is generally the easiest to solve.

Let us start with two genes each with two alleles, A/a and B/b, and extend the ideas to more genes later. We will use the notation of A,a etc., rather than A^+, A^- at this point because we wish to consider genes or loci in general, not just QTL. Also, the A,a notation simplifies the description of events. Consider the F_1 from a cross between two

Meiosis

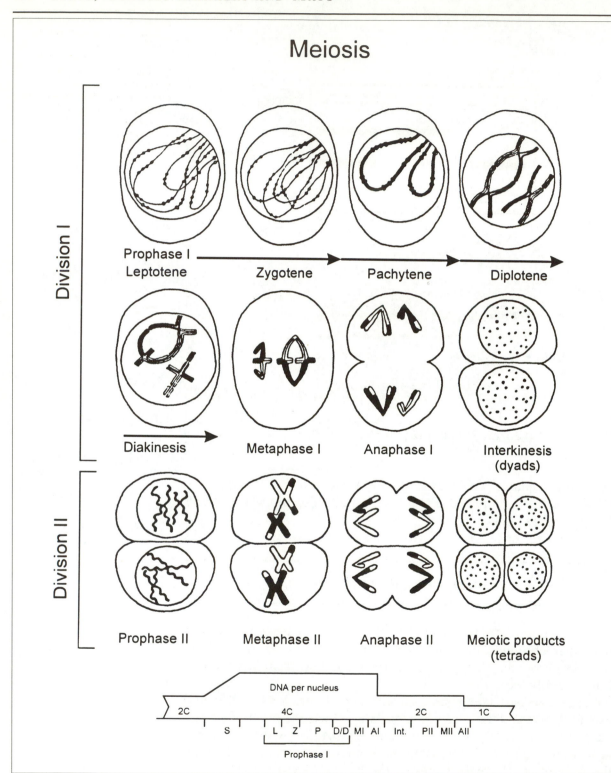

homozygous, true-breeding parents, P_1 (AABB) and P_2 (aabb). If these F_1 individuals are selfed or intercrossed, and if these genes are on different chromosomes, i.e. they are unlinked, then they will segregate independently in an F_2, as enshrined in Mendel's second law. However, if they are on the same chromosome, this will no longer be true and in order to see what happens at gamete formation in this case, we need to consider the events at meiosis as set out for two pairs of homologous chromosomes in Figure 6.5.

Immediately before meiosis the maternal and paternal chromosomes replicate to give two identical copies (chromatids) of each, but the chromatids of each chromosome stay closely attached to give a maternal pair and a paternal pair. At zygotene of meiosis, the maternal and paternal sets of each homologous pair come to lie very close together with almost exact gene for gene association. This process is known as **synapsis** and results in a structure known as a **bivalent**. This association is held together by a ladder-like backbone called the **synaptonemal complex** which is maintained through pachytene as the chromosomes shorten and thicken. After this, however, despite the chromatids staying together in maternal and paternal pairs, the homologous chromosomes repel each other and at the next stage, diplotene, the homologues would completely separate if it were not for sites along the chromosomes, called **chiasmata** (the singular is chiasma), where the paternal and maternal chromatids appear to be joined as shown in Figure 6.5.

The organization of the chiasmata is illustrated in Figure 6.6. The detailed molecular events leading to a chiasma are complex and not yet well understood, but the following simplified account describes the net effects of these events. During zygotene, pachytene or possibly earlier, there has been a break at effectively the same point in one of the maternal and one of the paternal chromatids (Figure 6.6(i)). This breakage has been 'repaired', but by connecting the maternal end of one to the paternal end of the other, reciprocally. This results in a **cross-over** in two of the four strands (chromatids) so that two of the chromatids have one end of maternal origin and the other of paternal origin as can be clearly seen in Figure 6.6(ii). The other two chromatids were not involved in the breakage and rejoining and, therefore, are entirely maternal or paternal. It is not in fact possible to tell which chromatids have actually crossed over by inspection of a diplotene chromosome, although it is commonly assumed that one can do so, largely because of the way chiasmata are represented

Figure 6.5 Diagrammatic representation of the principal stages of meiosis. (Reproduced by kind permission of Dr G.H. Jones, Plant Genetics Group, School of Biological Sciences, The University of Birmingham, UK.)

Figure 6.6 Stages in the formation of a cross-over to produce a chiasma. (i) Breakage of maternal and paternal chromatids; (ii) rejoining to produce recombinant chromatids; (iii) as for (ii) but chromatids observed in a different orientation.

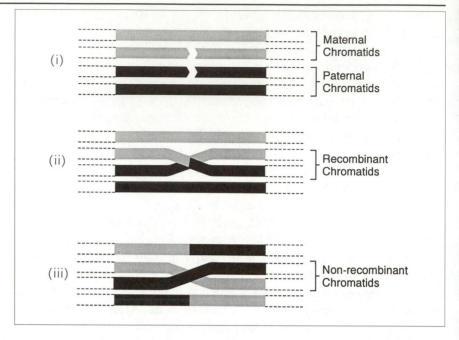

Figure 6.6 Stages in the formation of a cross-over to produce a chiasma. (i) Breakage of maternal and paternal chromatids; (ii) rejoining to produce recombinant chromatids; (iii) as for (ii) but chromatids observed in a different orientation.

in diagrams like Figure 6.6(ii). The bivalent is a three-dimensional structure which is compressed into two dimensions on the microscope slide and when represented on the printed page. In fact, either of the arrangements of chromatids shown in Figure 6.6(ii) or (iii) could reflect what is actually seen through the microscope. An actual diplotene is illustrated in Figure 6.7 to illustrate how crossing over at chiasmata can be seen and the chiasmata counted.

The number and location of chiasmata can be scored under the microscope and such studies in a wide range of animal and plant species have revealed certain critical features. Each bivalent normally has at least one chiasma and failure to form a chiasma typically results in non-disjunction of chromosomes at anaphase I. However, as with many rules in biology, there are some notable exceptions. For example, there are no chiasmata in male meiosis in *Drosophila* and several other species, but chromosomes assort regularly at anaphase I.

A chromosome may have two or more chiasmata, the number typical of a given chromosome being roughly proportional to its length, though rarely does any chromosome have more than five or six chiasmata. For any given chromosome in a species, the number and positions of the chiasma in different meioses within the same individual will vary. It is as though there were a very large number of sites where breakage and reunion of chromatids could occur, but that in any one meiosis, only a very few sites do actually form chiasmata.

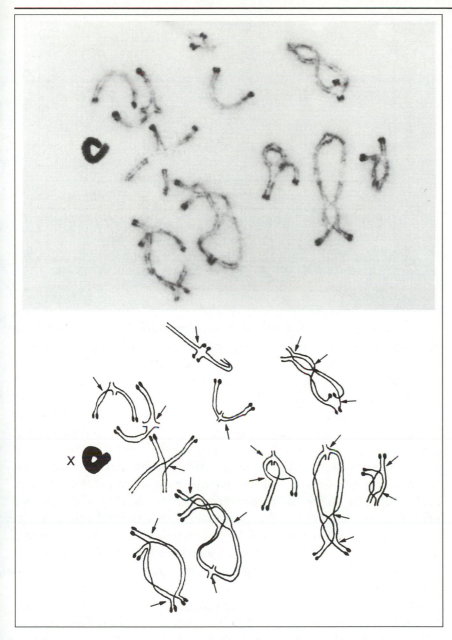

Figure 6.7 Diplotene in male meiosis of the locust, *Schistocerca gregaria*, which has 11 paired autosomes plus an unpaired X chromosome which stains darkly. Crossing over at chiasmata can be clearly seen and this particular nucleus has a total of 20 chiasmata indicated by arrows on the diagram. (Reproduced by kind permission of Dr G.H. Jones, Plant Genetics Group, School of Biological Sciences, The University of Birmingham, UK.)

Although chiasmata may occur anywhere along a chromosome, their distribution is by no means uniform along its length. There are often regions, particularly associated with heterochromatin around the centromeres and telomeres, in which chiasmata form less frequently, and there may be 'hot-spots' in other regions. In wheat, for example, chiasmata tend to form towards the ends of chromosome arms. This non-uniform distribution results in a distortion in the

Figure 6.8 Relationship between chiasma distribution along a chromosome and cross-over frequency. Genetic maps (g) are distortions of the true physical map (p) because of the non-uniform distribution of chiasmata along the chromosome. Genes in regions of high chiasmata frequency appear farther apart than those in regions where chiasmata seldom form. (C, centromere.)

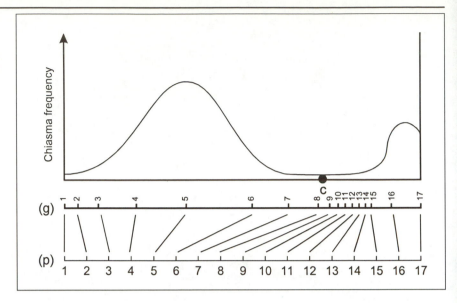

relationship between the physical map, representing the true gene positions along the DNA in kbp, and the genetical map as is shown in Figure 6.8. A further complication arises from the fact that chiasma frequency and distribution is often different in male and female meioses of the same species [4].

These chiasmata, or their genetical equivalents, cross-overs, are the simplest and most useful way we have at present to measure the distance between genes on the chromosome and to determine the gene order. What is called the **genetic distance** between any two genes is simply a function of the average number of cross-overs that occur between them. It is analogous to measuring the distance between two towns on a highway, not in kilometres but by the average number of lane diversions, due to road repairs, that are encountered over a large number of journeys along the highway. That length of chromosome which on average has one cross-over, as estimated from a large number of meioses, is defined as being 50 centimorgans (cM) long in honour of the distinguished geneticist, Thomas Hunt Morgan. It follows from this that if a length of chromosome has, on average, μ chiasmata then it will be 50μ cM long.

It is possible in some species to count the total number of chiasmata on all the chromosomes in a nucleus and then to obtain the average over a number of nuclei. It is, however, generally easier to obtain such counts from male meioses. Some examples are shown for various species in Table 6.3. Armed with these values of μ per nucleus, the total map length of the chromosomes is simply 50μ cM. In many species, some or all of the individual chromosomes can be distinguished,

Table 6.3 Relationship between chromosome number, size and chiasma frequency in different species

Species	Chromosome number per gamete	Total genome size (kbp)	Map length (cM)	Mean Xa frequency per nucleus	Expected total map length	kbp per cM
Man	23	$2.8 \times 10^{6*}$	2708	55	2750	1034
Drosophila	4	$1.8 \times 10^{5*}$	279	–	–	645
Bread wheat	21	$1.6 \times 10^{7\dagger}$	2575	55	2750	6214
Tomato	12	$1.3 \times 10^{6\dagger}$	1400	22	1100	929
Maize	10	$3.0 \times 10^{6\dagger}$	1350	27.05	1353	2222
Rice	10	$4.4 \times 10^{5\dagger}$	1575	–	–	279
Arabidopsis	5	$1.5 \times 10^{5\dagger}$	501	10	500	299
Brassica oleracea	9	$6.1 \times 10^{5\dagger}$	1080	22	1100	565

[*] Weaver, R.F. and Hedrick, P.W. (1989) *Genetics*, 2nd edn, W.C. Brown, Dubuque, USA.
[†] Arumaganathan, K. and Earle, E.D. (1991) Nuclear DNA content of some important plant species. *Plant Mol. Biol. Rep.*, **9**(3), 208–18.

and it is then possible to estimate, cytologically, the mean number of chiasmata, and hence the map length in cM, of particular chromosomes.

6.6 Chiasma frequency and recombination frequency

Cytological observations on chiasmata can, therefore, define the total map size of the whole genome or particular chromosomes. This is an important parameter because it places a limit to the total map length that one would expect to reach when all the individual genes are mapped. However, it is not possible to count the number of chiasmata between individual genes because, with few special exceptions, the individual genes cannot be seen under the microscope. It is necessary, therefore, to find some other way to estimate the relevant chiasma frequency and this involves looking at the results of the genetic exchange that are caused by the chiasma, i.e. the recombination of maternal and paternal alleles.

We must now explore how these chromosomal events, which result in chiasmata, affect the assortment of genes A/a and B/b. In Figure 6.9 we illustrate the F_1 at diplotene of meiosis. In Figure 6.9(i) there is a single cross-over outside the region A/a to B/b and in (ii) there is a single cross-over between A/a and B/b. These represent the two possible types of event at meiosis involving one cross-over. If we follow (i) through the events of meiosis we end up with four products, two of which contain chromosomes with AB and two with ab. That is,

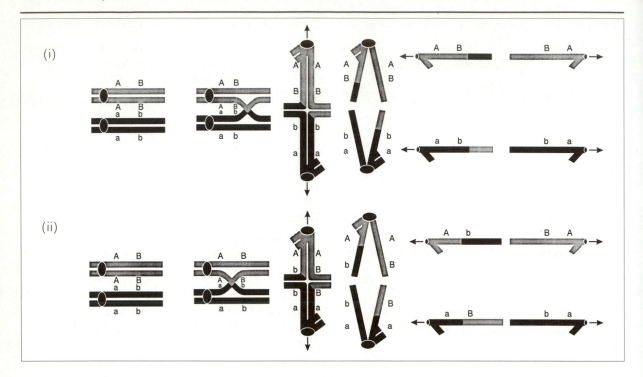

Figure 6.9 Consequences of crossing over. Meiosis with (i) chiasma outside the region A–B, so all resulting chromosomes are non-recombinant for AB; (ii) chiasma between A and B, so 50% of the chromosomes are recombinant for AB.

all the chromosomes contain one or other of the original, parental combinations. If this was in a male meiosis, all four products could become gametes, but in female meioses, only one of the four is destined to be included in a gamete. However, the net result from (i) is that all gametes are parental combinations of alleles, AB or ab. With Figure 6.8(ii) on the other hand, only half the products of meiosis are parental, *AB* and *ab*, while the other half are non-parental, Ab and aB. The latter are termed **recombinant** gametes as the maternal and paternal genes have been recombined. Gene mapping in plants and animals is almost entirely based at present on estimating the frequency of these recombinant gametes; the recombination frequency. We will now attempt to explain why and how this is done.

We have just shown that a single chiasma between two heterozygous genes in an F_1 will result in half the gametes being recombinant. It was also pointed out above that the number and positions of chiasmata vary between meioses, with the restriction that there is always at least one per chromosome pair. Now, if the two genes A/a B/b are very close together, then it is very unlikely that a chiasma will occur between them and hence only very rarely will recombinant gametes be produced. Conversely, if the two genes were at opposite ends of the chromosome, there would always be a chiasma between them and therefore 50% of the gametes produced will be recombinant as

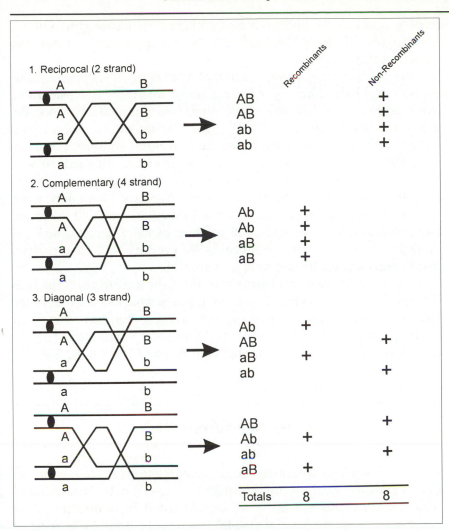

Figure 6.10 Double crossing over between genes. There are three types of double crossing over which can occur and their consequences in terms of the number of recombinant and non-recombinant gametes they produce are shown.

far as genes A and B are concerned. Therefore, the frequency of recombinant gametes is a function of the distance separating the two genes.

But what happens if the two genes are sufficiently far apart that two chiasmata occasionally form between them? This is illustrated in Figure 6.10. If we assume that any of the four chromatids can be involved in the recombinational events then there are four possible situations with two chiasmata. In Figure 6.10, the first event is set to have occurred between strands 2 and 3 in every case, while the second event occurs between various combinations of the four strands. Thus in (i) strands 2 and 3 are again involved and this results in reversing the first event producing all parental gametes. In (ii) the second event involves strands 1 and 4, resulting in all recombinant gametes. Since (i) and (ii) are equally likely possibilities, the net result is that half the gametes are recombinant. Situations (iii) and (iv) both involve three of the four strands and both result in half the gametes being recombinant.

Thus, we come to the conclusion that, providing all four chromatids are involved independently in the exchange, and there is considerable genetical evidence that this is true, then two chiasmata between the genes A/a, B/b will also lead to 50% recombinant gametes on average. Indeed, if the same argument is applied to meioses with three or more chiasmata between the two genes, the same result is obtained; 50% of the gametes are recombinant. This conclusion is based on an average over many meioses but this is what we are concerned with in practice since each female gamete comes from a different meiosis and, given the enormous difference between the number of male gametes and the number of progeny they produce, most progeny will have derived from male gametes from different meioses, too.

So, as far as the two genes are concerned, there are just two situations that need to be considered; first those meioses which have at least one chiasma between A/a and B/b and second those that have none. Let the frequencies of these two events be θ and $1 - \theta$ respectively. The frequency of recombinant gametes between A/a and B/b will be $50\theta\%$.

6.7 Estimation of recombination frequency

Having shown how chiasmata cause recombination as seen in recombinant gametes, we must now consider how to estimate the number of recombinant gametes. Clearly we cannot count them directly as the genotype of a gamete is not visible. It has to be done by looking at the progeny of the F_1 and inferring the gamete genotypes from these progeny.

The simplest way to see this is to consider a backcross of an F_1 to the double recessive (aabb) parent, as illustrated in Table 6.4. This type of cross, known as a test-cross, results in a one-to-one relationship between the phenotypes of the progeny and the genotypes of the gametes. Thus any individual which has a phenotype Ab must have come from a recombinant gamete, Ab, etc. Table 6.4 also shows data from a test-cross in which 135 progeny were scored of which seven were recombinant. The frequency of recombinant gametes is, therefore, approximately 5%.

What does this recombination frequency (RF) of 5% tell us about the relative position of genes A/a and B/b? Had the two genes been on opposite ends of the same chromosome there would always have

Table 6.4 The use of a test-cross to calculate the recombination frequency between two genes. (i) Allelic arrangement in cross; (ii) test-cross data; (iii) tests for single factor ratios; and (iv) estimates of RF

(i)

A	B		a	b

$\times$

a	b		a	b

F_1 $\quad\quad\quad$ P_2

(ii)

		Progeny genotypes		Progeny phenotypes	Number
F_1 gametes	Parental	AB	AaBb	AB	68
		ab	aabb	ab	60
	Recombinant	Ab	Aabb	Ab	4
		aB	aaBb	aB	3
				Total	135

(iii) Test for single factor ratios:

Number of A individuals $= 68 + 4 = 72$

Number of a individuals $= 60 + 3 = 63$

$\chi^2 = 0.6$; n.s.

Number of B individuals $= 68 + 3 = 71$

Number of b individuals $= 60 + 4 = 64$

$\chi^2 = 0.37$; n.s.

(iv) RF $= (4 + 3)/135 = 0.0518$ or 5.18%

been a chiasma between them and the recombination frequency would have been 50%. The fact that it is only 5% tells us that they are quite close and compatible with just one chiasma occurring between them in every 10 meioses or a mean chiasma frequency of $\mu = 0.1$.

Referring back we see that $\mu = 0.1$ is equal to 5 cM, i.e. the recombination frequency and the cM distance are equal. In fact, cM were so defined that 1% recombination is equivalent to 1 cM. Unfortunately, this simple relationship breaks down over longer intervals of chromosome as a simple illustration will show. Suppose the two genes were at opposite ends of a chromosome which had a mean chiasma frequency of 2.5, i.e. it is $2.5 \times 50 = 125$ cM long. Since there is always at least one chiasma between these two terminal genes, the recombination frequency is 50%. Recombination frequency between two genes can never be more than 50%, which is also the recombination frequency expected of (unlinked) genes on different chromosomes. Whereas

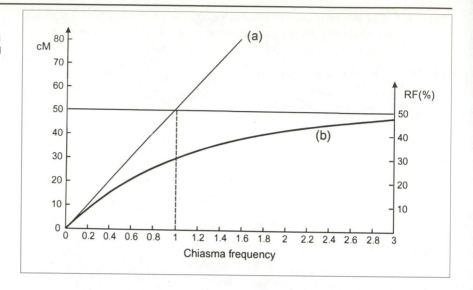

cM distance will increase linearly with the number of chiasmata, recombination frequency will plateau at 50%, as shown in Figure 6.11. This problem is readily resolved, as will become clear later.

So far we have simply considered the estimation of the distance between two genes. Having established the principles, let us now consider three genes. As before, suppose we start with an F_1 which was produced from a cross between two parents, AABBCC and aabbcc, and we wish to determine the order of the genes and their distances apart; the order could be ABC, ACB or BAC. If this F_1 is test-crossed to the triple recessive parent, aabbcc, we might obtain the results illustrated in Table 6.5. The steps in the process are as follows.

Firstly, check that, at all three loci, the alleles are segregating in the 1 : 1 ratio expected of a backcross. This is important because disturbed segregation may upset our estimate of recombination frequency. Disturbed segregation can arise from two different causes, differential survival of the two alleles or misclassification of the genotypes, and the problems associated with these will be discussed later in this chapter. In the example in Table 6.5, χ^2 tests show the ratios not to be significantly disturbed, so we can proceed to the next stage.

Secondly, calculate the recombination frequencies between each pair of genes, A–B, A–C and B–C in the same way as before (Table 6.4). This results in the following values:

$$A\text{–}B = 0.1925 \text{ or } 19.25\%$$

$$A\text{–}C = 0.3025 \text{ or } 30.25\%$$

$$B\text{–}C = 0.3900 \text{ or } 39.00\%$$

Table 6.5 Determination of recombination frequencies and gene order with three genes in coupling linkage. (i) The cross; (ii) the data; (iii) tests of single factor ratios; and (iv) estimates of recombination frequencies

(i)　A　　B　　C　　　　a　　b　　c

———————————　×　———————————

　　a　　b　　c　　　　a　　b　　c

(ii)

	Class	Progeny phenotypes $\equiv F_1$ gametes	Number of progeny
F_1 gametes	1	ABC	103
		abc	120
	2	ABc	51
		abC	49
	3	Abc	14
		aBC	7
	4	AbC	24
		aBc	32
		Total	400

(iii)

Number of A:a individuals = 192 : 208

$\chi^2 = 0.64$; n.s.

Number of B:b individuals = 193 : 207

$\chi^2 = 0.49$; n.s.

Number of C:c individuals = 183 : 217

$\chi^2 = 2.89$; n.s.

(iv)

$R_{A-B} = (14 + 7 + 24 + 32)/400 = 0.1925$

$R_{A-C} = (51 + 49 + 14 + 7)/400 = 0.3025$

$R_{B-C} = (51 + 49 + 24 + 32)/400 = 0.3900$

Since B–C is the largest RF, genes B and C must be the farthest apart while A is in-between, i.e. the sequence is,

$\longleftarrow$————————— 39.00% —————————$\longrightarrow$

══B════════════A════════════C══

$\longleftarrow$—— 19.25% ——$\longrightarrow$ $\longleftarrow$—— 30.25% ——$\longrightarrow$

It is noticeable that the RF B–C is less than the sum of the intervening RFs, B–A and A–C, and this is due to the fact that recombination

Table 6.6 Derivation of expected recombination frequencies in a three marker cross (to be read in conjunction with Table 6.5)

	B	A	C
	I	II	
	b	a	c

Assume recombination frequency in region I $= p$
and recombination frequency in region   II $= q$

Class	Gametes, in gene order	Recombination in interval		Probability	
		I	II		
1	BAC bac	No	No	$f_1 = (1-p)(1-q) = 1-p-q+pq$	
2	BAc baC	No	Yes	$f_2 = (1-p)q$	$= q - pq$
3	bAc BaC	Yes	Yes	$f_3 = pq$	$= pq$
4	bAC Bac	Yes	No	$f_4 = p(1-q)$	$= p - pq$

Recombination frequency in region:

$B-A = f_3 + f_4 = p$
$A-C = f_2 + f_3 = q$
$B-C = f_2 + f_4 = p + q - 2pq$

frequencies, unlike cM distances, are not additive over large distances as was explained earlier. This is basically due to recombinational events occurring in both intervals, B–A and A–C, so that as far as B–C is concerned the second recombinational event cancels the effect of the first. This can be shown formally as follows.

Consider the situation as set out in Table 6.6 where the recombination frequencies in region I (B–A) is p and that in interval II (A–C) is q, with the corresponding frequencies of non-recombination being $1 - p$ and $1 - q$ respectively. Providing the probability of crossing over in regions I and II is independent, and this can be tested, the probability of obtaining each type of gamete is simply the product of the probabilities of the relevant events in each interval. For example, the probability of producing a completely non-recombinant gamete, BAC or bac is $(1 - p) \times (1 - q)$, and the expected frequencies of this and the other three classes are given in the Table 6.6. If we combine these as we would combine the actual numbers to obtain the recombination frequencies in the intervals B–A, A–C and B–C, we obtain the values p, q and $p + q - 2pq$. Thus the distance between the genes farthest apart is less than the sum of the two intervals by exactly twice their product. In the example of Table 6.5, $p = 0.1925$ and $q = 0.3025$ so we would expect the recombination frequency B–C to be 0.3785 which is close to its observed value of 0.39.

We stated above that this relationship depends on the probability of crossing over, and hence chiasma formation, being independent in the two intervals. This can easily be tested by comparing the observed and expected frequencies of double crossing over, i.e. recombination in both B–A and A–C simultaneously. The expected number of gametes with double recombination is $p \times q \times N$, i.e. 22.8 while the observed number is 21. The ratio of these, the observed divided by the expected number of recombinants, is called the **coefficient of coincidence**, CoC. If crossing over in the two regions is independent then CoC = 1. Departure from CoC = 1 is called **chiasma or cross-over interference**, **I** (**I = 1 − CoC**). If this is positive, i.e. there are fewer cross-overs than expected, it suggests that the presence of a chiasma in one interval hinders the formation of a second chiasma in the other; negative interference would suggest synergism. In our example in Table 6.5, CoC = 21/22.8 = 0.92, which is not significantly different from unity and hence we can assume independence.

There is published evidence for almost complete chiasma interference up to 15 cM in *Drosophila*, while our own data using molecular data in *Brassica* indicate the same to be true in plants. Thus, we should not expect two cross-overs closer than 15 cM.

We have thus seen how, using a test-cross, we can measure the genetical distance between three genes and determine their order. As new genes are identified, the same procedures can be used to map them with respect to the others and slowly, as more genes become available for study, a complete genetic map can be constructed. Up until the mid-1980s, most of the genes that were being mapped were associated with clear-cut phenotypic effects which meant that they involved major mutants. Thus mapping generally progressed by looking at progeny from crosses between carriers of the new mutation with genotypes carrying one, two or three previously mapped genes. With the advent of RFLP and other molecular markers, one is commonly in the position of having a very large number of recognizable loci segregating simultaneously in the same cross. In order to handle these, a variety of computer software has been developed which establish the overall map that best fits the combined data [5,6].

The genes in the examples we have used so far have been arranged in such a way that one chromosome contains AB while the other contains ab. An F_1 in this configuration is said to be in **coupling**. Had the arrangement been Ab/aB it is said to be in **repulsion**. If an individual which is A/a B/b is taken from a population it would be impossible to know if it was in coupling or repulsion and this would have to be inferred from the frequencies of the progeny. If it was in repulsion, then the recombinant gametes would be A–B and a–b. Were we wrongly to assume the genes to be in coupling and estimate the recombination frequency from the number of A–b and a–B progeny, then the estimates would exceed 50%, and this should immediately suggest an error. Sampling variation apart, recombination frequencies must be less than or equal to 50%.

Although the essential principles of gene mapping are conveniently illustrated using test-crosses, i.e. backcrosses, they are not the only generation that can be used; in fact any generation derived from a double heterozygote will provide information, such as F_2, F_3, F_∞ or doubled haploid lines. Appropriate formulae for calculating recombination frequencies for these are given in Table 6.7 for the simple cases of either complete dominance or co-dominance. Where one is attempting to map genes in natural populations, particularly with outbreeders, there are particular problems. If the material can be self-fertilized, then any particular parent can be considered as being

Table 6.7 Formulae for estimating recombination frequencies (R) for data from various generations assuming A and B in coupling and co-dominance (i). The different genotypes to which the observed frequencies, a–i refer are illustrated in (ii). R for F_2 is estimated as that value which satisfies the equation and can be obtained by the method of scores (see Bailey, 1961) or by evaluating the equation

(i)

Mapping population	Estimation equation
Backcross	$R = (f + h)/(e + f + h + i)$
Doubled haploid lines	$R = (c + g)/(a + c + g + i)$
Recombinant inbred lines (SSD lines)	$R = (c + g)/\{2(a + i)\}$
F_2	$(c + g)\{2/R\} + (b + d + f + h)$ $\times \{(1 - 2R)/R(1 - R)\} - (a + i)\{2/(1 - R)\}$ $+ e\{2(2R - 1)/(1 - 2R + 2R^2)\} = 0$

(ii)

Genotype	AA	Aa	aa
BB	a	b	c
Bb	d	e	f
bb	g	h	i

equivalent to an F_1 and its progeny to an F_2. The F_2 model can therefore be used to map the loci which were heterozygous in the parent, although, because there are no equivalents to P_1 and P_2, the linkage phase of any particular pair of genes, i.e. whether they are in coupling or repulsion, has to be deduced from the genotypes of the F_2 progeny. Different parents will be heterozygous for different loci and so the map information from separate families will have to be combined in order to build up a complete linkage map.

Alternatively, individuals can be taken from the population, genotyped and mated in pairs to yield a number of full-sib (FS) families. In a particular FS family, any pair of segregating loci will represent either an F_2 or a backcross. For example, if the parents in a particular cross were,

$$\frac{\text{A b c d}}{\text{a B c d}} \quad \times \quad \frac{\text{a B C D}}{\text{A b c d}}$$

then the progeny would represent an F_2 for AB and a backcross for CD, while for BC it is neither. Not only would the linkage phase be unknown, but for the AB pair the linkage phase could be the same or opposite in the two parents, and has to be deduced from the progeny. Human families are of this sort but, with the added problem that the family sizes are very small, it is necessary to combine data from many families as well as from other relatives in order to construct linkage maps. Again, special computer software is available to achieve this.

The reliable estimation of recombination frequencies requires large samples. Because the recombination frequency, R, is a proportion, its variance is $R(1 - R)/N$, from the binomial distribution. Thus an estimate of recombination frequency of 5%, i.e. $R = 0.05$, from 135 test-cross progeny (see Table 6.4) would have a standard error of $\sqrt{(0.05 \times 0.95/135)} = 0.019$, while to obtain a recombination frequency of 1% which is at least twice its standard error would require almost 400 progeny. Sample sizes, therefore, need to be quite large.

6.8 Mapping functions

We will now turn to the problem of converting recombination frequencies, which are not additive, into cM distances which are. This requires the application of a mapping function, the most widely used being that due to J.B.S. Haldane [7], and assumes no chiasma interference.

Suppose that the true mean chiasma frequency between our two genes A–B is μ. Given that the average number of chiasmata on a chromosome is generally small, typically between two and three, then μ is likely to be very small. It follows, therefore that the frequency distribution of meioses with 0, 1, 2, ... chiasmata in the interval A–B will follow a Poisson distribution with mean μ. It was demonstrated earlier that what is important for recombination is the proportion, θ, of meioses with at least one chiasma in the interval A–B. This, of course, is equal to 1 minus the proportion of meioses with no chiasma in this interval, i.e. $\theta = (1 - e^{-\mu})$. Hence the recombination frequency is,

$$R = \theta/2 = (1 - e^{-\mu})/2.$$

Turning this equation round, gives

$$e^{-\mu} = 1 - 2R.$$

Hence

$$\mu = -\ln(1 - 2R),$$

where $\ln = \log_e$. Thus the distance in cM is,

$$\mu \times 50 = (-\ln(1 - 2R)) \times 50 \qquad \text{[Eqn 6.1]}$$

$$(\text{conversely, } R = (1 - e^{-cM/50})/2).$$

If we apply this formula to the three gene case we examined earlier, we obtain

$$B-A = 24.31$$

$$A-C = 46.44$$

$$B-C = 75.71$$

and we see that B–C is very close to the sum of the other two intervals, 70.75, any difference being due to the sampling error of the estimates of R. The standard error of R_{B-C} is 0.024 $(= \sqrt{(0.39 \times 0.61/400)})$, and so its 95% confidence limits are 0.39 ± 0.048, i.e. 0.438 and 0.342. These correspond to map distances of 104.4 and 57.6, which emphasize the lack of precision when mapping over large distances. The corresponding range for B–A is 37.5 to 29.4 cM, which is much smaller, whereas had we mapped two genes with $R = 0.05$, then the confidence interval would be between 4.07 and 6.49. Thus although R has a minimum standard error at 0.5, the cM distance is most precise when R is small.

An example of an RFLP map of a chromosome of *Brassica oleracea* is shown in Figure 6.12. This map was obtained from doubled haploid lines derived from an F_1 and well illustrates the

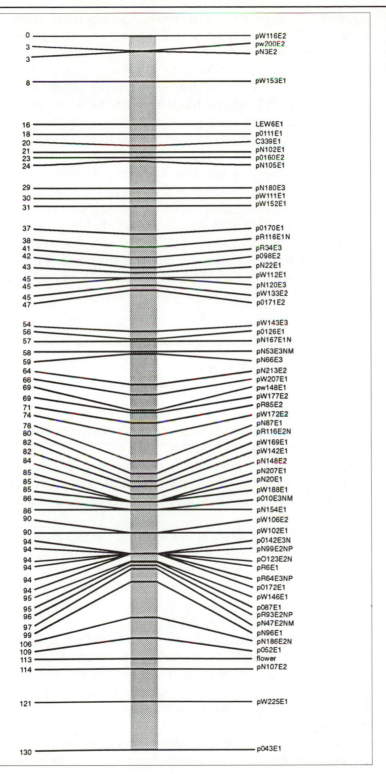

Figure 6.12 A typical molecular marker map for a single chromosome. The example is taken from chromosome 3 of *Brassica oleracea*.

detailed coverage that can easily be obtained with molecular markers. The markers on this map are not true genes and hence are of little interest in their own right. However, they provide an essential framework for locating other genes of interest and hence such maps are known as **framework maps**. Their use in locating and studying QTL will be discussed in Chapter 8 but they are also invaluable for comparing genome organization in different related genera and also as markers for identifying and following important genes in marker-aided selection programmes.

Haldane's mapping function assumes that there is no cross-over interference, but as we have said, this is not true over distances of at least 15 cM. A more realistic function would incorporate this complete interference distance with the random component of Haldane. Such a function, which is commonly used, is that of Kosambi [8]. The effect of this interference can be visualized from Figure 6.11. The RF curve will exactly follow the straight line (a) up to 15 cM and will then follow a line approximately parallel to (b) from then on.

6.9 Segregation distortion

It is not uncommon to find that when progeny from a test-cross, F_2 or other mapping population are scored, some loci show distorted segregation. As we said earlier, this may arise from one of two causes; misclassification of some genotypes or differential survival of different genotypes. Their consequences are quite different for mapping, however, and it is therefore necessary to distinguish between them.

Misclassification arises either because the genotype is ambiguous or from scoring errors. With molecular markers, autoradiographs may be difficult to read, for example, and a faint band may be missed. Changes in methylation occur spontaneously and a restriction site may be hidden or revealed, so mimicking mutation. It can be shown that such errors will inflate the estimate of RF, depending on the scale of misclassification. It therefore behoves the gene mapper to be absolutely scrupulous in checking for and removing dubious observations. We said earlier that cross-over interference is often complete for distances of around 15 cM, therefore data which indicate double cross-overs within this distance should be treated with suspicion.

Selection can occur at any stage from meiosis to scoring. A particular polymorphism in a chromosome may have differential viability and this may be particularly apparent when inbred mapping populations are used. DH lines produced by anther culture are particularly prone to selection for alleles that facilitate selection *in vitro* or at regeneration. Selection distortion of a single locus does not affect estimates of RF. With two or more linked loci showing independent selection, alternative methods of estimating RF are available.

Distortion due to misclassification is normally restricted to the locus concerned and will not affect neighbouring loci. Selective distortion will affect all linked loci in inverse proportion to their distance from the affected locus. The reader should consult Bailey (1961) for further details.

Summary

1. Differences between individuals in their DNA can be recognized at various levels; gross morphological differences, protein differences or base changes in the DNA.
2. Such differences can arise by mutation at the base level; they may have no effect on the individual that carries them through to being lethal.
3. The mapping of genes or loci on chromosomes normally relies on the frequency of chiasmata formation and crossing over between the genes.
4. A length of chromosome which has an average of μ chiasmata or cross-overs is defined to have a genetic length of 50μ cM.
5. It is normally not possible to count chiasmata between individual genes and so recombination frequency is used to estimate the chiasmata frequency.
6. There is growing evidence of considerable cross-over interference extending for a distance of about 15 cM in some species.
7. Because recombination frequency and chiasmata frequency are not linearly related, a mapping function is necessary to construct the map derived from recombination frequency data.
8. Segregation distortion of alleles at particular loci will normally only affect recombination frequency if it arises from misclassification of genotypes. Distortion due to selection does not generally create problems for mapping.

References

1. Griffiths, A.J.F., Miller, J.H., Gebhart, W.H., Lewontin, R.C. and Suzuki, D.T. (1996) *An Introduction to Genetic Analysis*, 6th edn. W.H. Freeman & Co., New York.
2. Sajanta, A., Puomilahti, S., Johnsson, V. and Ehnholm, C. (1992) Amplification of reproducible allele markers for amplified fragment length polymorphism analysis. *Biotechniques*, **12**, 16.
3. Tanksley, S.D., Young, N.D., Paterson, A.H. and Bonierbale, M.W. (1989) RFLP mapping in plant breeding: new tools for an old science. *Biotechnology*, **7**, 257–64.
4. Lagercrantz, U. and Lydiate, D.J. (1995) RFLP mapping in *Brassica nigra* indicates differing recombination rates in male and female meiosis. *Genome*, **38**, 255–64.
5. Lander, E.S., Green, P., Abrahamson, J., Barlow, A., Daly, M.J., Lincoln, S.E. and Newburg, L. (1987) MAPMAKER; an interactive computer package for constructing primary genetic linkage maps of experimental and natural populations. *Genomics*, **1**, 174–81.
6. Stam, P. (1993) Construction of integrated genetic linkage maps by means of a computer package: Joinmap. *The Plant Journal*, **5**, 739–44.
7. Haldane, J.B.S. (1919) The combination of linkage values and the calculation of distances between the loci of linked factors. *J. Genet.*, **8**, 299–309.
8. Kosambi, D.D. (1943) The estimation of map units from recombination values. *Ann. Eugen. Lond.*, **12**, 172–5.

Further reading

Bailey, N.T.J. (1961) *Introduction to the Mathematical Theory of Linkage*. Oxford University Press.
Mather, K. (1938) *The Measurement of Linkage in Heredity*. Methuen and Co., London.

Gene counting and location

7

We have assumed in previous chapters that the quantitative traits are controlled by many underlying genes whose individual effects are small and cumulative. We have further assumed that these genes have additive and dominance effects, that they may interact and that they show all the standard segregation patterns of normal genes on chromosomes. In fact, the whole subject of quantitative genetics has been built around a series of assumptions, which, though plausible and amenable to testing, are based on almost no actual knowledge of the individual genes themselves. We do not know what these genes do, how many there are, nor where they are located on the chromosomes.

It would seem likely that for some traits, such as grain yield in cereals, a very large number of genes are potentially involved. Some will be directly concerned with the number of grains per spike, the number of spikes per tiller and the number of tillers per plant. Some may affect the production of photosynthetic products and their storage in the grain, while yet others may affect root growth, disease susceptibility or germination time of the seed. Thus, a whole army of genes may affect yield, many of which are far removed from the actual trait itself, but all may show allelic variation in populations which we observe as genetical variation in our analyses and which will respond to selection. Clearly, allelic effects of some genes may be large, those of many more may be small, while the vast majority probably have negligible effects, although their combined effect could be considerable.

7.1 Reasons for locating QTL

One reason for wishing to locate QTL is to provide a means of answering some fundamental questions about the genetic control of quantitative traits. How many genes are there? Are there many as was suggested for yield, or are there just a few important genes segregating while the rest are of little or no significance? How are these genes distributed across the genome? Are there 'hot-spots' on particular chromosomes for particular traits or is there a relatively random distribution? In considering these questions we must always bear in mind that we are solely concerned with those genes that are segregating in the crosses or populations under study. Although there are potentially hundreds of genes which might affect yield, only a small proportion of these may actually be segregating in any given instance. Quantitative genetics can say nothing about the monomorphic loci because they do not contribute to variation.

Knowing where particular genes are on the chromosome can help to distinguish between the effects of linkage and pleiotropy, which is of particular concern to breeders (pleiotropy being the situation where the same gene(s) control two or more traits). It would also be possible to look at the additive and dominance effects of individual loci to see to what extent dominance is uni-directional and what size dominance ratios actually are at various loci. Moreover, one could start to analyse interactions between genes more precisely and possibly identify the mode of gene action and interaction in a few cases.

Turning to a more immediately practical level, a knowledge of the location of useful QTL might open up various opportunities to improve selection efficiency or medical diagnosis. This would depend on being able to identify other easily recognizable genes, marker loci, linked to the QTL which could be used to 'flag-up' the presence of the QTL. Such genes would allow for marker assisted selection (MAS) which would be particularly useful in alien gene transfer via backcrossing, improving traits with low heritability or penetrance and improving efficiency by selecting individuals at a juvenile stage. Ultimately, precise location of useful or important QTL could open up the possibilities of gene cloning and transformation into animals and plants of economic interest or for ameliorating genetic deficiencies in humans.

7.2 Background to methodology

Standard methods of gene mapping in eukaryotes depend on scoring the frequency of recombination between genes, the basic principles of which are set out in Chapter 6, and rely on a number of premises.

Perhaps the most important of these is that the genotypes at the relevant loci can be unambiguously identified, either directly in the individuals themselves or by simple progeny tests. Linkage analysis is also adversely affected by segregation distortion, particularly if caused by misclassification but also by viability disturbance.

Mapping quantitative traits, therefore, creates special problems, because the genotype is never unambiguously determined from the phenotype. With several genes controlling a single trait, a given phenotype could be produced by many different genotypes while the same genotype could, because of environmental variation, produce many phenotypes. On their own, therefore, quantitative traits are not particularly amenable to conventional linkage analysis and hence supplementary information is required. This information has to come from the joint study of easily recognizable, single gene marker loci.

The first account of linkage of a quantitative trait to a major gene locus was provided by Sax in 1923 [1]. He was studying the inheritance of seed weight and seed colour in an F_2 of a cross between two lines of the bean, *Phaseolus vulgaris*. Seed weight behaved like a typical quantitative trait, with a continuous distribution, whereas seed colour involved the segregation of a single gene, P/p. Among an F_2 of 166 plants, he observed that the mean weights of the three seed colour genotypes differed as is shown in Table 7.1. Individuals which are PP had heavier seeds than those that were pp, while the heterozygote was almost exactly intermediate. Either the alleles for seed colour were having a pleiotropic effect on seed weight or there was a gene (or genes) affecting seed weight which was closely linked to P/p.

Attempts to use major gene mutants to identify chromosomes or parts of chromosomes which affected quantitative traits continued throughout the following 60 years. These studies were most successful in *Drosophila* and wheat, where the availability of lines containing special chromosomes plus major markers, made it possible to construct genotypes which contained precisely delineated sections of chromosome from particular parental lines [2–5]. Such genotypes enabled quantitative differences to be attributed to specific chromosome regions (Chapter 8).

Table 7.1 Bean weight and colour in an F_2 of *Phaseolus vulgaris* [1]

Number of plants	Average weight of 10 beans (g)	Colour genotype
41	2.64	pp
80	2.83	Pp
45	3.07	PP

Major gene mutants, however, have a number of disadvantages for the analysis of quantitative traits, particularly if one is interested in natural or breeding populations. Firstly, there are few of them and they are unlikely to exist in the populations under study. Given that the QTL may occur throughout the genome, a large number of markers are required, at least one per chromosome and preferably more, and these would have to be introduced into the population from different sources before the study could begin. Such alien introductions could also introduce other genetic material including the very QTL under study.

Another problem with major mutants is that they are very likely to have pleiotropic effects on the quantitative trait, e.g. dwarfing genes in cereals have side effects on yield. They are also likely to affect the fitness of the individuals that carry them: major mutants are normally rare in populations because they are disadvantageous. This would result in the marker locus showing segregation distortion, which, in turn, could affect the measurement of recombination frequency. Finally, major mutants are generally recessive, so it would be necessary to progeny test some phenotypes in order to ascertain their genotype.

These factors have mitigated against the widespread use of single gene markers in hunting for associations with quantitative traits except in very circumscribed laboratory exercises. However, the discovery of naturally occurring, single locus polymorphisms at the DNA level in the 1980s has opened up the possibility of analysing crosses and populations of plants and animals including man [6]. Molecular markers have all the advantages for quantitative genetic analysis that major mutants lacked.

They are common and any individual in an outbreeding species or an F_1 is likely to be heterozygous for hundreds of sites. There are some exceptions however. Some members of the Solanaceae, such as *Nicotiana*, seem to have relatively few, as does wheat, although the latter may be due to the relatively narrow genetic base of cultivated wheats. Molecular markers are mainly outside coding regions and hence most are unlikely to have any effect on the phenotype. Thus pleiotropy with the quantitative trait is not likely to be a problem. Moreover, since they are also common and naturally occurring, they are unlikely to affect fitness and so will not distort segregation ratios in linkage studies. Finally, the majority of molecular markers, though not all (e.g. RAPDs), are co-dominant and so all genotypes are unambiguously identifiable.

There are essentially two approaches for using genetic markers to locate and study QTL. The first, described below, relies on the association between the quantitative traits and marker loci in segregating populations such as F_2s or natural populations. The second, described in Chapter 8, uses markers to engineer chromosomes of particular genetical constitution in order to confine genetical differences to defined chromosomal regions. However, a variety of methods have also been described which use relationships between the biometrical parameters of means and variances to estimate the number of QTL segregating in a population and these will be critically evaluated towards the end of this chapter.

7.3 QTL and marker loci in segregating populations

In order to illustrate the principles of locating and estimating the effects of individual QTL, we will, for simplicity, use a population of doubled haploid lines obtained from an F_1. Such lines, it will be recalled from Chapter 4, can be obtained by regenerating individual plants from haploid pollen grains or ovules and thus represent completely homozygous individuals, each from a different gamete. On selfing, such homozygous plants produce true breeding, DH lines.

Consider a set of DH lines produced from an F_1 derived from a cross between two homozygous parents which differ at four molecular marker loci, A, B, C and D. We will refer to the marker alleles from P_1 as A_1, B_1, etc. and those from P_2 as A_2, B_2, etc. Table 7.2 illustrates data from a sample of 25 such lines which have been scored both for a quantitative trait and their genotypes at these four marker loci. Such a small sample of 25 lines and four markers would be too small for reliable analysis, but is used simply to illustrate the approach. Standard linkage analysis (see Chapter 6) of the three markers A, B, C in these data shows them to be linked with recombination frequencies A–B = 0.16, B–C = 0.32, A–C = 0.4, from which, by Equation 6.1, we can deduce the map, in cM (Haldane), to be:

A————————B————————————————————————C

19.28 cM   51.08 cM

Marker D is unlinked to the other three.

The trait data are the means of four replicates (blocks) and the appropriate ANOVA of the 100 original data items (25 lines × 4 replicates) is shown in Table 7.3(i). From this we see that the lines differ significantly for the quantitative trait and that the additive genetical variance, V_A^*, as estimated from $\frac{1}{2}\sigma_L^2$, is 13.16. We now wish to see whether any of this variation can be associated with the marker loci A, B, C or D.

Table 7.2 Genotype at four marker loci (A–D) and trait mean for 25 doubled haploid lines from an F_1. 1 and 2 refer to *11* and *22* homozygotes respectively

DH line	Marker locus				Mean trait value
	A	B	C	D	
1	1	2	2	2	39.8711
2	1	2	2	1	31.9503
3	1	1	1	1	43.9023
4	2	2	2	1	37.9634
5	2	2	2	2	41.1388
6	1	2	1	2	39.1713
7	1	1	2	1	41.9898
8	1	1	1	1	43.5941
9	2	2	2	1	33.4596
10	2	2	2	2	39.1542
11	2	2	1	1	33.3379
12	1	1	1	1	43.2697
13	2	2	1	1	32.6693
14	2	2	2	1	32.2222
15	1	1	1	1	42.7623
16	2	1	1	2	51.2106
17	2	2	2	1	33.6086
18	2	2	2	2	42.1210
19	1	1	2	1	31.2853
20	1	1	2	1	41.8986
21	2	2	2	2	41.1959
22	2	2	2	1	44.4769
23	2	2	1	2	41.4460
24	1	1	2	1	42.5452
25	2	2	2	1	31.9783

Marker means

$\overline{M_1M_1} = \bar{X}_1$	40.2036	42.4953	41.2626	37.8185
n_1	11	9	9	17
$\overline{M_2M_2} = \bar{X}_2$	38.2845	37.2353	37.9287	41.9136
n_2	14	16	16	8
$\delta_i = (\overline{M_1M_1} - \overline{M_2M_2})/2$	0.9596	2.6300	1.6670	−2.0476

This can be achieved by comparing the mean trait values of all those DH lines which are *11* at a particular locus with those that are *22*, and these means, calculated from the data, are shown at the bottom of Table 7.2. It appears that at loci A, B and C the alleles from parent 1 are associated with an increasing effect on the trait, because the means for the *11* genotypes are largest, whereas parent 1 has a

Table 7.3 Analysis of variance of trait data on 25 DH lines. Data in Table 7.2. (i) Full ANOVA; (ii) ANOVAS for individual marker loci

(i)

Source	df	SS	MS	F	P	ems
Lines (L)	24	2597.9112	108.2463	36.7	***	$\sigma^2_{LB} + 4\sigma^2_L$
Blocks (B)	3	2.6193	0.8731	<1.0	n.s.	$\sigma^2_{LB} + 25\sigma^2_B$
L × B	72	212.5800	2.9525	–		σ^2_{LB}
Total	99	2813.1105				

Estimate of $\sigma^2_L = 26.32$, $V^*_A = 13.16$

(ii)

Source	df	MS(A)	MS(B)	MS(C)	MS(D)
$\overline{M_1 M_1}$ versus $\overline{M_2 M_2}$	1	90.7478	637.4615**	256.0870	364.9118*
Between lines, within marker classes	23	109.007	85.2369	101.8184	97.0870

decreasing effect at the unlinked locus, D. In order to test whether these differences are significant we need to compare them with the variation between lines, within marker classes. If we consider marker B, for example, the SS between the classes is

$$\{(\bar{X}_1 \times n_1)^2/n_1 + (\bar{X}_2 \times n_2)^2/n_2 - [(\bar{X}_1 \times n_1) + (\bar{X}_2 \times n_2)]^2/(n_1 + n_2)\}4$$
$$= 637.4615$$

As the SS between the 25 lines is 2597.9112 (Table 7.3(i)), the SS 'between lines within marker classes' is, by difference, 1960.4497 (i.e. MS = 85.2369, for 23 df), resulting in the revised ANOVA shown in Table 7.3(ii). This confirms that the difference between the means at marker B is significant, and similar analyses (Table 7.3(ii)) confirm this to be true of marker D, also, but not A or C. Although the four tests are not strictly orthogonal, they strongly support the view that parent 1 has at least one increasing QTL on the chromosome carrying linked markers A, B and C, while it carries a decreasing QTL on the chromosome carrying marker D.

It is now necessary to develop a genetical model to help us interpret these observations. Consider first a single marker locus, M, which is linked to a QTL, Q, with a recombination frequency, R, Table 7.4(i). As before, we use the subscripts 1 and 2 to indicate the parental source of each allele. The F_1 will produce four types of gamete, and hence four types of doubled haploid line, in the frequencies shown in Table 7.4(ii). Using the now familiar model for the QTL genotype means of,

$$Q_1Q_1 = m + a$$

$$Q_2Q_2 = m - a$$

the expected means of the marker genotypes can easily be obtained as shown in Table 7.4(iii). Half the difference between the two means, $\delta\{= (1 - 2R)a\}$, clearly depends on the size and direction of the QTL effect, a, and the distance of the QTL from the marker, R. If the marker and the QTL are unlinked, i.e. $R = 0.5$, then $1 - 2R$, and hence δ, are expected to be zero. On the other hand, if the QTL is so close to the marker as not to recombine with it, i.e. $R = 0.0$, then δ is a, as expected since the marker is effectively allelic to the QTL.

The expected relationship between δ for markers at various distances in recombination frequency from a QTL are shown in Figure 7.1(i) and rescaled in terms of cM in Figure 7.1(ii). A particular QTL will exert an effect over the whole chromosome and associations

Table 7.4 Derivation of the model for a single marker locus and a QTL in doubled haploid lines from an F_1. (i) The F_1 constitution; (ii) DH genotypes, frequencies and genetic values; (iii) expectations of marker means and differences

(i)

M_1	Q_1

M_2	Q_2

$$\longleftarrow R \longrightarrow$$

(ii)

F_1 gametes	Frequency	DH genotype	Genetic value
$M_1 Q_1$	$\frac{1}{2}(1 - R)$	$M_1M_1 Q_1Q_1$	$m + a$
$M_1 Q_2$	$\frac{1}{2}(R)$	$M_1M_1 Q_2Q_2$	$m - a$
$M_2 Q_1$	$\frac{1}{2}(R)$	$M_2M_2 Q_1Q_1$	$m + a$
$M_2 Q_2$	$\frac{1}{2}(1 - R)$	$M_2M_2 Q_2Q_2$	$m - a$

(iii)

$$\overline{M_1M_1} = \{\tfrac{1}{2}(1 - R)(m + a) + \tfrac{1}{2}(R)(m - a)\}/\tfrac{1}{2} = m + (1 - 2R)a$$

$$\overline{M_2M_2} = \{\tfrac{1}{2}(R)(m + a) + \tfrac{1}{2}(1 - R)(m - a)\}/\tfrac{1}{2} = m - (1 - 2R)a$$

$$\delta = (\overline{M_1M_1} - \overline{M_2M_2})/2 = (1 - 2R)a$$

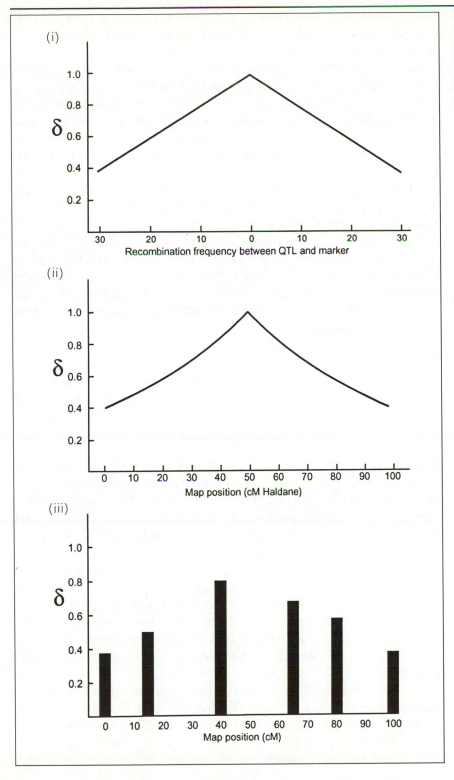

Figure 7.1 QTL mapping. Expected additive effects, $\delta = (1 - 2R)a$, associated with markers at various genetic distances from a QTL of additive effect $a = 1$. (i) Abscissa in recombination frequency; (ii) abscissa in cM; (iii) observed values of δ associated with six marker loci.

with different linked markers may be reflections of the same QTL. In practice, there will be data on a limited number of markers and typical results are shown in Figure 7.1(iii).

Because the difference between the marker means is a function of both a and R, it is not possible from information on a single marker to determine both parameters. At the very least, it is necessary to have information on two markers which are known to be on either side of the QTL, flanking markers, since a will be the same for both and the two recombination frequencies, i.e. that between marker 1 and the QTL and marker 2 and the QTL, will have to combine to give the recombination frequency between the markers which can be independently estimated. The problem then becomes that of identifying the flanking markers.

Based on the relationships set out above, several statistical approaches have been proposed to locate individual QTL and to estimate their effects. Although they involve different degrees of statistical sophistication, they all have similar reliability and we will consider just two of the most straightforward, '**marker regression**' and '**interval mapping**'.

7.3.1 Marker regression

This relies on the linear relationship between the size of the differences in the marker means and the recombination frequency between the QTL and the individual markers as illustrated in Figure 7.1(i),

$$\text{i.e. } \delta_i = a(1 - 2R_i),$$

where δ_i and R_i are, respectively, half the difference between the means at the ith marker and the recombination frequency between the QTL and the ith marker [7]. At the true position of the QTL, there should be a linear regression of δ_i on $(1 - 2R_i)$ with slope a which passes through the origin (see Figure 7.2). That is, it is equivalent to the normal linear regression, $y = c + bx$, where $y = \delta_i$, $c = 0$, $b = a$ and $x = (1 - 2R_i)$. However, the true location of the QTL is not known and so this regression has to be carried out at regular intervals, e.g. every 2 cM, along the chromosome, the most likely position of the QTL being that at which the residual SS of the regression is minimized. The reader should note that, when a linear regression through the origin is to be fitted, the uncorrected SSx, SSy and SPxy are used, and the residual SS has n − 1 and not n − 2 df.

The procedure is illustrated in Table 7.5(i) for a possible QTL at 0.0 cM using the data of Table 7.2 for the three linked markers, A, B and C. The positions of the three markers are given together with the corresponding values of δ_i, from Table 7.2, which correspond to the y_i values. As we are starting with the presumed

Table 7.5 Marker regression analysis of data in Table 7.2. (i) Basic data for QTL at 0.0 cM; (ii) regression ANOVA of data in (i); (iii) basic data for QTL at 38 cM; (iv) regression ANOVA of data in (iii)

(i)

Marker (i)	Marker position (cM)	δ_i	Distance of QTL from marker (cM)	R_i	$(1 - 2R_i)$
A	0.0	0.9596	0.0	0.0	1.0
B	19.28	2.6300	19.28	0.16	0.68
C	70.36	1.6670	70.36	0.38	0.24

$\sum \delta_i^2 = 10.6166$; $\sum (1 - 2R_i)^2 = 1.52$

$\sum \delta_i(1 - 2R_i) = 3.1481$; $a = b = 3.1481/1.52 = 2.07$

(ii)

Source	df	SS	MS	F	P
Regression	1	6.5200	6.52	3.20	n.s.
Residual	2	4.0966	2.04		
Total	3	10.6166			

(iii)

Marker (i)	Marker position (cM)	δ_i	Distance of QTL from marker (cM)	R_i	$(1 - 2R_i)$
A	0.0	0.9596	38.00	0.266	0.468
B	19.28	2.6300	18.72	0.156	0.688
C	70.36	1.6670	32.36	0.238	0.524

$\sum \delta_i^2 = 10.6166$; $\sum (1 - 2R_i)^2 = 0.9669$

$\sum \delta_i(1 - 2R_i) = 3.1320$; $a = b = 3.1320/0.9669 = 3.24$

(iv)

Source	df	SS	MS	F	P
Regression	1	10.1452	10.15	15.3	***
Residual	1	0.4714	0.47	<1	n.s.
Error	22	15.3625	0.6983		

QTL location being at the beginning of the linkage group, i.e. at marker A, the distances of the QTL from the three markers, in cM, will be as in column 4. Use of Equation 6.2 converts these cM distances to R_i and, hence, $1 - 2R_i$. The $(1 - 2R_i.)$ correspond to the x_i values.

Linear regression of the y_i onto the x_i at this point results in a residual SS of 4.0966, while a, as estimated from the regression

Figure 7.2 QTL mapping. Relationship between the marker-associated effect, δ and $1 - 2R$.

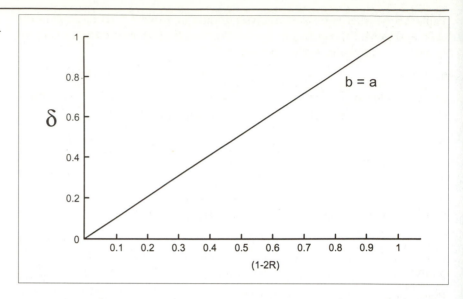

slope, b, is 2.07. Repeating this analysis at 2 cM intervals across the whole region between markers A and C yields a range of residual SS as shown in Figure 7.3. The most likely position of the QTL from Figure 7.3 is where the residual SS is at a minimum, i.e. at 38.0 cM, at which point the QTL effect a is estimated to be 3.242.

The error SS for this analysis is essentially a function of the 'Lines' SS in Table 7.3(i) less the SS due to the regression and the residual. Since the regression analysis was based on the δ_i obtained by averaging over four replicates and 25 lines, the appropriate total SS for the regression analysis is the 'Lines' SS ($= 2597.9112$, see Table 7.3(ii)) divided by 100. After removing the regression and residual SS we are left with a SS of 15.3625, for 22 df which is used as the error. At the optimum QTL location, significance of the regression confirms that a is not zero, while the non-significance of the residual MS indicates that the model is adequate to explain the observed results. The data are thus consistent with a single QTL with $a = 3.24$ at 38.0 cM on this linkage group, and the resulting map is as follows,

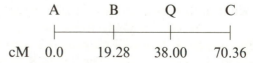

The process has been illustrated for one set of linked markers but it can be repeated for all other sets of linked markers in order to identify and characterize the effects of QTL for the trait on other

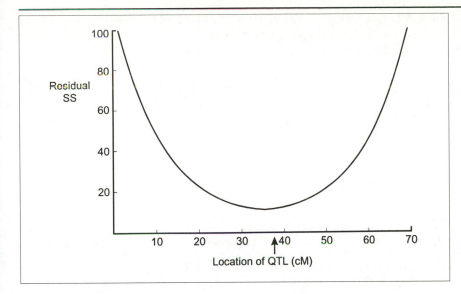

Figure 7.3 QTL mapping. Change in residual SS with putative QTL position along the chromosome using marker regression. The arrow indicates the likely location of the QTL where the SS is at a minimum. (Based on data from Tables 7.2 and 7.3.)

chromosomes. In this way, an overall picture of the number, locations and sizes of significant QTL which are segregating in the cross can be identified.

It was estimated earlier that, for the set of 25 DH lines used for illustration, the additive genetic variation due to all QTL, $2V_A^*(=\sum a^2)$ was 26.32 (Table 7.3(i)). Therefore, the QTL identified at 38.0 cM with $a = 3.24$ contributes 40% to this variance (i.e. $3.24^2 \times 100/26.32$). The QTL linked to marker 'D' must contribute at least another 16% ($2.04^2 \times 100/26.32$) and thus more than half the genetical variation for this trait can be explained by these two QTL alone. The remaining variation must be due to QTL on other chromosomes.

7.3.2 'Interval mapping' by multiple regression

Another approach is to use flanking markers for 'interval mapping' using individual observations rather than marker means. Two different statistical procedures have been proposed to accomplish this, the one using maximum likelihood [8–10] and the other using weighted least squares [11,12]. Since they give essentially identical results, we will illustrate the latter because it involves standard regression procedures which readers may find easier to use and interpret.

This approach can be explained in terms of a set of DH lines as before. Consider a QTL between two flanking marker loci, A and B, i.e. the F_1 is,

$$\frac{A_1 \quad Q_1 \quad B_1}{A_2 \quad Q_2 \quad B_2} \quad \text{(see Table 7.6(i))}$$

Table 7.6 Derivation of model for flanking marker regression analysis of doubled haploid lines. (i) Constitution of F_1; (ii) frequencies and genetic values of DH lines; (iii) marker genotype means

(i)

$$
\begin{array}{ccc}
A_1 & Q_1 & B_1 \\
\hline\hline
A_1 & Q_2 & B_2
\end{array}
$$

$$\longleftarrow R_A \longrightarrow \longleftarrow R_B \longrightarrow$$
$$\longleftarrow \text{—————} R \text{—————} \longrightarrow$$

(ii)

DH genotype	Frequency		Genetic value
$A_1A_1Q_1Q_1B_1B_1$	$\frac{1}{2}(1-R_A)(1-R_B)$	$\Big\}\frac{1}{2}(1-R)$	$m+a$
$A_1A_1Q_2Q_2B_1B_1$	$\frac{1}{2}(R_A)(R_B)$		$m-a$
$A_1A_1Q_1Q_1B_2B_2$	$\frac{1}{2}(1-R_A)(R_B)$	$\Big\}\frac{1}{2}(R)$	$m+a$
$A_1A_1Q_2Q_2B_2B_2$	$\frac{1}{2}(R_A)(1-R_B)$		$m-a$
$A_2A_2Q_1Q_1B_1B_1$	$\frac{1}{2}(R_A)(1-R_B)$	$\Big\}\frac{1}{2}(R)$	$m+a$
$A_2A_2Q_2Q_2B_1B_1$	$\frac{1}{2}(1-R_A)(R_B)$		$m-a$
$A_2A_2Q_1Q_1B_2B_2$	$\frac{1}{2}(R_A)(R_B)$	$\Big\}\frac{1}{2}(1-R)$	$m+a$
$A_2A_2Q_2Q_2B_2B_2$	$\frac{1}{2}(1-R_A)(1-R_B)$		$m-a$

(iii)

$$A_1A_1B_1B_1 = \{\tfrac{1}{2}(1-R_A)(1-R_B)(m+a)+\tfrac{1}{2}(R_A)(R_B)(m-a)\}/\tfrac{1}{2}(1-R)$$
$$= m+a(1-R_A-R_B)/(1-R)$$
$$A_1A_1B_2B_2 = \{\tfrac{1}{2}(1-R_A)(R_B)(m+a)+\tfrac{1}{2}(R_A)(1-R_B)(m-a)\}/\tfrac{1}{2}(R)$$
$$= m-a(R_A-R_B)/(R)$$
$$A_2A_2B_1B_1 = \{\tfrac{1}{2}(R_A)(1-R_B)(m+a)+\tfrac{1}{2}(1-R_A)(R_B)(m-a)\}/\tfrac{1}{2}(R)$$
$$= m+a(R_A-R_B)/(R)$$
$$A_2A_2B_2B_2 = \{\tfrac{1}{2}(R_A)(R_B)(m+a)+\tfrac{1}{2}(1-R_A)(1-R_B)(m-a)\}/\tfrac{1}{2}(1-R)$$
$$= m-a(1-R_A-R_B)/(1-R)$$

Let the recombination frequency between A and B be R, which can be estimated as described in Chapter 6, and let R_A and R_B be the recombination frequencies between the QTL and markers A and B respectively. The F_1 will produce eight gametes, and hence eight unique DH lines, the expected frequencies and genetic values of which are as shown in Table 7.6(ii). However, only the marker genotypes are identifiable, of which there are just four, A_1B_1, A_1B_2, A_2B_1, and A_2B_2. The expected means of each of these are easily obtained from Table 7.6(iii). For example, A_1B_1 comprises $A_1Q_1B_1$ and $A_1Q_2B_1$, thus the mean is,

$$A_1B_1 = m + a(\tfrac{1}{2})[(1 - R_A)(1 - R_B) - (R_A - R_B)]/(\tfrac{1}{2})(1 - R)$$

$$= m + a[(1 - R_A)(1 - R_B)]/(1 - R). \qquad \text{[Eqn 7.1]}$$

Similarly, the other three are,

$$A_1B_2 = m - a[(R_A - R_B)]/R \qquad \text{[Eqn 7.2]}$$

$$A_2B_1 = m + a[(R_A - R_B)]/R \qquad \text{[Eqn 7.3]}$$

$$A_2B_2 = m - a[(1 - R_A)(1 - R_B)]/(1 - R). \qquad \text{[Eqn 7.4]}$$

Clearly, R, R_A and R_B are related, and one can either assume that recombination in the two intervals is independent (i.e. $R = R_A + R_B - 2R_AR_B$) or that there is complete interference (i.e. $R = R_A + R_B$). Either way, since R is known, it is only necessary to estimate R_A to locate the QTL. As with the previous approach, this method involves estimating m and a at a range of possible QTL positions between A and B by regression and locating the QTL at that position where the residual SS is at a minimum. This will correspond to the position where the variance ratio which tests the regression is at a maximum. In this case, one is regressing the trait score (y_i) of the genotype, i.e. a DH line, on to the coefficient of a (x_i). This is illustrated in Table 7.7 using the data in Table 7.2 for a QTL at an arbitrary position 2 cM from A, and assumes no interference. As we have shown, the markers A and B are 19.28 cM apart, so in the illustration, the QTL is 2 cM from A and 17.28 cM from B. These distances translate into the following recombination frequencies,

$$R = 0.1600; \quad R_A = 0.0196; \quad R_B = 0.1461.$$

Using these values, the coefficients of a in Equations 7.1–7.4 give the following values of x for DH lines of the relevant marker genotype,

Genotype at locus		Coefficient of a	$= x$
A	B		
1	1	$(1 - R_A - R_B)/(1 - R)$	$= 0.9932$
1	2	$-(R_A - R_B)/R$	$= 0.7906$
2	1	$(R_A - R_B)/R$	$= -0.7906$
2	2	$-(1 - R_A - R_B)/(1 - R)$	$= -0.9932$

These values of x are shown with the corresponding genotypes in Table 7.7, and normal linear regression analysis yields the SS and estimates shown in Table 7.7(ii). It can be seen that at 2 cM, the residual SS is 616.9039, a is estimated to be 1.1963 while F(regression) is 1.2 and hence not significant.

Repeating this procedure at 2 cM intervals results in the residual SS declining to 497.0128 at 18 cM, i.e. just before marker B, at which point a is estimated to be 2.6320. The whole procedure is then repeated to explore the interval B–C. Over the whole range A–C, the residual SS reaches a minimum at 35.28 cM, i.e. 16 cM to the right of marker B (see Figure 7.4). At this point, the estimate of QTL effect, a, is 3.68, while the value of F(regression) is maximized at 8.7 and is highly significant.

As with 'marker regression', method (i), the regression and remainder MS test the significance of the parameter estimates and the adequacy of the model respectively, as shown by the ANOVAS in Table 7.8 at the two putative QTL positions. In both ANOVA, the SS being partitioned by the regression are identical because it reflects the SS between line means. This is the same as the 'Lines' SS in Table 7.3 apart from a factor of four, because we are using the means of four blocks in the regression analysis. The ANOVAS show that at 2 cM (Table 7.8(i)) the residual MS is significant but the regression is not. At the optimum position for the QTL, i.e. 35.28 cM, the regression is highly significant, confirming a QTL at this point (Table 7.8(ii)). The residual MS is still significant, however, due to the presence of other QTL such as that which we showed earlier to be associated with the unlinked marker, D.

The two methods clearly lead to similar but not identical conclusions. Thus 'marker regression' indicated a QTL at 38 cM with $a = 3.24$, while 'flanking marker regression' places the QTL at 35.28 cM with $a = 3.68$. Computer simulation indicates that the distribution of the estimates of QTL position and effect are almost identical by the two methods and reasonably correlated [7,13]. However, the former attempts to test the consistency of the single QTL model with the data from all linked markers simultaneously, while the latter does not, at least not explicitly.

We illustrate the use of flanking marker and marker regression analyses to locate QTL controlling grain yield in barley in Figure 7.5. The data were obtained from plot trials of 100 DH lines derived from a cross between two commercial varieties, and clearly illustrate the location of a single QTL on chromosome 5.

In order to facilitate an understanding of the methodology, we have used DH lines since the genetic models are simple, both in terms of the genetical parameters themselves, i.e. only m and a, and

Table 7.7 Data for carrying out flanking marker regression based on information in Table 7.2. (i) Genotypes at three marker loci (A–C), x_i values and trait means for 25 doubled haploid lines from an F_1. 1 and 2 refer to *11* and *22* homozygotes respectively. The x_i values are calculated from the formulae in Equations 7.1–7.4 in the text, assuming the QTL to be located 2 cM from A and 17.28 cM from B, the distance A to B being 19.28 cM; (ii) SS, SP and estimate of *a*

(i)

DH line	Marker locus				Mean trait value
	A	B	C	x_i	
1	1	2	2	0.7906	39.8711
2	1	2	2	0.7906	31.9503
3	1	1	1	0.9932	43.9023
4	2	2	2	−0.9932	37.9634
5	2	2	2	−0.9932	41.1388
6	1	2	1	0.7906	39.1713
7	1	1	2	0.9932	41.9898
8	1	1	1	0.9932	43.5941
9	2	2	2	−0.9932	33.4596
10	2	2	2	−0.9932	39.1542
11	2	2	1	−0.9932	33.3379
12	1	1	1	0.9932	43.2697
13	2	2	1	−0.9932	32.6693
14	2	2	2	−0.9932	32.2222
15	1	1	1	0.9932	42.7623
16	2	1	1	−0.7906	51.2106
17	2	2	2	−0.9932	33.6086
18	2	2	2	−0.9932	42.1210
19	1	1	2	0.9932	31.2853
20	1	1	2	0.9932	41.8986
21	2	2	2	−0.9932	41.1959
22	2	2	2	−0.9932	44.4769
23	2	2	1	−0.9932	41.4460
24	1	1	2	0.9932	42.5452
25	2	2	2	−0.9932	31.9783

(ii)

$$SS_x = 22.7573$$
$$SS_y = 649.4771$$
$$SP_{xy} = 27.2264$$
$$SS \text{ regression} = 32.5732$$
$$SS \text{ residual} = 616.9039$$
$$b = a = 1.1963$$

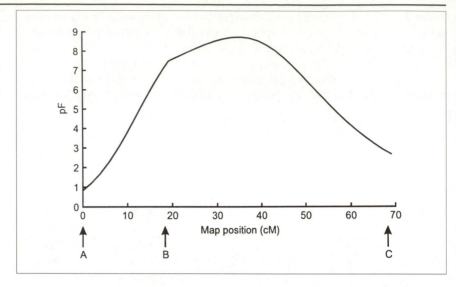

Figure 7.4 QTL mapping. Change in the variance ratio (pF) testing the significance of the regression in flanking marker regression mapping with putative QTL along the chromosome. (Data from Tables 7.7 and 7.8.) In the present case with DH lines, regression relates to just one parameter, i.e. $p = 1$ in pF. In general, $p > 1$ and pF becomes χ^2. A, B and C are marker loci.

their coefficients. It is, however, straightforward to derive the expectations for other populations such as backcrosses, F_2s, SSD lines, etc. and some of these are shown for the two approaches illustrated above in Table 7.9. With method (i), evidence suggests that it is better to locate the QTL by regression on to the additive difference between means and then to estimate the dominance effect at this

Table 7.8 ANOVA for interval mapping by regression. (i) ANOVA for a putative QTL at 2 cM from marker A; (ii) corresponding ANOVA at the most likely QTL position, 35.28 cM, as judged by the minimum residual SS

(i)

Source	df	SS	MS	F
Regression	1	32.5732	32.3732	1.2 n.s.
Residual	23	616.9039	26.8219	36.3***
(Line means)	24	649.4771	27.0616	36.7***
L × B	72	53.145	0.7381	

(ii)

Source	df	SS	MS	F
Regression	1	178.2746	178.2746	8.7**
Residual	23	471.2032	20.4871	27.7***
(Line means)	24	649.4778	27.0616	
L × B	72	53.145	0.7381	

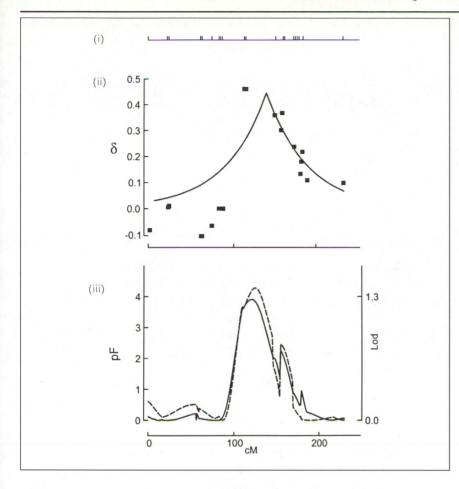

Figure 7.5 Mapping for yield QTL in barley. The locations of genetic markers on chromosome 5 are shown in (i) while the likely positions of the QTL are indicated by the peaks for marker regression mapping (ii) and for flanking marker mapping by regression (pF) and log-likelihood (Lod) (iii). (Reproduced by kind permission of Dr D. Laurie, The John Innes Centre, Norwich, UK and Dr J. Bezant, Plant Genetics Group, School of Biological Sciences, The University of Birmingham, UK.)

position. With method (ii), multiple regression can be used to estimate a and d jointly. There is no test of the significance of the residual MS with method (ii) unless replicated genotypes are used, such as doubled haploid or recombinant inbred lines.

7.4 Handling more than one QTL on a chromosome

So far we have considered the simple case where there is one QTL controlling the trait per linkage group and we can explore each linkage group in turn for evidence of a QTL, locate it and estimate its genetic effects. The existence of QTL on other chromosomes will affect the variability of the estimates but will not bias them. But what if there are more than one QTL per chromosome?

Table 7.9 Genetic models appropriate to (i) marker regression and (ii) flanking marker regression for various generations

(i)

Generation	$\overline{M_1M_1} - \overline{M_2M_2}$	$\overline{M_1M_1} - \overline{M_1M_2}$	$\overline{M_1M_2} - \frac{1}{2}(\overline{M_1M_1} + \overline{M_2M_2})$
F_2	$(1-2R)a$	–	$(1-2R)^2 d$
Bc	–	$\frac{1}{2}(1-2R)(a \pm d)$	–
SSD	$\{(1-2R)/(1+2R)\}a$	–	–
DH	$(1-2R)a$	–	–

(ii)

Genotype	F_2	
	a	d
$A_1A_1B_1B_1$	$[(1-R_A)^2(1-R_B)^2 - R_A^2 R_B^2]/(1-R)^2$	$[2R_A(1-R_A)R_B(1-R_B)]/(1-R)^2$
$A_1A_1B_1B_2$	$[(1-R_A)^2 R_B(1-R_B) - R_A^2 R_B(1-R_B)]/$ $R(1-R)$	$[R_A(1-R_A)(1-R_B)^2 + R_A(1-R_A)R_B^2]/$ $R(1-R)$
$A_1A_1B_2B_2$	$[(1-R_A)^2 R_B^2 - R_A^2(1-R_B)^2]/R^2$	$[2R_A(1-R_A)R_B(1-R_B)]/R^2$
$A_1A_2B_1B_1$	$[R_A(1-R_A)(1-R_B)^2 - R_A(1-R_A)R_B^2]/$ $R(1-R)$	$[(1-R_A)^2 R_B(1-R_B) + R_A^2 R_B(1-R_B)]/$ $R(1-R)$
$A_1A_2B_1B_2$	0	$[R_A^2 R_B^2 + R_A^2(1-R_B)^2 + (1-R_A)^2 R_B^2$ $+(1-R_A)^2(1-R_B)^2]/[R^2 + (1-R)^2]$
$A_1A_2B_2B_2$	$[R_A(1-R_A)R_B^2 - R_A(1-R_A)(1-R_B)^2]/$ $R(1-R)$	$[(1-R_A)^2 R_B(1-R_B) + R_A^2 R_B(1-R_B)]/$ $R(1-R)$
$A_2A_2B_1B_1$	$[R_A^2(1-R_B)^2 - (1-R_A)^2 R_B^2]/R^2$	$[2R_A(1-R_A)R_B(1-R_B)]/R^2$
$A_2A_2B_1B_2$	$[R_A^2 R_B(1-R_B) - (1-R_A)^2 R_B(1-R_B)]/$ $R(1-R)$	$[R_A(1-R_A)(1-R_B)^2 + R_A(1-R_A)R_B^2]/$ $R(1-R)$
$A_2A_2B_2B_2$	$[R_A^2 R_B^2 - (1-R_A)^2(1-R_B)^2]/(1-R)^2$	$[2R_A(1-R_A)R_B(1-R_B)]/(1-R)^2$

Common sense suggests that two QTL which are very close together will behave very much like one combined QTL, while two which are a long way apart will behave as if on two linkage groups. So what happens in between?

It is easy to extend the single marker model to two or more QTL, and for DH lines the difference between marker means is simply

$$(1-2R_1)a_1 + (1-2R_2)a_2$$

where R_1 and R_2 are the recombination frequencies between the marker and QTL_1 and QTL_2 respectively. This confirms that if the two QTL are very close, i.e. $R_1 = R_2 = R$, then the difference in means is $(1-2R)(a_1 + a_2)$, the sum of the two effects. The two QTL could be linked in coupling, i.e. a_1 and a_2 are the same sign or

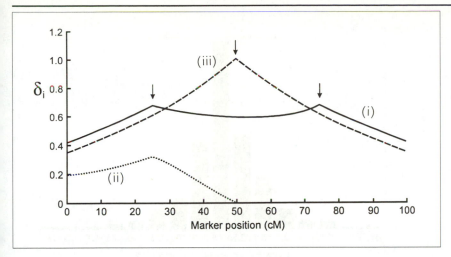

Figure 7.6 Consequences for marker effects (δ_i) with two QTL; (i) in association; (ii) in repulsion. QTL positions are indicated by arrows. The single QTL case (iii) is given for comparison. QTL effects are $a = 1.0$ for (iii) and $a = 0.5$ for (i) and (ii).

they could be in repulsion and cancel each other out. The effect of two QTL, which are 40 cM apart, on the difference in marker means for DH lines, is shown in Figure 7.6(i) for coupling and in Figure 7.6(ii) for repulsion. The single QTL curve is shown for comparison in Figure 7.6(iii). In order to identify two, rather than one QTL, it is necessary to be able to distinguish graphs (i) or (ii) from (iii) and with few or poorly situated markers and small population sizes this can be difficult. A common result found by simulation, is that a spurious, 'ghost', QTL is identified somewhere between the two true QTL [12]. It is possible to fit models involving two QTL to such data but these require examining all possible positions of the two QTL, estimating their effects at each set of positions and identifying those positions which minimize the residual SS as before. It is then necessary to show that there has been a significant reduction in the residual SS as a result of fitting two rather than one QTL. Simulation suggests that the power of such tests is not high unless the QTL are well separated, i.e. by more than 20 cM, and then only with large sample sizes. Because of the difficulty of resolving linked genes from individual genes controlling quantitative traits, Mather coined the term **'effective factor'** to describe the linked set of genes identified by these methods in order to distinguish them from individual genes. These effective factors are largely inherited intact, but will occasionally recombine.

This difficulty in separating two linked QTL from one is hardly surprising when the confidence limits of the estimated QTL locations are examined. For example, computer simulation has shown that with an F_2 population of 300 individuals, five well-spaced markers and a QTL with heritability of 20%, the 95% confidence interval spans

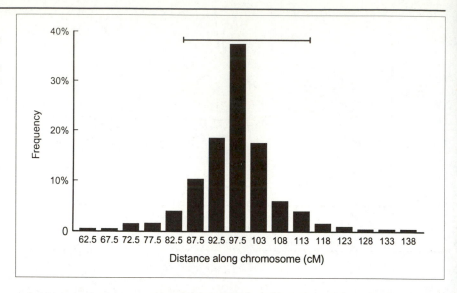

Figure 7.7 The distribution of QTL locations obtained from 1000 simulated F_2 populations of 300 individuals and using five well-spaced markers. Simulations were based on a single QTL with a heritability of 20% situated at 97.5 cM. The 95% confidence interval is indicated by the bar above the histogram.

30 cM, i.e. 30% of a typical chromosome (Figure 7.7) [13,14]. In order to reduce this interval significantly, populations of several thousand individuals need to be genotyped and scored.

Another difficulty associated with the flanking marker techniques is that a large number of statistical tests are carried out and so it is necessary to avoid Type I errors, i.e. detecting effects as significant by chance. Various approaches have been suggested to minimize this problem while not missing genuine QTL [8]. It is often assumed that the more markers there are on a chromosome the more precisely any QTL will be located. Simulation has shown this not to be generally true; **four or five well-spaced markers are quite adequate**, while further markers scarcely affect the precision [13,14]. Figure 7.8(i) illustrates the close correlation of QTL locations obtained with markers 5 cM and 20 cM apart while Figure 7.8(ii) shows the corresponding correlation for the estimates of the QTL effects. Clearly a four-fold increase in marker density has made no perceptible improvement on the estimate of QTL position or effect.

The use of multiple markers thus allows approximate QTL locations to be obtained but, unless the heritability of the individual QTL is high and very large populations are used, these locations will merely indicate the general region of the chromosome such as a particular chromosome arm. Such accuracy is, however, often adequate for marker-assisted selection studies and to identify chromosome regions worthy of more detailed investigation. It is possible to quickly identify key marker loci associated with some traits in a segregating population by looking at the marker genotypes of individuals which are extreme for the quantitative trait. Thus, DNA

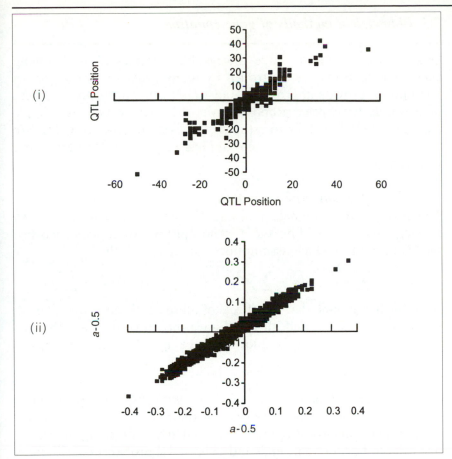

(i)

(ii)

Figure 7.8 Comparison of the dense map (40 markers 5 cM apart, X-axis) and sparse map (10 markers, 20 cM apart, Y-axis) with respect to estimates of (i) QTL position and (ii) QTL effect (*a*) based on 1000 simulations.

can be sampled from a group of very tall individuals and bulked, while a similar bulk can be made of DNA from the shortest individuals. Those markers which show extreme allelic differences between the two bulks are likely to be closely linked to key height QTL. This approach is called bulked segregant analysis.

Given this degree of imprecision in QTL location by such methods, how is it possible without recourse to inordinately large experiments, to locate them more precisely and accurately? The most attractive approach is to use markers to construct chromosomes containing short, prescribed regions from different sources in such a way that one can be sure that any genetic differences between two lines must be due to genes in these short regions. Such 'genetically engineered' chromosomes provide much more precise QTL locations and hence more realistic opportunities for gene cloning and transformation, which in turn may lead to a more detailed analysis of the nature of the QTL themselves. This approach will be explained more fully in the next chapter.

7.5 *Biometrical methods of gene counting*

Quantitative geneticists have always been interested in knowing whether their traits are controlled by many genes or just a few. In attempts to provide approximate answers to this problem, a variety of methods have been proposed over the years which use simple biometrical relationships to estimate the likely number of QTL. We will consider some of these, emphasizing their strengths and weaknesses.

7.5.1 Use of means and variances

This involves the parameters $[a]$ and V_A^*, or $[d]$ and V_D^* [15]. As we showed earlier, providing there is no epistasis or linkage between the QTL, both of which assumptions can be tested, then

$$[a] = r_a \sum a_i; \quad 2V_A^* = \sum a_i^2; \quad [d] = \sum d_i; \quad 4V_D^* = \sum d_i^2.$$

If we further assume that the effects of all the genes are equal, i.e. all the a_is are the same and all the d_is are the same and that the two parents are extreme high and low lines, i.e. $r_a = 1$, then with k QTL

$$[a] = ka; \quad 2V_A^* = ka^2; \quad [d] = kd; \quad 4V_D^* = kd^2.$$

Thus one can estimate k as $[a]^2/2V_A^*$ or $[d]^2/4V_D^*$. The precision of such estimates will depend largely on the reliability of V_A^* and V_D^* which is generally low. In practice, particularly in breeding situations, one will not have extreme lines and so $[a] < ka$, probably considerably less, and thus the estimate of k will be reduced. Similarly, ambi-directional dominance will reduce the estimates obtained using $[d]$.

If the two parents are known to be in association, for example they may be extreme high and low selections from a population, then it is necessary to consider the effects of linkage and variable gene effects on estimates of k.

In the absence of epistasis, linkage affects the variance but not the mean, as is shown in Chapter 10. Since we are assuming that the parents are in association, any linkage will be in coupling which will inflate the additive variance, V_A^*. The effect of this is clearly to cause k to be underestimated.

With variable additive gene effects at different loci we can redefine a_i as follows:

$$a_i = a + \delta_i$$

where

$$\sum \delta_i = 0$$

and

$$a = \sum a_i/k.$$

Thus $\sum a_i = ka$

$$a_i^2 = (a + \delta_i)^2$$

$$= a^2 + \delta_i^2 + 2a\delta_i$$

$$2V_A^* = \sum a_i^2 = ka^2 + \sum \delta_i^2 + 2a \sum \delta_i$$

$$= ka^2 + k\sigma_\delta^2 + 0$$

$$= k(a^2 + \sigma_\delta^2)$$

where $\sigma_\delta^2 = \epsilon\delta_i^2$ from which it follows that:

$$\hat{k} = [ka]^2/k[a + \sigma_\delta^2].$$

Again, k will be an underestimate of the true number of genes given any variation in the a_is. To date, there is insufficient information on any particular system to guide us as to the true size and shape of the distribution of a_i's but whatever evidence there is suggests that the distribution of a_i would be very skewed with a few large, several intermediate and very many small gene effects. To take a simple example, if there were 50 genes of which two had additive effects $a = 10$, 10 genes had $a = 5$ and 38 genes had $a = 1$, then,

$$\sum a_i = (2 \times 10) + (10 \times 5) + (38 \times 1)$$

$$= 108$$

$$\sum a_i^2 = (2 \times 100) + (10 \times 25) + (38 \times 1)$$

$$= 488$$

$$\hat{k} = \left[\sum a_i\right]^2 / \sum a_i^2$$

$$= 23.9$$

while

$$\hat{a} = \sum a_i/\hat{k} = 108/23.9$$

$$= 4.52$$

i.e. it would appear that there are approximately 24 genes with an average effect of 4.52 units while, in fact, there are 50 genes with varying effects but averaging 2.16 units. Thus, the estimate is biased towards the few loci of large effect. The population derived from the cross will behave in fact like an equivalent population containing 24 genes of equal effect ($a = 4.52$ units). Similar arguments apply to estimates from dominance parameters also.

7.5.2 The use of extreme SSD lines

We have seen that the problems of gene dispersion between the original parents of a cross cause $\sum a$, and hence k, to be underestimated. One method that has been employed to avoid this difficulty is to produce a large number of true breeding lines from a cross and to use the extreme lines to substitute for the parents. These should represent a greater degree of association.

The value of $\sum a$ can be estimated as half the difference between the highest and lowest lines, while $2V_A^*$ can be estimated from the genetical variation between the inbred lines. However, the likely range of the highest and lowest lines is, itself, a function of the variance of the population (as is $2V_A^*$) and the number of lines sampled. Therefore, the estimate of $[a]$ is $(P_H - P_L)/2$, which will be some function, f, of σ, while $2V_A^* = \sigma^2$. Thus,

$$\hat{k} = (f\sigma)^2/\sigma^2 = f^2$$

and is a function of the Normal distribution, not k. Given any typical experiment of this sort with, say, 100 derived lines, the likely estimate of $\sum a$ will be between 2σ and 3σ and hence k ($= f^2$) will be between 4 and 9. Exceptionally, particularly with skewed distributions, larger estimates might be obtained, but scoring more lines has little effect. Such methods are, therefore, of no real value as far as the accuracy of the estimates is concerned.

7.5.3 The number of independent genotypes

A further method that has been proposed involves identifying the number of independent lines that can be derived from a cross. If the two parents used differ at a single locus, then by inbreeding from their F_2, only two distinct true breeding lines can be produced, which will, of course, be identical to the original parents, A^+A^+ and A^-A^-. With two loci, there are four distinct homozygous genotypes ($A^+A^+B^+B^+$, $A^+A^+B^-B^-$, $A^-A^-B^+B^+$, $A^-A^-B^-B^-$) and with three loci, eight genotypes, etc. Thus, if one could show that there were four and only four distinct genotypes among the derived lines then one could deduce that there were only two genes segregating, while with eight genotypes there should be three genes, etc.

This approach, however, involves two assumptions: first, that each genotype has a different phenotype and second, that genuine differences in phenotype can be identified experimentally. If all gene effects are equal then the number of distinct phenotypes is simply

$k + 1$ while with unequal gene effects the potential maximum is 2^k. It would, for instance, be impossible to tell whether eight different phenotypes indicated seven genes of equal effect or three with very unequal effects such that each of the eight genotypes had a recognizable phenotype. The more loci there are, the less likely it is that all genotypes would be distinguishable even with unequal gene effects. One therefore arrives at two extreme estimates of k, the higher being based on the assumption of equality of gene effects.

The second problem mentioned above was that of actually distinguishing two genotypes even when their expected phenotypes are different. This will depend on the size of the standard error of the mean of each line relative to the size of individual gene effects and hence is determined by the number of individuals of each genotype that are scored. Thus, the more individuals scored, the smaller the difference that can be detected between two means and hence the more phenotypes one can recognize as being significantly different, i.e. $\hat{k}$ will depend on the size of the experiment.

7.5.4 Genotype assay

Yet another method of estimating k is based on calculating the proportion of individuals in the F_n generation of inbreeding that are still heterozygous at one or more loci controlling the character in question. This has been called 'genotype assay' [16]. For example, with one locus, the proportion heterozygous would be $\frac{1}{2}$ in the F_2, $\frac{1}{4}$ in the F_3, $\frac{1}{8}$ in the F_4, etc. With two loci, on the other hand, the corresponding proportions would be $\frac{3}{4}$, $\frac{7}{16}$, $\frac{15}{64}$, etc. The general formula for the proportion of individuals heterozygous at least at one locus, in the F_n generation with k genes is,

$$P(\text{Het}) = 1 - \{(2^{n-1} - 1)/2^{n-1}\}^k.$$

The values of $P(\text{Het})$ for a range of values of k and n are shown in Table 7.10. Thus if we know the proportion of individuals still heterozygous at generation F_n, we can calculate k.

Table 7.10 Proportion of individuals in F_n which are heterozygous for at least one QTL when there are k QTL segregating

Generation	k					
	1	2	3	4	5	10
F_2	0.5	0.75	0.875	0.938	0.969	0.999
F_3	0.25	0.438	0.578	0.684	0.763	0.944
F_4	0.125	0.234	0.330	0.414	0.487	0.737
F_5	0.063	0.121	0.176	0.228	0.275	0.476

The problem is therefore to determine which individuals are heterozygous at one or more loci or, alternatively, which are not heterozygous at any loci. This is best achieved by selfing or test-crossing every individual; any individual which is heterozygous at one or more loci will show segregation among its progeny and this can be determined by progeny testing. However, the sensitivity of the test will obviously depend on the family sizes and the number of families used to progeny test each F_n individual. As can be seen from Table 7.10, the more loci there are, the further one should inbreed before progeny testing, because if the value of P is high, the discrimination is poor, so it is better to adjust the level of inbreeding so that P is approximately 0.5. For example:

With 1 locus, 0.5 are heterozygous in F_2

with 5 loci, 0.487 are heterozygous in F_4

with 10 loci, 0.476 are heterozygous in F_5.

Thus, if one found that 48% of the individuals in the F_5 were still heterozygous, this would indicate that at least 10 QTL were involved. Also, because of continued recombination, k increases with generations of inbreeding.

None of these biometrical methods of gene counting is very reliable and all lead to underestimates under most conditions. Although genotype assay is probably the least unreliable, the amount of experimental work involved in obtaining the estimate of k would be large relative to the value of the information gained. In practice, it would be useful to know whether the trait of interest is controlled by just one or two loci, in which case they can be handled like major genes, or if there are several to many loci, in which case one needs to use biometrical approaches to study them. The above theory does, however, lead one to be cautious about statements that a particular trait is controlled by just one or two genes.

7.6 General conclusions on QTL counting and locating

All attempts to locate and count the number of QTL in segregating populations, whether by means of molecular markers or biometrical parameters, are limited by the size of the experiment and the magnitudes of the individual gene effects. They will always underestimate the true number of genes and focus attention on the few loci of largest effect. Biometrical methods would yield upper limits of around 9–12

Table 7.11 Estimates of numbers of QTL obtained using genetic markers in segregating populations

Barley[*]		Tomato[†]		Maize[‡]	
Trait	Nos. QTL	Trait	Nos. QTL	Trait	Nos. QTL
Yield	6	Mass per fruit	11	Corn borer tunnel length	7
Lodging	6	Fruit pH	9	Height	3
Height	10	Soluble solids	7	–	–
Heading	9	–	–	–	–
Grain protein	6	–	–	–	–
α-Amylase	9	–	–	–	–
Diastatic power	9	–	–	–	–
Malt extract	7	–	–	–	–

[*] See reference [5]; [†], see reference [15]; [‡], see reference [17].

QTL and power calculations indicate that marker location methods will have similar limits.

There is a growing body of data on the number and effects of QTL from trials in several plants [17–19] and animals where molecular marker data have been used in association with quantitative trait data, a few examples of which are shown in Table 7.11. It is apparent from these examples that the estimates of QTL numbers for any given trait are entirely consistent with those theoretically expected. The range of sizes of gene effects emerging from these data are of considerable interest. For example, in a species cross in tomato [18], the number of QTL detected was inversely proportional to the size of their effects, as might be expected. Extrapolating the curve, which is always a risky procedure, would, nevertheless, suggest that there are probably very many more QTL whose individual effects are too small to measure.

Expressing gene effects as a proportion of the phenotypic variation is an appropriate guide to short-term selection response. It does, however, exaggerate the relative importance of the gene in terms of the ultimate response that might be achievable. To take a specific example, if a trait is segregating for 1 QTL with $a = 9$ and 19 QTL each with $a = 1$, then the first QTL will contribute 81% to the genetic variance $(9^2 / \sum a^2)$ while the other 19 contribute 1% each. Any attempt to count the QTL segregating in this population would identify just the one gene. However, of the combined, ultimate response to selection, $\sum a$, the large QTL contributes less than one-third $\left(\frac{9}{28}\right)$.

QTL location and association with particular markers could be a valuable aid to selection for short-term response, which is the principal aim of commercial breeding. It may, however, limit long-term response.

The subject of marker-assisted selection will be discussed more fully in Chapter 15.

Summary

1. Genes contributing to the variation in crosses and populations can be counted and located using numerous techniques.
2. Biometrical methods provide estimates of the overall numbers of genes controlling a particular trait in the source population but shed little light on the relative contribution or location of individual genes.
3. Marker-based methods, on the other hand, when applied to segregating populations provide a means of locating QTLs to chromosomal regions and measuring their additive and non-additive effects.
4. There is little difference in the precision and accuracy of the major methods of QTL location currently available, i.e. Marker Regression, Multiple Regression and Maximum Likelihood. For typical quantitative traits with heritabilities of less than 40% on an individual basis, the 95% confidence intervals of estimated QTL position can be 30 cM or more.
5. The precision of QTL location depends more on sample size than on the density of markers and no great increase in precision is obtained with more than five well-spaced markers per chromosome.
6. The efficiency of all methods depends mainly on the size of the population and the heritability of the trait. Statistical considerations alone dictate the number of QTL that might be estimated and these suggest that it will be difficult to identify more than about 12 QTLs for a particular trait in a given population sample.

References

1. Sax, K. (1923) The association of size differences with seed-coat pattern and pigmentation in *Phaseolus vulgaris*. *Genetics*, **8**, 552–60.

2. Breese, E.L and Mather, K. (1957) The organisation of polygenic variation within a chromosome in *Drosophila*. I. Hair characters. *Heredity*, **11**, 373–95.

3. Law, C., Snape, J.W. and Worland, A.J. (1983) Aneuploidy in wheat and its use in genetical analysis, in *Wheat Breeding and its Scientific Basis* (ed. F.G.H. Lupton), Chapman & Hall, London, pp. 71–108.

4. Mather, K. and Harrison, B.J. (1949) The manifold effect of selection II. *Heredity*, **3**, 131–62.

5. Thoday, J.M. (1961) Location of polygenes. *Nature*, **191**, 368–70.

6. Beckmann, J.S. and Soller, M. (1983) Restriction fragment length polymorphisms in genetic improvement: methodologies, mapping and costs. *Theor Appl. Genet.* **67**, 35–43.

7. Kearsey, M.J. and Hyne, V. (1994) QTL analysis: a simple 'marker regression' approach. *Theor. Appl. Genet.*, **89**, 698–702.

8. Jansen, R.C. (1993) Interval mapping of multiple quantitative trait loci. *Genetics*, **135**, 205–11.

9. Knapp, S.J., Bridges, W.C. and Birkes, D. (1990) Mapping quantitative trait loci using molecular marker linkage maps. *Theor. Appl. Genet.*, **79**, 583–92.

10. Lander, E.S. and Botstein, D. (1989) Mapping Mendelian factors underlying quantitative traits using RFLP linkage maps. *Genetics*, **121**, 185–99.

11. Haley, C.S. and Knott, S.A. (1992) A simple regression method for mapping quantitative trait loci in line crosses using flanking markers. *Heredity*, **69**, 315–24.

12. Martinez, O. and Curnow, R.N.C. (1993) Estimating the locations and sizes of the effects of quantitative trait loci using flanking markers. *Theor. Appl. Genet.*, **85**, 480–8.

13. Hyne, V., Kearsey, M.J., Pike, D.J. and Snape, J.W. (1995) QTL analysis: unreliability and bias in estimation procedures. *Molecular Breeding*, **1**, 273–82.

14. Darvasi, A., Weintreb, A., Minke, V., Weller, J. and Soller, M. (1993) Detecting marker-QTL linkage and estimating QTL gene effect and map location using a saturated genetic map. *Genetics*, **134**, 943–51.

15. Wright, S. (1934). The results of crosses between inbred strains of guinea pigs, differing in the number of digits. *Genetics*, **19**, 537–51.

16. Jinks, J.L and Towey, P. (1976) Estimating the number of genes in a polygenic system by genotype assay. *Heredity*, **37,** 69–81.

17. Hayes, P.M., Liu, B.H., Knapp, S.J. *et al.* (1993) Quantitative trait locus effects and environmental interaction in a sample of North American barley germ plasm. *Theor. Appl. Genet.*, **87**, 392–401.

18. Paterson, A.H., Damon, S., Hewitt, J.D. *et al.* (1991) Mendelian factors underlying quantitative traits in tomato: comparison across species, generations and environments. *Genetics*, **127**, 181–97.

19. Schön, C.C., Lee, M., Melchinger, A.E., Guthrie, W.D. and Woodman, W.L. (1993) Mapping and characterisation of quantitative trait loci affecting resistance against second generation European corn borer in maize with the aid of RFLPs. *Heredity*, **70**, 648–59.

Further reading

Beckmann, J.S. and Soller, M. (1986) Restriction fragment length polymorphisms in plant genetic improvement. *Oxford Surveys of Plant Molecular and Cell Biology*, **83**, 196–250.

Dudley, J.W. (1993) Molecular markers in plant improvement: manipulation of genes affecting quantitative traits. *Crop Science*, **33**, 660–8.

Mulitze, D.K and Baker, R.F. (1985) Evaluation of biometrical methods for estimating the number of genes. I. Effects of sample size. *Theor. Appl. Genet.*, **69**, 553–8.

Tanksley, S.D. (1993) Mapping polygenes. *Annu. Rev. Genet.*, **27**, 205–33.

Designer chromosomes 8

We saw in the previous chapter that, although it was possible to use a combination of quantitative trait data and marker information from individuals in segregating populations to locate QTL, the precision and reliability of the locations were poor while it was particularly difficult to separate linked QTL. These difficulties are due mainly to the relatively low frequency with which chiasma occur along a chromosome and the correspondingly low level of recombination. Although the precision may be adequate for marker aided selection it would be far too low for detailed genetical analysis, chromosome walking or map-based cloning.

Table 8.1 contains some data on chiasma frequencies in rye, *Secale cereale*, from which, in conjunction with Figure 8.1, we can calculate that 22% of chromosomes in gametes will not have been involved in a single recombinational event while only about 1% of chromosomes will have been involved in more than two. Recombination frequencies between markers and QTL depend on these infrequent events and given a small finite sample of gametes the same event will contribute recombination information to several markers. If the chiasma distribution in the meioses that produced the gametes in the mapping population departed from that expected, even if only because of chance sampling, then all the map positions will be correspondingly biased. More precise mapping of QTL can be achieved only by specifically selecting for recombinational events in particular regions and this essentially requires constructing chromosomes of defined constitution. The various ways in which this can be done is the subject of this chapter.

8.1 Chromosome engineering

The basic principle underlying these approaches is to create two genotypes which are identical apart from a defined region on a particular

Table 8.1 Chiasma frequency in chromosomes of *Secale cereale*

No. of chiasmata per chromosome	Frequency	Unrecombined chromosomes (%)
1	0.267	50.0
2	0.716	25.0
3 or more	0.017	12.5
Mean	–	22.0

chromosome. One can then be sure that any genetical differences in phenotype between these two genotypes must be due to genes in this defined region. The smaller the region, the more precisely the positions of these genes will be known. There will, of course be a very large number of regions which can be examined in this way depending on their size, so the results of earlier marker-based studies on segregating populations (Chapter 7) would be used to focus attention on particular chromosomal sections.

Figure 8.1 Bivalents normally have at least one chiasma, frequently two and sometimes three or more. Nevertheless, chromosomes often come through meiosis without being involved in any recombination. The reasons for this are illustrated for one or two chiasmata to show why so little recombination actually occurs.

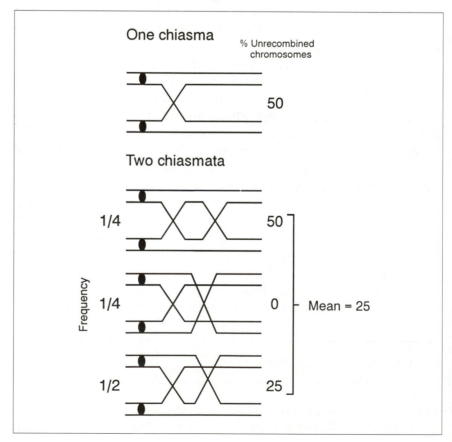

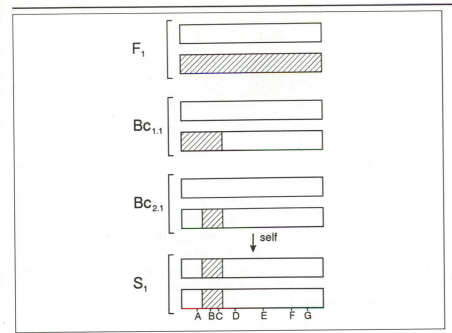

Figure 8.2 Stages in the construction of backcross substitution lines using molecular markers. The shaded region represents an introgressed donor region which is finally made homozygous by selfing.

Although there are various ways of achieving this end, the procedures essentially involve starting from an F_1 and repeatedly backcrossing to a recurrent parent in order to introgress the defined region from the donor parent into the recipient using alleles at marker loci to identify the required genotypes as illustrated in Figure 8.2. At each generation, individuals are selected which are heterozygous for the particular region to be introgressed, i.e. B–C in Figure 8.2, and homozygous for the recurrent parent's alleles at all other loci, i.e. A, D, E, F and G. The use of co-dominant markers such as RFLPs facilitates this process; if dominant markers, such as the majority of RAPDs are used, then it is important to have the dominant alleles in the non-recurrent parent in order to identify the relevant genotypes which are heterozygous for the donor region. It may take two or three generations of backcrossing before the desired genotype is attained, but every generation should result in one getting closer to the objective as more distant markers are removed by recombination. Simultaneous selection for individuals homozygous for alleles from the recurrent parent on all other chromosomes should be practised at the same time in order to accelerate the return to the recurrent parent's genotype. Once an individual of the desired marker genotype is found, the chromosome with the introgressed region has to be made homozygous either by intercrossing two identical genotypes or by selfing, and selecting those individuals which are homozygous for the donated region from among the progeny. Such homozygotes can be maintained as true breeding, **substitution lines** (Figure 8.2).

Figure 8.3 Detail of a chromosome with substituted region involving four molecular markers. The cross-overs to introduce the donor region (darkly shaded) must have occurred in the regions A–B and C–D. The precise length of the donor region can be determined by identifying more markers in the lightly shaded regions.

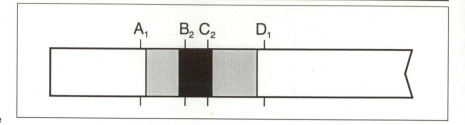

Such a substitution line can be put into a replicated and randomized trial together with the recurrent parent and scored for all traits of interest; any differences detected between the two lines must be due to a QTL (or QTLs) in the substituted region. The precise length of the substituted region is circumscribed by the distance between the flanking markers and the selected markers. Thus in the example in Figure 8.2, recombination has occurred somewhere between A and B and between C and D (Figure 8.3). If the distances between B or C and the flanking markers are large, then the actual length of the segment will not be well defined. However, the QTL will have to be located between A and D, so the limits to its position are defined by the distance A to D. Should it be necessary to define the length more precisely, it would always be possible to find additional polymorphic markers in the intervals A–B and C–D, and to use these in further backcrosses to reduce the distance further. Potential markers for this purpose would be suggested from other available framework maps.

8.2 Locating a QTL within the substituted region

One of the problems with trying to locate QTL in populations was the wide confidence interval of the estimate, often in excess of 30 cM. Substitution lines reduce this confidence interval to the distance between the homozygous flanking markers, A and D as shown by the example in Figure 8.3. However, given that a significant effect has been associated with a particular substitution the location of the putative QTL can be estimated much more precisely by an appropriate backcross or F_2 as follows.

Consider an F_1 between the substitution line and the recurrent parent which has been backcrossed to the recurrent parent, and assume that the QTL is between marker loci, B and C, as shown in Table 8.2. Let the recombination frequencies between B and the QTL (Q) be R_B and that between C and the QTL be R_C respectively.

Table 8.2 Mapping a QTL within a substituted chromosome in a backcross. B and C are the flanking markers and Q is the QTL

	F_1		Recurrent parent		

B1　　　Q1　　　C1　　　　　　　　B1　　　Q1　　　C1

$$\equiv | \equiv | \equiv | \equiv$$
$$\equiv | \equiv | \equiv | \equiv \qquad \times \qquad \equiv | \equiv | \equiv | \equiv$$

B2　　　Q2　　　C2　　　　　　　　B1　　　Q1　　　C1

$|- R_B -|- R_C -|$

$|- R_{BC} -|$

↓

		Frequency	Genetic value

B1　　　Q1　　　C1

$\frac{1}{2}(1 - R_{BC})$ 　　　$m + \alpha \quad (= \epsilon Y_1)$

B1　　　Q1　　　C1

B2　　　Q2　　　C2

$\frac{1}{2}(1 - R_{BC})$ 　　　$m - \alpha \quad (= \epsilon Y_2)$

B1　　　Q1　　　C1

B1　　　Q1　　　C2

$\frac{1}{2} R_C$ 　　　$m + \alpha$

B1　　　Q1　　　C1

$= \frac{1}{2} R_{BC}$ 　　　$m + \alpha(R_C - R_B)/R_{BC} \quad (= \epsilon Y_3)$

B1　　　Q2　　　C2

$\frac{1}{2} R_B$ 　　　$m - \alpha$

B1　　　Q1　　　C1

B2　　　Q1　　　C1

$\frac{1}{2} R_B$ 　　　$m + \alpha$

B1　　　Q1　　　C1

$= \frac{1}{2} R_{BC}$ 　　　$m - \alpha(R_C - R_B)/R_{BC} \quad (= \epsilon Y_4)$

B2　　　Q2　　　C1

$\frac{1}{2} R_C$ 　　　$m - \alpha$

B1　　　Q1　　　C1

Providing the distance B to C is fairly short we can ignore double recombination and hence the recombination frequency between B and C, $R_{BC} = R_B + R_C$; R_{BC} will be known from earlier map information or can be estimated from the backcross in the usual way (Chapter 6). As shown in Table 8.2, the non-recombinant classes have genetic values;

$$B_1-C_1/B_1-C_1 \quad m + \alpha \quad (=Y_1)$$
$$B_2-C_2/B_1-C_1 \quad m - \alpha \quad (=Y_2)$$

where α is half the difference between the phenotypes of the homozygous and the heterozygous substitutions. The recombinant genotypes B_1-C_2 and B_2-C_1 will contain a mixture of genotypes whose relative frequencies will depend on whether recombination occurred to the left or right of the QTL. As shown in Table 8.2, the genetic values of these genotypes will be,

$$B_1-C_2/B_1-C_1 \quad m + \alpha(R_B - R_C)/R_{BC} \quad (=Y_3)$$
$$B_2-C_1/B_1-C_1 \quad m - \alpha(R_B - R_C)/R_{BC} \quad (=Y_4)$$

α can be estimated as $(Y_1 - Y_2)/2$, while $(Y_3 - Y_4)/2 = \alpha(R_B - R_C)/R_{BC}$. Therefore R_A can be estimated as

$$[1 + \{(Y_3 - Y_4)/(Y_1 - Y_2)\}]R_{BC}/2$$

This is illustrated using some data on bristle number in *Drosophila* in Table 8.3.

8.3 Detecting substitution effects

So far we have considered the production and analysis of a single substituted region but there is no limit to the number of independent substitutions that can be made. In practice, a single initial backcross can be used to initiate very many different substitutions and a library of lines can be created which cover the whole or at least a large proportion of the genome (Figure 8.4). Such a library constitutes a very valuable resource because it can be maintained indefinitely and studied for any number of traits over locations and years [3]. As with any study to locate and measure the effects of QTL, positive identification will depend on the size of the effect and the degree of replication; the larger the trial the more QTL will be found.

In order to be certain that a difference between a substitution line and its recurrent parent is due to the introgressed region, it is necessary

Table 8.3 Estimating the location of a QTL controlling bristle number in *Drosophila melanogaster* [1]. The recombination frequency between the mutant markers cp (clipped wing) and Sb (stubble bristles) is known to be 0.129 [2]

	F_1			Recurrent parent	
cp	Q_1	Sb	cp	Q_1	Sb

$\times$

cp^+	Q_1	Sb^+	cp	Q_1	Sb

$\longleftarrow R_B \longrightarrow \longleftarrow R_C \longrightarrow$

Backcross progeny		Mean	
cp	Sb	18.00	(Y_1)
cp^+	Sb^+	20.62	(Y_2)
cp	Sb^+	18.95	(Y_3)
cp^+	Sb	19.19	(Y_4)

$$\alpha = \tfrac{1}{2}(Y_1 - Y_2) = -1.31$$

$$R_B = \tfrac{1}{2} R_{BC}\{1 + \tfrac{1}{2}(Y_3 - Y_4)/(Y_1 - Y_2)\}$$

$$= 0.07$$

$$R_C = R_{BC} - R_B$$

$$= 0.129 - 0.07 = 0.059$$

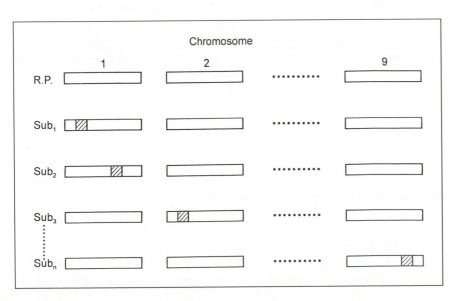

Chromosome

1   2   9

R.P.

Sub₁

Sub₂

Sub₃

Subₙ

Figure 8.4 The recurrent parent (R.P.) and a library of single substitution lines. Substituted regions are shaded.

to be sure that there are no other regions introgressed elsewhere in the genome. It is therefore essential to have markers that provide a good coverage of the genome so that the risk of such 'unseen' substitutions is low. There are several ways of coping with this possibility, however. One is to make replicate, independent substitutions of a particular region on the assumption that it is unlikely that the same unseen substitution would occur in both. If the replicated lines are consistent in their trait scores, then one can be confident that the QTL is in that region. Inconsistency, on the other hand is ambiguous. Either a second region is involved or the same region is involved but with slightly different substitutions which do not both include the QTL.

Another approach is to produce F_2 or backcross progeny from the F_1 between the recurrent parent and the substituted line, and to score these progeny both for the trait and for the genotype of the markers within the introgressed tract, i.e. B–C in our example. Only differences attributable to the region B–C will be detected, while other, unwanted regions should segregate independently. This approach has two other advantages. It is possible to look at dominance (d) as well as additive (a) effects of the QTL and, because all the progeny come from the same parents, it is possible to eliminate non-genetic causes of differences such as maternal effects, seed age, etc. The latter is a very important consideration when studying the individual effects of QTL which may well be small compared with extraneous non-genetic factors which can cause large differences between families. The F_2/ backcross approach does however have the disadvantage with plants that the progeny cannot be raised in single genotype plots in order to simulate agronomic conditions.

Not only does the technique of substituting small defined regions provide more precise QTL locations but it also provides a means of handling and manipulating the QTL just like any major gene. The F_1 between any two different substitution lines, each of which has been shown to carry a QTL, allows the interaction of two different QTL to be examined in the subsequent F_2 or backcross generations simply by comparing the trait means of the nine classes of marker genotype as illustrated in Table 8.4.

8.4 Manipulating whole chromosomes

Although it may be relatively simple to construct substitutions in individual chromosomes, handling many chromosomes simultaneously may be technically demanding in resources. An obvious solution is

Table 8.4 The analysis of an F_2 between two substitution lines on different chromosomes, each known to contain a QTL. Parameters *aa, ad* and *dd* represent the genetic interactions of genes of chromosomes I and II, etc.

$$F_1 \quad \begin{matrix} Q_{11} \\ \blacksquare \\ \square \\ Q_{12} \end{matrix} \quad \begin{matrix} Q_{22} \\ \square \\ \blacksquare \\ Q_{21} \end{matrix}$$

	$Q_{11}Q_{11}$	$Q_{11}Q_{12}$	$Q_{12}Q_{12}$
$Q_{21}Q_{21}$	$m + a_1 + a_2 + aa$	$m + d_1 + a_2 + ad$	$m - a_1 + a_2 - aa$
$Q_{21}Q_{22}$	$m + a_1 + d_2 + ad$	$m + d_1 + d_2 + dd$	$m - a_1 + d_2 - ad$
$Q_{22}Q_{22}$	$m + a_1 - a_2 - aa$	$m + d_1 - a_2 - ad$	$m - a_1 - a_2 + aa$

to create lines which contain whole or half chromosome substitutions initially and then to use these to engineer smaller substitutions as described above. The construction of whole chromosome substitutions is facilitated when there are methods available to enable whole chromosomes to be transmitted from generation to generation without risk of recombination. A few systems of this sort that can be exploited are described below, but they are restricted to a limited range of organisms.

8.5 Chromosome substitution methods in different species

8.5.1 Fungi

Fungi have many advantages as organisms for genetical analysis, which is why they have been extensively used since the 1940s for understanding gene action and recombination. They are also important commercially in fermentation, baking, as a food and, of course, as pests and saprophytes. They can be raised as haploids, hence simplifying genetical analysis, vegetatively reproduced, and raised under highly controlled environments. It is not surprising, therefore, that they offer useful features for studying quantitative traits.

In some haploid organisms, for example the fungus *Aspergillus nidulans*, recombination suppression can be achieved by passing from a diploid spore to a haploid without going through meiosis. Such a process may occur naturally but can be induced by certain drugs such as *p*-fluorophenylalanine or Benlate [4, 5]. Since meiosis is not involved, the chromosomes do not recombine and the net result is that each haploid spore resembles a gamete in that it contains

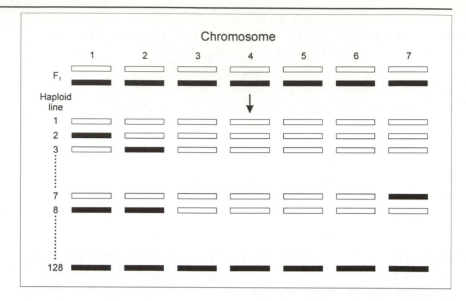

one homologue at random of each of the 2n chromosomes. Providing markers are available for each of the n chromosomes, it is then possible to produce a culture of the fungus from each haploid spore and identify its marker constitution. This can prove the parental origin of each of the n chromosomes in the haploid culture, as illustrated in Figure 8.5. With seven chromosomes as in *Aspergillus*, there will be 2^7 (=128) different types of spore from any diploid parent and given a suitably large sample of spores, it should be possible to obtain a representative of every one, a suitably large sample being ~400 in order to be 95% certain of obtaining them all. Once produced, such a set of lines can be used to locate genes on chromosomes as well as to provide source material for individual chromosome mapping studies as will be illustrated later.

8.5.2 *Drosophila melanogaster*

The fruit fly, *Drosophila melanogaster*, has played an important part in the study of quantitative traits for the same reasons that have contributed to its extensive wider use in genetical analyses. Its value is as a model for studying the genetic architecture of traits and the response to artificial and natural selection. It also provides several complementary approaches for whole chromosome manipulation. Another important advantage is that it has a haploid karyotype of only four chromosomes, one of which, the fourth, is very small and contributes little to genetical variation.

The first feature of *Drosophila* which aids whole chromosome manipulation is the achiasmate meiosis in males. Thus, the male gametes of an F_1 consist of all haploid combinations of the four

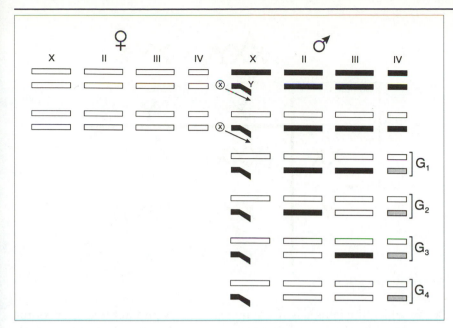

Figure 8.6 Construction of chromosome substitution lines in *Drosophila*, relying on achiasmate male meiosis to suppress recombination. G_1–G_4 are the four genotypes representing all possible combinations of chromosomes II and III.

pairs of chromosomes (i.e. $2^4 = 16$), completely unrecombined. The males are therefore an ideal vehicle for 'cloning' whole chromosomes and passing them on to the next generation. Moreover, a particular set of chromosomes can be maintained in an unrecombined state indefinitely by maintaining them solely through the male line, although they may mutate slowly in the process. In the female, on the other hand, meiosis is normal except that there is no recombination on chromosome IV.

Using these features, it is very easy to substitute single whole chromosomes from one strain into another by backcrossing using the female as the recurrent, homozygous parent as shown in Figure 8.6 because, ignoring the relatively inert fourth chromosome, there are just four different male genotypes. All that is needed are co-dominant markers on the chromosomes in order to distinguish the parental source of individual chromosomes both in the homozygous and heterozygous state. If these males are scored for some quantitative trait, then the contribution of each chromosome to the trait can be estimated. For example, the difference in phenotype between G_2 and G_4 flies (Figure 8.6) must reflect the difference between chromosome II in the heterozygous and homozygous state. The method of analysis will be illustrated more fully later. Unfortunately, it is not possible to transmit the X chromosome from father to son using normal females because *Drosophila* has an X/Y sex chromosome system in which the Y is passed from father to son and the X from father to daughter. It is, however, possible to use a special stock of *Drosophila* called 'Attached

Figure 8.7 Progeny genotypes obtained from 'attached X' female *Drosophila*. XX represents the 'attached X' chromosomes. Progeny in brackets normally fail to reach maturity and, as a result, the surviving male progeny receive their X chromosome from their father and **not** their mother.

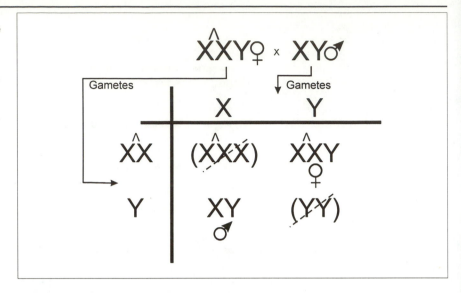

X', in which the female is XXY instead of XX and her two X chromosomes are attached and inherited together. The result of crossing an attached X female to a normal male is illustrated in Figure 8.7 which shows that the sons all inherit the X from their father and their Y from their mother, i.e. the reverse of normal sex-linked inheritance. Attached X female stocks provide a means of maintaining X chromosomes free of recombination and thus permit their effective multiplication and maintenance. These can be used for whole chromosome substitution by creating an XXY stock with the same autosomes as the recurrent parent.

A further important and useful feature available with *Drosophila* is the presence of stocks which contain what are commonly called 'Balancer' chromosomes. These are chromosomes which contain several overlapping inversions and are labelled by readily recognizable morphological markers which are generally a mixture of both dominant and recessive alleles [2]. These inversions effectively prevent recombination between chromosomes in the female when the balancer chromosome is heterozygous with a chromosome of another sequence such as a normal or different balancer chromosome. Not only do the complex inversions make it difficult, physically for the two chromosomes to pair but chiasmata within an inverted segment lead to chromosome bridges and fragments at anaphase so that effectively only unrecombined chromosomes pass into the gametes. Normally, such stocks are maintained with the chromosome in the heterozygous state against a different balancer chromosome and, because they also contain recessive lethal alleles, the balancer chromosome is prevented from becoming homozygous. A large number of such stocks exist with balancing chromosomes for all three major chromosomes [2].

These balancer chromosomes enable the geneticist to pass chromosomes of interest through a female *Drosophila* and to 'collect' them in her gametes intact and without recombination, so achieving in the female what was possible by achiasmate meiosis in the male. How these features can be used is best illustrated by an example. Consider two inbred, true breeding lines of *Drosophila*, A and B, which differ in some quantitative trait. Let us suppose that we wish to find out which of the three major chromosomes, X, II and III, contribute to differences in phenotype and measure the size of the chromosome effects. For simplicity, the three pairs of chromosomes will be illustrated as:

$$\text{Line A} = \frac{A\ A\ A}{A\ A\ A} = A\ A\ A$$

$$\text{Line B} = \frac{B\ B\ B}{B\ B\ B} = B\ B\ B$$

We now need to construct six other lines, so that, together with the two original lines, we have all eight possible homozygous combinations of chromosomes, i.e.

Line	X	II	III
1	A	A	A
2	A	A	B
3	A	B	A
4	A	B	B
5	B	A	A
6	B	A	B
7	B	B	A
8	B	B	B

The first step is to cross both lines, A and B to the balancer stock and then to backcross the selected progeny to lines A and B as illustrated in Figure 8.8. The eight chromosome sets are identified by the presence or absence of the markers in the balancer chromosomes. As a result of this mating scheme the A and B chromosomes are prevented from unwanted recombination by ensuring that when they are heterozygous, they are either in the male or balanced against a recombination suppressing chromosome in the female. The resulting set of eight substitution lines can be maintained indefinitely.

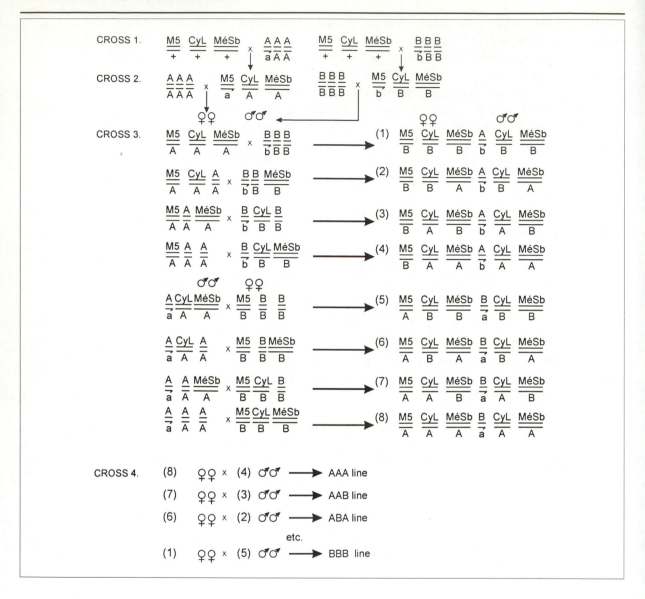

Figure 8.8 Crossing scheme to produce all eight possible homozygous substitution lines for the three main chromosomes in *Drosophila* using marked, balancer chromosomes. M5, CyL, MéSb are balancer chromosomes identified by the dominant markers. A, B represent wild-type chromosomes from lines A and B, while a and b represent Y chromosomes.

8.5.3 *Triticum aestivum*

The final method to be considered has been widely used for genetical analysis in wheat but, in principle, it is possible in any allopolyploid. It avoids recombination of a particular chromosome during passage through several generations of backcrossing by preventing the chromosome in question from pairing with its homologue. This is achieved by removing the homologue and keeping the chromosome to be manipulated in a monosomic, as opposed to the normal disomic, condition.

Wheat is an **allohexaploid**, with $2n = 6x = 42$ chromosomes, having been derived from three ancestral diploid progenitors which provide the so-called A, B and D genomes. These progenitors are thought to be:

A genome – *Triticum monoccoccum*

B genome – *Aegilops speltoides*

D genome – *Aegilops squarrosa*.

Each genome individually consists of seven homologous pairs of chromosomes, i.e.

1A, 2A, 3A, . . . 7A

1B, 2B, 3B, . . . 7B

1D, 2D, 3D, . . . 7D.

Chromosomes with the same number, e.g. 1A, 1B, 1D, are called a **homoeologous** group. They share considerable sequence homology indicating a common origin and will pair with each other in certain genetic backgrounds; a single gene on chromosome 5B is known to prevent such pairings normally. Because the genome is present in six copies it is possible for plants which lack one or two complete chromosomes to survive and to reproduce, although they may be somewhat abnormal in appearance due to unbalanced dosage effects of certain critical genes. In 1953 [6] all 21 sets of **nullisomic** and **monosomic** lines were produced in the spring wheat variety, Chinese Spring. Nullisomics completely lack one homologous pair, e.g. they may have no representative of chromosome 5A, and hence only have 40 chromosomes. Monosomics lack one of a particular homologous pair, e.g. they have just one 5A, and hence have 41 chromosomes. As might be expected, the monosomics are more viable than the corresponding nullisomics and are therefore more useful for genetical analysis. Although these aneuploids were first produced in Chinese Spring, they are now available in a range of modern commercial wheats.

Monosomics are much more viable but they also give disturbed segregation ratios which are capitalized on for breeding. Thus, if we use the notation X– to indicate a monosomic individual, where X indicates a chromosome is present and – indicates its absence, we do not expect an equal number of gametes which are X or –. Among female gametes only 25% carry the chromosome, i.e. are X while the remaining 75% are –. This is because univalent chromosomes fail to orientate properly at metaphase I of meiosis. Among male gametes, on the other hand, 96% contain the chromosome and this is almost certainly the result of natural selection for those pollen grains which have the normal 'haploid' condition. Thus, on random union of gametes from selfing an X– individual, progeny with the following ploidy are obtained:

Ploidy	Chromosome No.	%
Disomic (XX)	21 pairs	$24 = 96\% \times 25\%$
Monosomic (X−)	20 pairs + 1	$73 = (96\% \times 75\%) + (4\% \times 25\%)$
Nullisomic (−)	20 pairs	$3 = 4\% \times 75\%$

Providing nullisomic and/or monosomic lines are available in a **recipient** wheat variety it is possible to substitute the given nulli- or monosomic chromosome with chromosomes from another, **donor** variety. This is illustrated in Figure 8.9 for a case in which a chromosome 7A from a donor variety, Hope, is to be substituted for the Chinese Spring 7A. The donor is first crossed as male parent to the Chinese Spring line which is nullisomic for chromosome 7A to produce an F_1 which is monosomic for the donor 7A but disomic and heterozygous for all the other 20 pairs of chromosomes. Because 7A is monosomic it is unable to recombine with the Chinese Spring

Figure 8.9 Crossing scheme using nullisomic lines to introduce chromosome 7A from a donor wheat variety into a recipient, Chinese Spring.

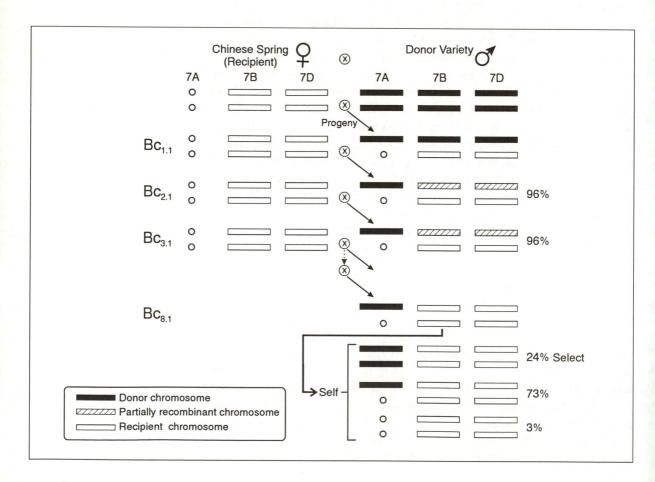

7A; normal recombination will still occur among the other 20 pairs. The F_1 is then backcrossed as male to the Chinese Spring nullisomic and because of the distorted segregation in monosomic male meioses, 96% of the gametes and hence 96% of the progeny will be monosomic for the donor 7A. Such plants have to be identified cytologically, i.e. by scoring root tip mitoses, so it is important to use the distorted segregation to increase their frequency and so make it easier to find them. Once identified, they are again backcrossed as the pollen parent to the Chinese Spring line which is nullisomic for 7A and the process repeated for several backcross generations. All the time, the donor 7A is protected from recombination while the repeated back-crossing results in the other chromosomes becoming homozygous for the Chinese Spring genotype. The speed of approach to the recurrent parent's genotype can be accelerated by selection for molecular markers. Selfing a selected monosomic after five or six generations of backcrossing results in approximately 24% of the progeny being disomic and these can be identified cytologically.

Figure 8.10 Crossing scheme using monotelocentric lines to introduce chromosome 7A from a donor wheat variety into a recipient, Chinese Spring.

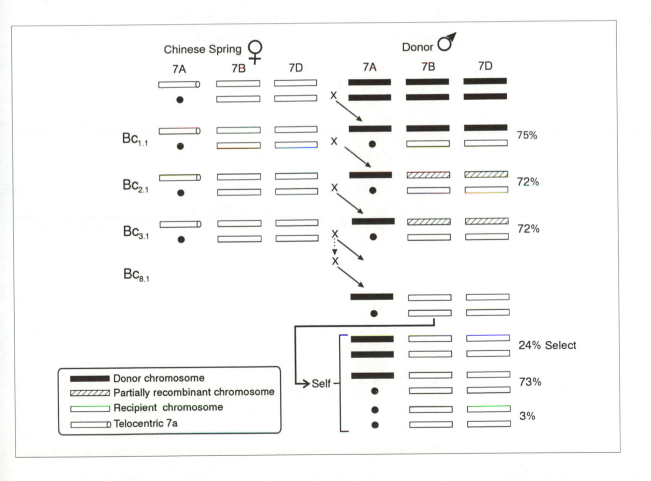

Such progeny are homozygous for the 7A from the donor variety, Hope, and virtually homozygous for all other chromosomes from Chinese Spring. The same process can, of course, be used to substitute any other chromosome.

However, because of the poor viability of nullisomics, crossing schemes usually employ monosomic rather than nullisomic lines. Their use, however, requires that the donor and recipient chromosomes are distinguishable. In order to 'label' them in such a way that they can be identified under the microscope, recipient lines carrying di-telocentric chromosomes have to be used. Otherwise, the crossing scheme is very similar to that described for nullisomics and is illustrated in Figure 8.10 [7].

8.6 Use of chromosome substitution lines

In the three situations described above it has been possible to construct new, true-breeding lines which contain defined combinations of chromosomes from two parental lines. In the case of wheat, it was possible to substitute one chromosome in a given line whereas in *Aspergillus* and *Drosophila* all possible combinations of chromosomes could be created. Once produced, there are a very wide range of uses to which they can be put both for genetical analysis and, in the case of wheat, for breeding purposes. Let us consider the *Drosophila* substitution lines first as they offer the widest range of analytical possibilities.

The set of eight substitution lines in *Drosophila* can be scored for some trait, in a suitably randomized and replicated trial. An example of such data is illustrated in Table 8.5(i) using body weight, and the corresponding ANOVA is presented in Table 8.5(ii). From this ANOVA we see that there are highly significant differences between the lines for bristle number and hence the different chromosomes are having differential effects on the trait. The eight lines constitute an orthogonal set of all possible combinations of chromosomes, therefore the SS for differences between lines, with 7 df, can be partitioned into seven independent items, each with 1 df, testing the various chromosomal effects and their interactions. These are: the three additive effects, one for each chromosome, the three pairwise chromosomal interactions (X × II, X × III and II × III) and the three-way interaction (X × II × III). All significant effects can be estimated using orthogonal comparisons as shown in Table 8.5(iii). These data show that there are genes from A on the X chromosome which increase body weight

Table 8.5 The analysis of eight true-breeding substitution lines of *Drosophila*. The data are the mean body weights (mg) of 10 females and the experiment was raised on four separate occasions. (i) Data and model; (ii) ANOVA; (iii) estimates of chromosomal effects and interactions

(i)

Substitution line			Occasion					Model (orthogonal comparisons)							
X	II	III	1	2	3	4	Total	m	a_1	a_2	a_3	aa_{12}	aa_{13}	aa_{23}	aaa_{123}
A	A	A	1.129	1.160	1.305	1.266	4.860	1	1	1	1	1	1	1	1
A	A	B	1.199	1.234	1.436	1.760	5.629	1	1	1	−1	1	−1	−1	−1
A	B	A	1.260	1.361	1.730	1.480	5.831	1	1	−1	1	−1	1	−1	−1
A	B	B	1.392	1.425	1.400	1.421	5.637	1	1	−1	−1	−1	−1	1	1
B	A	A	1.156	1.126	1.170	1.235	4.687	1	−1	1	1	−1	−1	1	−1
B	A	B	1.218	1.247	1.338	1.309	5.112	1	−1	1	−1	−1	1	−1	1
B	B	A	1.215	1.234	1.335	1.344	5.128	1	−1	−1	1	1	−1	−1	1
B	B	B	1.280	1.156	1.336	1.368	5.140	1	−1	−1	−1	1	1	1	−1
Totals			9.849	9.943	11.05	11.182	42.024								

(ii)

Source	df	SS	MS	F	P
Lines (L)	7	0.2865	0.0409	4.09	**
Occasions (O)	3	0.1879	0.0626	6.26	**
L × O	21	0.2110	0.0100		

(iii)

$a_1 = 0.059$; ** $aa_{12} = -0.016$ n.s. $aaa_{123} = -0.017$ n.s.

$a_2 = -0.045$; * $aa_{13} = -0.004$ n.s.

$a_3 = -0.032$ n.s. $aa_{23} = -0.043$ *

($a_1 = 0.059$) while on chromosomes II and III genes from A decreased body weight ($a_2 = -0.045, a_3 = -0.032$) though a_3 is not significant. It also appears that genes on chromosomes II and III interact. A similar analysis could be applied to the set of fungal lines.

Because the eight *Drosophila* lines are all homozygous, only the additive effects and their interaction between chromosomes can be estimated. If these lines were to be intercrossed, however, all possible homozygous and heterozygous combinations of chromosomes can be produced, 27 female and 18 male. Scoring these should allow a fuller dissection of the dominance effects of individual chromosomes together with their interactions [8].

Where the experimental material consists of lines with just single chromosomes substituted, as in wheat, the analytical possibilities are somewhat restricted. Producing progeny by crossing the original line with the substituted lines allows the additive and dominance effects associated with each substitution to be measured, while crosses between substitution lines allow digenic epistatic effects to be tested and estimated, as was illustrated in Table 8.4. It is also possible to treat crosses between substitution lines and their recurrent parents just like any other cross between two inbred parents and analyse the six basic generations as described in Chapters 2 and 3.

Such substitution lines have, however, proved an extremely useful tool in investigating the genetical control of a wide range of traits in wheat [7]. They have also allowed the breeder the possibilities of identifying chromosomes or chromosome regions in different lines which contain useful alleles, with the potential for combining them in an improved variety.

The QTL associated with substituted chromosomes or chromosome regions constitute a useful resource for breeders as they can easily be transferred into breeding lines.

Summary

1. Molecular markers allow small, defined regions of a chromosome from a donor line to be introgressed into a recipient variety. Such substitution lines provide a valuable tool for precisely locating QTL and measuring their effects.
2. The special features of the biology of wheat and *Drosophila* and some fungi allow whole chromosomes from one line to be substituted into another line without recombination.
3. Such lines provide useful material for analysing the genetic effects associated with particular whole chromosomes.
4. Substitution lines, whether of small chromosomal segments or of whole chromosomes are a useful source of identifiable QTL for breeding purposes.

References

1. Wolstenholme, D.R. and Thoday, J.M. (1963) Effects of disruptive selection: VII A third chromosome polymorphism. *Heredity*, **18**, 413–31.
2. Lindsley, D.L. and Grell, E.H. (1967) Genetic variations of *Drosophila melanogaster*. Carnegie Institute of Washington, Publication No. 627.

3. Tanksley, S.D. (1993) Mapping polygenes. *Annu. Rev. Genet.*, **27**, 205–33.

4. Morpurgo, G. (1961) Somatic segregation induced by *p*-fluorphenyl-alanine. *Aspergillus Newsletter*, **2**, 10.

5. Varga, J. and Croft, J.H. (1994) Assignment of RFLP, RAPD and iso-zyme markers to *Aspergillus nidulans* chromosomes using chromosome substituted segregants of a hybrid of *A. nidulans* and *A. quadrilineatus*. *Curr. Genet.*, **25**, 311–17.

6. Sears, E.R. (1953) Nullisomic analysis in common wheat. *Am. Nat.*, **87**, 245–53.

7. Law, C., Snape, J.W. and Worland, A.J. (1983) Aneuploidy in wheat and its use in genetical analysis, in *Wheat Breeding and its Scientific Basis*, (ed. F.G.H. Lupton), Chapman & Hall, London, pp. 71–108.

8. Kearsey, M.J. and Kojima, K. (1967) The genetic architecture of body weight and egg hatchability in *Drosophila melanogaster*. *Genetics*, **56**, 23–37.

9 Populations

9.1 Problems of studying populations

We have developed the basic theory of quantitative genetics through the use of populations of lines or individuals obtained from an F_1, either by selfing, sib-mating or backcrossing. As stated before, this approach has been used to simplify the algebra required to derive the expectations of the various generation means, variances and covariances. However, a very large proportion of populations which researchers wish to study and analyse are clearly not of this type. Thus, most animal populations, including domestic mammals and, in particular, man together with many outbreeding plants cannot be considered to have been derived from an F_1. Even where inbreeding is possible and true-breeding lines are available, as for example in maize, wheat, mice and even chicken, the lines used for genetical analysis are most often derived from quite different populations and only rarely from the same F_1.

The genetical modelling of F_2-derived populations is simplified because we know that there are only two alleles segregating at every locus and that these alleles have predictable frequencies in the F_2 and derived populations. Linkage and epistasis create some difficulties but their effects and consequences can be quite easily handled. The extension of the genetical analysis of quantitative traits to other situations involves many more unknown parameters with, generally, less statistics available to study them.

Let us briefly consider what some of these problems are. First, we will not know how many alleles there are segregating at each polymorphic locus and this number will almost certainly vary over loci. Second, we will have no idea of the frequencies of these alleles and these frequencies may vary over loci. Third, we cannot assume that the individuals in our population mate at random unless we force them to do so, and so there may well be correlations between relatives for non- or quasi-genetical reasons. The fourth difficulty concerns the non-independence of the genotypes at different loci. With F_2-derived populations such effects can safely be assumed to arise solely from linkage but in populations of ill-defined ancestry such an assumption would be invalid. The fifth and final major problem concerns the types of family that can be used for genetical analysis. The valuable families of the basic generations, parents, F_1, F_2 and backcrosses, are not available and so all information has to be derived from FS and HS analyses supplemented by covariances with other relatives such as parents. When studying such populations, therefore, we are faced with more unknown parameters while at the same time having fewer and far less precise analytical tools at our disposal.

We cannot expect to cope with all these problems simultaneously on top of those raised previously, so we are forced to simplify. It is well known from studies of genetic polymorphisms at identifiable gene loci such as isozymes that more than two alleles can coexist in natural populations and, by extrapolation, this is probably also true of QTL. However, it is common in such situations for two alleles to predominate while we cannot be sure that all alleles will affect the trait differentially. Therefore, it is probably not too unrealistic to assume that there are no more than two alleles at each locus in terms of their effects on the trait. It would be quite unreasonable to assume that allele frequencies are equal, however, and so this possibility has to be built into the models.

9.2 Solutions

The mating structure of the population is important for the proper interpretation of our analyses. We have considered both random mating and selfing so far, and these two extremes can be handled with our populations. Mixed inbreeding and random mating would be very difficult to cope with. Therefore, in so far as it is possible with experimental organisms, attempts are normally made to enforce a defined mating system on the material; natural mating is too ill-defined. In studies of natural populations, among which our own species is included, it is necessary to attempt to measure any departure from random mating as far as the trait is concerned and build this into the models. Otherwise we will assume that the population has been deliberately randomly mated or selfed.

Let us first consider a population polymorphic at k QTL with the frequency of the increasing A^+ and decreasing A^- alleles being p

and q, where $p + q = 1$. Using a single gene model, as before, then the frequencies and genetic values of the three genotypes in the population, with random mating, will be,

Genotype	A^+A^+	A^+A^-	A^-A^-
Frequency	p^2	$2pq$	q^2
Genetic value	$m + a$	$m + d$	$m - a$

The mean and variance of this population can be derived as before. Thus,

$$\text{mean} = m + p^2 a + 2pqd + q^2(-a)$$

Because $(p^2 - q^2)a = (p + q)(p - q)a$, and $(p + q) = 1$,

$$\text{mean} = m + (p - q)a + 2pqd. \qquad \text{[Eqn 9.1]}$$

This simplifies in the special case of $p = q = \frac{1}{2}$ to $m + \frac{1}{2}d$, which is the mean of an F_2.

The variance is also obtained in the same way as before.

$$V_G = p^2 a^2 + 2pqd^2 + q^2(-a)^2 - [(p - q)a + 2pqd]^2$$

$$= 2pq[a + d(q - p)]^2 + 4p^2 q^2 d^2. \qquad \text{[Eqn 9.2]}$$

Again, if we put $p = q = \frac{1}{2}$ into this formula we obtain the familiar genetical variance of an F_2 at a single locus of,

$$\tfrac{1}{2}a^2 + \tfrac{1}{4}d^2.$$

The formulae can be extended to cover k **independent** loci in the usual way. Thus the population mean is,

$$\text{mean} = m + \sum (p_i - q_i)a_i + \sum 2p_i q_i d_i \qquad \text{[Eqn 9.3]}$$

and the variance,

$$V_G = \underline{\sum 2p_i q_i [a_i + d_i(q_i - p_i)]^2} + \underline{\sum 4p_i^2 q_i^2 d_i^2}. \qquad \text{[Eqn 9.4]}$$

The two components underlined in Equation 9.4 recur in all the derivations of FS and HS variances as well as parent–offspring covariances, therefore, by defining

$$V_A = \sum 2p_i q_i [a_i + d_i(q_i - p_i)]^2$$

and

$$V_D = \sum 4p_i^2 q_i^2 d_i^2$$

we obtain,

$$V_G = V_A + V_D$$

as before for an F_2.

Table 9.1 Expectations of variance components (σ^2s) from the FS and HS experimental designs for randomly mating populations where the allele frequencies are unknown

σ^2	Model
σ_F^2	$\frac{1}{4}V_A + V_{EM}$
σ_M^2	$\frac{1}{4}V_A$
σ_{MF}^2	$\frac{1}{4}V_D + V_{EC} - V_{EM}$
$\sigma_{F/M}^2$	$\frac{1}{4}V_A + \frac{1}{4}V_D + V_{EC}$
σ_B^2	$\frac{1}{2}V_A + \frac{1}{4}V_D + V_{EC}$
σ_W^2	$\frac{1}{2}V_A + \frac{3}{4}V_D + V_E$

$$h_n^2 = \frac{V_A}{V_A + V_D + V_E}$$

Fortunately, all the other expectations derived in Chapter 5 for FS and HS families derived from an F_2 translate into identical components by substituting V_A and V_D for V_A^* and V_D^*, as shown in Table 9.1. Those who enjoy tedious algebra may wish to prove this for themselves and the appropriate table, corresponding to Table 4.8 for an F_2, is shown in Table 9.2. Therefore all the BIPs, NCI and NCII experiments can be performed and analysed as described in Chapter 5, only the interpretations of the components change.

9.3 Consequences of definitions of V_A and V_D

It is worth noting several features which emerge from these definitions. First, although V_D is purely dominance variation, V_A includes additive and dominance effects; V_A is purely additive variation only when $p = q$. Nonetheless, V_A is referred to as the additive genetical variance, as before, although it is additive only in the statistical sense, being a measure of GCA. Similarly, V_D is still called the non-additive variance. In both cases, however, the size of the component is affected by the allele frequencies and the sign of d_i. The result of this is that such parameters cannot be used to estimate the dominance ratio because the ratio $\sqrt{(4V_D/2V_A)}$ will vary with p_i as shown in Figure 9.1. This shows that the dominance ratio is greatest when the dominant allele is the most frequent and, given that alleles favoured by natural selection are likely to be both dominant and most frequent, such a situation is likely to be very common in natural populations. It

Table 9.2 Sib-mating a population with unequal gene frequencies. (i) Structure of families; (ii) summary of means and variances; (iii) derivation of covariance of offspring onto mid-parental values

(i)

Female parent	Frequency		Male parent: A^+A^+ p^2	A^+A^- $2pq$	A^-A^- q^2	Row means (= HS means)
A^+A^+	p^2	Family	P_1	$Bc_{1.1}$	F_1	
		(Freq)	(p^4)	$(2p^3q)$	(p^2q^2)	
		Family mean	$m+a$	$m+\frac{1}{2}(a+d)$	$m+d$	$m+pa+qd$
		σ_W^2	V_E	$\frac{1}{4}(a-d)^2+V_E$	V_E	
A^+A^-	$2pq$	Family	$Bc_{1.1}$	F_2	$Bc_{1.2}$	
		(Freq)	$(2p^3q)$	$(4p^2q^2)$	$(2pq^3)$	
		Family mean	$m+\frac{1}{2}(a+d)$	$m+\frac{1}{2}d$	$m+\frac{1}{2}(-a+d)$	$m+(p-q)a+\frac{1}{2}d$
		σ_W^2	$\frac{1}{4}(a-d)^2+V_E$	$\frac{1}{2}a^2+\frac{1}{4}d^2+V_E$	$\frac{1}{4}(a+d)^2+V_E$	
A^-A^-	q^2	Family	F_1	$Bc_{1.2}$	P_2	
		(Freq)	(p^2q^2)	$(2pq^3)$	(q^4)	
		Family mean	$m+d$	$m+\frac{1}{2}(-a+d)$	$m-a$	$m-qa+pd$
		σ_W^2	V_E	$\frac{1}{4}(a+d)^2+V_E$	V_E	
Column means (=HS means)			$m+pa+qd$	$m+(p-q)a+\frac{1}{2}d$	$m-qa+pd$	$m+(p-q)a+\frac{1}{2}pqd$

(ii)

Type of family	P_1	$Bc_{1.1}$	F_1	F_2	$Bc_{1.2}$	P_2	Average
Frequency	(p^4)	$(4p^3q)$	$(2p^2q^2)$	$(4p^2q^2)$	$(4pq^3)$	(q^4)	1.0
Family mean (y_i)	$m+a$	$m+\frac{1}{2}(a+d)$	$m+d$	$m+\frac{1}{2}d$	$m+\frac{1}{2}(-a+d)$	$m-a$	$m+(p-q)a+\frac{1}{2}pqd$
σ_W^2	V_E	$\frac{1}{4}(a-d)^2+V_E$	V_E	$\frac{1}{2}a^2+\frac{1}{4}d^2+V_E$	$\frac{1}{4}(a+d)^2+V_E$	V_E	$pq\{a+d(q-p)\}^2+3p^2q^2d^2+V_E$
Mid-parent (x_i)	$m+a$	$m+\frac{1}{2}(a+d)$	m	$m+d$	$m+\frac{1}{2}(-a+d)$	$m-a$	$m+(p-q)a+\frac{1}{2}pqd$

(iii)

$$\text{Covariance } (y_i, x_i) = p^4\{a\cdot a\} + 4p^3q\{\tfrac{1}{2}(a+d)\cdot\tfrac{1}{2}(a+d)\} + 2p^2q^2\{d\cdot 0\} + 4p^2q^2\{\tfrac{1}{2}d\cdot d\}$$
$$+ 4pq^3\{\tfrac{1}{2}(-a+d)\cdot\tfrac{1}{2}(-a+d)\} + q^4\{(-a)\cdot(-a)\} - \{(p-q)a+\tfrac{1}{2}pqd\}^2$$
$$= pq\{a+d(q-p)\}^2$$

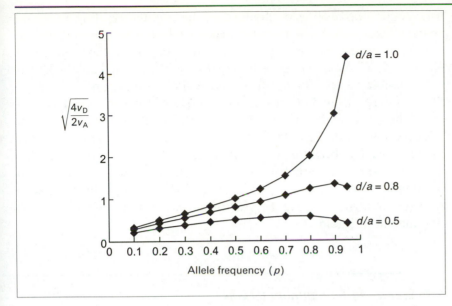

Figure 9.1 Dominance ratios should not be estimated using V_A and V_D from populations in which the allele frequencies of the QTL are unknown. The figure shows that such an approach would result in the estimated dominance ratio varying with the allele frequency and it would invariably be biased, often falsely, indicating considerable overdominance. The true level of dominance for each situation is given as d/a.

is also typical of the situation found in many heterotic traits by breeders. Also, the higher the degree of dominance the greater is the bias, with overdominance being indicated when it does not in fact exist. One should, therefore, never attempt to judge the degree of dominance from estimates of variance components in populations in which the allele frequencies are unknown.

Because the sign of d affects V_A, and both V_A and V_D depend on p, it follows that the size of the narrow and broad heritability will also depend on these factors. It is still however the case that regression of offspring on mid-parent is V_A/V_P, and this is the appropriate value of the narrow heritability for predicting selection response.

It does not generally make sense to consider an NCIII or TTC for studying such populations because the necessary parental, inbred testers would not exist. The best that one could do, would be to use extreme high and low selection lines from the population, on the grounds that they should differ at most of the loci segregating in the population for the selected trait. Unfortunately, such lines are only of use for the trait selected, and cannot be used for other traits as they would not necessarily differ at many of the segregating loci. One possible situation where the NCIII and TTC might be useful, however, is where an F_2 has been created but then allowed to evolve over a number of generations in response to natural selection. Providing the evolutionary period is not sufficiently great for new mutations to have accumulated and contamination can be ruled out, the original parents and F_1 may still be used as testers because they will differ at all genes segregating in the population. The allele frequencies may well have changed in the population, however, and some alleles may have become fixed, and so the expectations of the variance components will have changed from those given in Chapter 5. This can easily be handled as is shown in Table 9.3. It is still possible to test for epistasis, since the comparisons $\bar{L}_{1i} + \bar{L}_{2i} - 2\bar{L}_{3i} = 0.0$ for all sets of crosses irrespective of the genotype frequencies. The epistasis can no longer be partitioned into the additive × additive and additive × dominance + dominance × dominance types as with the F_2, because the mean is not purely a function of the additive × additive type of epistasis. The σ^2s for additive and dominance effects are still purely a^2 and d^2 respectively, each with the same coefficient; hence the previous tests for dominance and over-dominance still hold.

9.4 Studies of human populations

As we have seen, the analyses of non-F_2 populations have a number of problems, not least of which is the restricted range of experimental structures that can be applied to them. Human populations are particularly difficult, since experimental protocols are not possible and it is necessary to use the families which are naturally available. With such families, the normal assumptions of random mating and random allocation of genotypes to environments will not generally hold. Progeny stay together with their parents for something like 20% of their lives, so magnifying common environmental effects, while for many traits the possibility of cultural transmission is very strong.

We are particularly interested in the genetics of our own species, however, and it is ironic that we should pose so many difficulties for the geneticist. If these practical problems were not enough, social and political attitudes also intrude. No one would be upset to hear that the heritability of fat content of cow's milk was 70%, but if a similar value were published, say, for intelligence or personality in humans the political correctness of the scientist would immediately be questioned, as would be the relevance of the character. This was particularly well illustrated by the so-called Burt affair in the UK, in which an eminent psychologist was accused of scientific fraud after his death, on very flimsy evidence [1]. It was suggested that the heritabilities that he had obtained for intelligence in humans were based on fictitious data, invented to exaggerate the importance of heredity in the variation in IQ. These accusations conveniently

Table 9.3 Expectations of family means for a triple test cross

			Testers				
					$\frac{1}{2}$ Sums	$\frac{1}{2}$ Difference	Epistasis
		P_1	P_2	F_1			
F_2 genotype	Frequency	A^+A^+ ($\bar{L}_{1i}$)	A^-A^- ($\bar{L}_{2i}$)	A^+A^- ($\bar{L}_{3i}$)	$\frac{1}{2}(\bar{L}_{1i} + \bar{L}_{2i})$	$\frac{1}{2}(\bar{L}_{2i} - \bar{L}_{1i})$	$(\bar{L}_{1i} + \bar{L}_{2i}) - 2\bar{L}_{3i}$
A^+A^+	p^2	$m+a$	$m+d$	$m+\frac{1}{2}a+\frac{1}{2}d$	$m+\frac{1}{2}a+\frac{1}{2}d$	$-\frac{1}{2}a+\frac{1}{2}d$	0
A^+A^-	$2pq$	$m+\frac{1}{2}a+\frac{1}{2}d$	$m-\frac{1}{2}a+\frac{1}{2}d$	$m+\frac{1}{2}d$	$m+\frac{1}{2}d$	$-\frac{1}{2}a$	0
A^-A^-	q^2	$m+d$	$m-a$	$m-\frac{1}{2}a+\frac{1}{2}d$	$m-\frac{1}{2}a+\frac{1}{2}d$	$-\frac{1}{2}a-\frac{1}{2}d$	0
Mean		$m+pa+qd$	$m-qa+pd$	$m+(p-q)a+pd$	$m+\frac{1}{2}(p-q)a+\frac{1}{2}d$	$-\frac{1}{2}a+\frac{1}{2}(p-q)d$	0

$\sigma_S^2 = p^2(\frac{1}{2}a)^2 + q^2(-\frac{1}{2}a)^2 - \{\frac{1}{2}(p-q)a\}^2 = \frac{1}{2}pqa^2$

$\sigma_D^2 = p^2(\frac{1}{2}d)^2 + q^2(-\frac{1}{2}d)^2 - \{\frac{1}{2}(p-q)d\}^2 = \frac{1}{2}pqd^2$

ignored the fact that larger well-documented trials around the world had produced very similar estimates to Burt [2,3]. Notwithstanding these difficulties, many are keen to try to extract genetic information about a whole range of human quantitative traits and we will attempt to illustrate some of the basic approaches that they use.

The most extensively used families are full-sibs because they are common. One could obviously use the FS intra-class correlation as described in Chapter 5 to estimate the heritability, but this approach assumes that V_{EC} is small. Given the long period of time that human offspring remain within the natural or extended family, it may be dangerous to assume that V_{EC} is small and it requires testing in individual cases. Some attempt can be made to estimate the effects of common environment effects by looking at families in which the FSs are raised in different families as a result of fostering – the equivalent of randomizing over families. Such cases in which families are split up in this way are, thankfully, few while attempts are made to match the foster home to the natural home, which vitiates the principle of randomization.

9.5 The use of twins

FS families in which the progeny are twins, are potentially more useful because the sibs are the same age and hence exposed to similar environments. The general ANOVA of a set of n FS families of size 2 is shown in Table 9.4(i), together with the expectations of the σ^2s in Table 9.4(ii). These expectations assume that the sibs are standard FS. If the families consisted of di-zygotic (DZ), i.e. non-identical twins, the expectations would stay the same apart from the possibility that the expected values of E_w and V_{EC} might be different because twins potentially share a more similar environment simply because they grow up together.

If the twins are mono-zygotic (MZ), identical, on the other hand, the genetical and environmental expectations of the σ^2s are quite different as shown in Table 9.4(iii). Because the twins in any family are genetically identical, i.e. they are natural clones, there can be no genetic variation between them apart from that introduced by rare mutation which can be safely ignored. Therefore, as there is no genetic variation within a family, all the genetic variation occurs between families [4,5].

It may appear at first glance from the combined ANOVA of DZ and MZ families that because there are four parameters (V_A, V_D, V_E and

Table 9.4 Analyses of FS families including MZ and DZ twins. (i) Basic FS ANOVA with σ^2s; (ii) expectations of σ^2 for DZ twins and normal FS; (iii) expectations of σ^2 for MZ twins

(i)

Source	df	MS	ems
Between-families	$n-1$	MS_B	$\sigma_W^2 + 2\sigma_B^2$
Within-families	n	MS_W	σ_W^2
Total	$2n-1$		

(ii) $\quad \sigma_B^2 = \frac{1}{2}V_A + \frac{1}{4}V_D + V_{EC}$
$\qquad \sigma_W^2 = \frac{1}{2}V_A + \frac{3}{4}V_D + V_E$

(iii) $\quad \sigma_B^2 = V_A + V_D + V_{EC}$
$\qquad \sigma_W^2 = V_E$

V_{EC}) and four statistics, all parameters are estimable but this is not so; V_{EC} is never independent of $\frac{1}{2}V_A + \frac{1}{4}V_D$. Some use has been made of data from families where the twins have been separated and raised in different families, but the rarity of such cases and the non-random allocation of children to foster homes means that their value is reduced [2].

How far can one progress using the set of three common FS families, non-twins, DZ twins and MZ twins, without involving unrealistic or untestable assumptions? First of all, it is reasonable to assume that the underlying total genetical variation between individuals of the three groups is the same; whether an individual is a normal sib, MZ or DZ twin is a matter of chance. On the other hand, it is a distinct possibility that the average environments of the three groups are different, which would imply that we could be assessing the genes in different environments. This should be detectable as a difference in the trait means of the three groups and could arise for various reasons. Twins clearly impose different demands on their parents in terms of time and resources than do sibs born separately and these may create a different environment relative to normal sibs. The inter-relationships between the twins themselves, cooperative or competitive, may be very different from the relationships between normal sibs. Such differences may even occur between MZ and DZ twins, competition being at its greatest between identical genotypes. On *a priori* grounds, it is likely that the overall environments of MZ and DZ twins would be the most similar, but this could be tested by comparing the trait means for the two groups.

The factors which we have just considered as possible determinants of overall environmental differences between MZ and DZ twins could also contribute to differences in the size of V_E for MZ and DZ families. It is, however, more difficult to imagine why the environmental variation between families, V_{EC}, should differ. This is a function of genetical and environmental differences between families in the types of nurture they provide, such as diet, intellectual enrichment, exercise, stress, encouragement, etc. MZ and DZ twins should experience these variables equally and hence V_{EC} could reasonably be expected to be constant for both groups.

We have argued that DZ and MZ twins should have the same total genetical variance and the same environmental variation between families; what may differ are the V_Es. This possibility can easily be tested by comparing the total phenotypic variances of the two groups based on a suitably large sample. If there is a difference, then, since V_E for MZ twins is directly estimable from σ_W^2, V_E for DZ can be obtained by adding or subtracting, as appropriate, the difference in the two phenotypic variances. Table 9.5 gives the ANOVA of same-sex MZ and DZ twins from a large Swedish study [6] for the personality trait, neuroticism. It is clear that the within-family variances are always smaller for MZs than DZs and the corresponding intra-class correlations are always larger for MZ twins, as expected on a genetic model. It is also clear from Table 9.6 that the total variances of the MZ and DZ twin populations are very similar and certainly do not differ significantly, despite the very large numbers of df. It would thus appear that there is no reason to invoke any differential environment effects for the two types of twins.

It is possible to use weighted least squares procedures to fit models to these data (Table 9.5), and they confirm what was inferred above, that a simple V_A, V_E model fits the data for each sex adequately.

Table 9.5 Example of within- and between-family mean squares for MZ and DZ twins for a measure of neuroticism. Data based on a Swedish twin study [6]

Type of twins	Sex	Source	df	MS	t_{FS}
MZ	M	Between-families	1052	6.83	0.53
		Within-families	1053	2.09	
DZ	M	Between-families	1751	5.88	0.24
		Within-families	1752	3.62	
MZ	F	Between-families	1239	9.44	0.62
		Within-families	1240	2.19	
DZ	F	Between-families	1797	7.11	0.29
		Within-families	1798	3.89	

Table 9.6 Total variances (i) and parameter estimates (ii) derived from the data in Table 9.5

(i)

Type of twins	Sex	df	Total variance
MZ	M	2105	8.92
DZ	M	3503	9.5
MZ	F	2479	11.63
DZ	F	3595	11

(ii)

Sex	V_A	V_E	h_n^2
M	2.47	2.19	0.53
F	3.41	2.18	0.61

Indeed, it would appear that common environment effects are nowhere near as important as intuition might suggest for many aspects of human behaviour. One's family and school seem to have little influence on one's degree of neuroticism nor indeed to many other dimensions of behaviour and intelligence [7].

If V_{EC} was a problem, it would be possible to estimate the genetical variance within DZ families from the difference between the within family MS of DZ and MZ twins, $\frac{1}{2}V_A + \frac{3}{4}V_D$, and this estimate can be used to obtain a value for the broad heritability of the trait as,

$$2(\tfrac{1}{2}V_A + \tfrac{3}{4}V_D)/V_P = (V_A + 1\tfrac{1}{2}V_D)/(V_A + V_D + V_{EC} + V_E).$$

Again, using the neuroticism data in Table 9.5, we obtain broad heritability estimates of 0.68 and 0.64 for males and females respectively. These estimates are slightly inflated because V_D has a coefficient of $1\frac{1}{2}$ instead of 1, but the effect would be difficult to detect. Moreover, assortative mating among the parents of the twins, which can be shown to occur for many traits like stature and IQ, will counteract this, because it will reduce the genetical variation within families and increase that between them.

9.6 Heritabilities of human traits

When studying the genetics of domesticated animals and crop plants, estimates of narrow heritability are the principal aim because they are used to predict selection response. In human studies, the broad heritability (h_b^2) is more useful because $(1 - h_b^2)$ provides a measure of the scope of environmental influence, such as by educational or dietary reform, to affect the trait. In so far as there are identifiable environmental factors contributing to the phenotypic variance of a trait, they are likely to be components of V_{EC}. The within-family environmental components are likely to be far less tangible and will, of course, contain a large component due to factors which are not strictly environmental variables at all such as developmental noise, measurement error, etc. (Chapter 3). Thus, given the current environmental heterogeneity, only that proportion of V_P arising from V_{EC} is potentially capable of being exploited for social or medical purposes.

Heritability is but a function of the population and environment currently existing. A high heritability indicates that in the current environment there is little room for environmental change altering the phenotype. Suppose, for example, that a population has a mean of 80, a variance of 100, h_b^2 of 0.8 and V_{EC} of 9 for some particular trait (i.e. $V_{EC} \approx \frac{1}{2}$ the non-genetical variance). It is just possible that one particular dietary factor contributes a major part of V_{EC}, say 4 units of variance. If every one were to adopt the best diet, the population mean might improve by $2 \times \sqrt{4}$ to 84. It is also possible that somebody might introduce a new breakfast cereal, never eaten before, which increases the mean trait by 50 units, as well as eliminating nearly all the genetical component of phenotypic variance. Such a factor is equivalent to the introduction of antibiotics or vaccination. It changes the environment so completely that the original phenotypic variation in susceptibility to infection is removed. A high heritability, therefore, does not preclude the possibility of trait enhancement through environmental change, it merely suggests that the desired factor is not a major component in the present environment. Currently, the evidence for IQ is that its heritability is high, V_{EC} is small and no particular factor in the environment has yet been identified which can improve a person's IQ potential very much.

9.7 Genotype–environment correlation

Another factor to consider in natural populations and one that is particularly relevant to humans, is that of genotype–environment correlation. This implies that genotypes which enhance the trait are to be found in environments that do the same and *vice versa*. For example, consider genes for intelligence. Individuals who have more alleles for high intelligence are more likely to go to college and obtain well-paid employment. They are also likely to marry someone from the same socioeconomic class who will also have more + alleles for intelligence, i.e. there will be assortative mating. Their children will thus have the advantage not only of their parents' alleles for high ability but also the advantages of a 'good' home background which could stimulate and encourage intellectual development. The converse would occur at the other end of the intellectual spectrum and overall this results in a correlation between the genotype and the environment of the children. The environment is in one sense an extension of the phenotype produced by the genes, although it is the parents' genes that have created the environment. Such correlations may not be high but they are inevitable. Their effect will be to inflate the variation between families. Instead of being just $V_G + V_{EC}$ they will also contain twice the covariance between genotype and environment. This is another reason for estimating the heritability from the within family variances.

The progeny of two identical twins are related as HS since, genetically, they have one parent in common as shown in Figure 9.2. It was shown in Chapter 5 that the use of HS in experimental situations means that we have a measure of additive genetical variation that is not associated with dominance or common environment effects [8]. With twin families, however, the progeny will be kept together with their male parent and to some extent the advantage of HS is lost. The ANOVA of such a set of twin families is shown in Table 9.7.

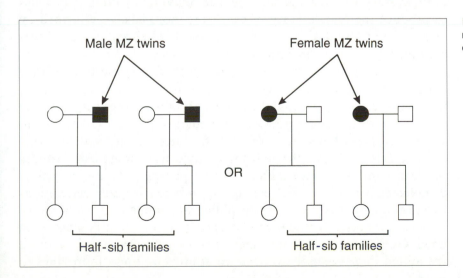

Figure 9.2 The families of monozygotic, identical twins are effectively half-sib families.

Table 9.7 Generalized ANOVA of the HS families of monozygotic twins described in Figure 9.2. (i) ANOVA; (ii) components

(i)

Source	df	MS	ems
Between HS groups	$n-1$	MS1	$\sigma_W^2 + r\sigma_{FS/HS}^2 + 2r\sigma_{HS}^2$
Between FS/HS groups	$n(2-1)$	MS2	$\sigma_W^2 + r\sigma_{FS/HS}^2$
Within FS families	$2n(r-1)$	MS3	σ_W^2

(ii)

$$\sigma_W^2 = \tfrac{1}{2}V_A + \tfrac{3}{4}V_D + V_E$$

$$\sigma_{FS/HS}^2 = \tfrac{1}{4}V_A + \tfrac{1}{4}V_D + V_{EC}$$

$$\sigma_{HS}^2 = \tfrac{1}{4}V_A$$

9.8 Diallel crosses

We described the analysis and interpretation of diallel crosses in Chapter 5 but indicated that they are used mainly to explore the genetics of crosses between inbred lines. For example, a breeder may have a number of promising inbred lines of some plant species such as maize or wheat, or relatively inbred populations such as breeds of cattle, and is interested to know how best to use these in crosses for future improvement. Because the lines are inbred it is not necessary to cross the same female individual to different males to produce the female HS groups; this is more easily performed by using several females from the same line to different males. There are thus few biological restrictions on their use.

As we stated earlier there is little reliable information to be gained from attempting to estimate genetical and environmental components from a diallel ANOVA because the df are normally so low as to make the estimates too unreliable for general predictive purposes. Their main value is in identifying pairs of parental lines which yield heterotic crosses. Not only can the hybrids be used as commercial varieties or strains but they also indicate which parents contain matching, complementary alleles that can be recombined in later generations by selection. If it is necessary to obtain an estimate of the additive genetical variance for selection purposes or to identify lines with good GCA, this can more easily be achieved by simply scoring the inbred lines themselves. Such an approach allows more lines to be included or, alternatively, a much higher level of replication to be employed than is possible with a diallel. The study of many hundreds of diallel crosses involving a wide variety of organisms and traits has confirmed that the line with the highest GCA is invariably the line with the highest mean score, even in those diallels where non-additive variation plays a major role. In the few cases where non-additive effects have such a major effect as to contradict this general rule then the actual GCA obtained from the HS line means is only relevant to those specific lines.

One further method of analysing inbred diallel crosses is the so-called Wr/Vr analysis [9]. This is a graphical approach which is designed to investigate the relative number of dominant alleles in the various lines making up the parents of the diallel, but is only relevant where the character studied displays non-additive variation as indicated by the 'b' items in the appropriate Hayman ANOVA as described in Chapter 5. It relies on the fact that the more recessive alleles a particular parent line has, the more variable will be the means of its various FS families when it is crossed to other lines. A parent which contains all the dominant alleles, on the other hand, will produce families which all have very similar means. These differences are measured as the variance of the means of the families in any given HS group, Vr, as shown in Table 9.8, using an 8×8 diallel for illustration. Similarly, the co-variance of the family means from any common parent with the means of the non-recurrent parent, Wr, will show the same relationship as the Vr. Plotting the Wr values against the Vr values should, in the absence of epistasis and linkage disequilibrium, produce a straight line with a slope of $b = 1.0$ (Figure 9.3). It is then possible to identify those parents with the most or fewest dominant alleles from their position on the graph. The intercept of the graph indicates the level of dominance. A negative, zero or positive intercept indicating overdominance, complete dominance and partial dominance respectively. In the example in Figure 9.3, dominance is clearly partial while parent V14 has the most dominant alleles and parent V7 the fewest.

9.9 Inbreeding in a population

Inbreeding can occur in a population either by design of the experiment or due to the natural mating of close relatives. It has consequences both to the mean and the variance, although the effect on the former is easier to model and to interpret. As with inbreeding

Table 9.8 Mean flowering time of the parents and F_1s in an 8×8 diallel of *Nicotiana rustica* averaged over 3 years. Entries in bold type are duplicated from the top half of the table and the parental scores in the last row are taken from the leading diagonal of the table. (i) Data; (ii) Wr/Vr values

(i)

	V2	V5	V7	V12	V14	V29	V38	V41
V2	34.2	21.7	35.0	27.4	21.1	30.0	21.0	21.9
V5	**21.7**	22.3	23.5	22.4	20.3	21.8	15.2	18.0
V7	**35.0**	**23.5**	38.6	25.5	21.0	32.8	18.8	22.2
V12	**27.4**	**22.4**	**25.5**	27.4	20.1	25.0	18.2	19.5
V14	**21.1**	**20.3**	**21.0**	**20.1**	20.9	23.2	15.6	20.1
V29	**30.0**	**21.8**	**32.8**	**25.0**	**23.2**	29.6	16.1	19.3
V38	**21.0**	**15.2**	**18.8**	**18.2**	**15.6**	**16.1**	12.3	14.6
V41	**21.9**	**18.0**	**22.2**	**19.5**	**20.1**	**19.3**	**14.6**	19.6
Parents	34.2	22.3	38.6	27.4	20.9	29.6	12.3	19.6

(ii)

Vr is the variance of the rth row, e.g.

$$Vr(V2) = \{34.2^2 + 21.7^2 + \ldots + 21.9^2 - (34.2 + 21.7 + \ldots + 21.9)^2/8\}/7$$
$$= 35.46$$

Wr is the covariance of the rth row with the parental means, e.g.

$$Wr(V12) = \{(34.2 \times 27.4 + 22.3 \times 22.4 + \ldots + 19.6 \times 19.5)$$
$$- [(34.2 + 22.3 + \ldots + 19.6) \times (27.4 + 22.4 + \ldots + 19.5)/8]\}/7$$
$$= 26.67$$

From the above diallel we get:

Array	Vr	Wr
V2	35.46	47.76
V5	7.60	19.77
V7	53.25	59.67
V12	13.25	26.67
V14	4.58	12.59
V29	33.08	47.77
V38	7.48	20.83
V41	5.66	17.38

from an F_2 as discussed in Chapters 2 and 4, inbreeding in a population decreases the frequency of heterozygotes and, given that there is some directional dominance, this decline in heterozygotes will cause the mean to change also; in a heterotic trait, this results in

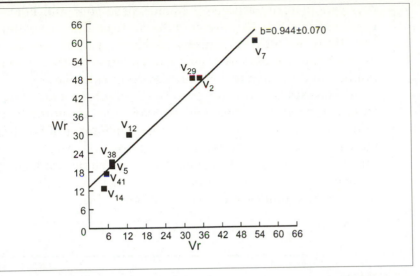

inbreeding depression. The extent of inbreeding in a population can be determined from the frequencies of genetic markers such as blood groups, isozymes or molecular markers. Significant deviations from Hardy–Weinberg equilibrium can, in some cases, indicate that inbreeding is occurring through a deficiency of heterozygotes. Differences in the trait mean for sub-populations with different amounts of inbreeding can indicate the direction of dominance. For example, many human populations which allow marriage between close relatives have reduced IQ which indicates that alleles for high IQ are dominant.

Summary

1. The study of natural or experimental populations other than F_2s involves many more unknown factors such as the mating system, allele frequencies and numbers of alleles.
2. Providing the population mates at random, all the methods described in Chapters 4 and 5 can be used and the interpretation of the ANOVA is the same. The parameter estimates, too, are identical, but their definitions are different.
3. Human populations can be analysed by these methods and normal FS as well as twin families can be used to study the inheritance of polygenic traits as they are all examples of bi-parental families.

continued

4. Analyses of many aspects of human behaviour and inheritance indicate that they are no different from other quantitative traits in other organisms. Additive genetic variance predominates, most environmental variation is that within families and not due to common environmental factors.

5. Diallel crosses are normally performed on inbred lines drawn from some relevant collection. As such they are normally a non-random, selected sample. Because the number of families increases geometrically with the number of parents, the number of parents used is generally small. These two factors mitigate against estimating parameters such as V_A and V_D, because the estimates are biased and unreliable.

References

1. Joynson, R.B. (1989) *The Burt Affair*. Routledge, London.
2. Bouchard, T.J. Jr., Lykken, D.T., McGue, M., Segal, N.L. and Tellegen, A. (1990) Sources of human psychological differences. The Minnesota study of twins reared apart. *Science*, **250**, 223–8.
3. Chipuer, H.M., Rovine, M.J. and Plomin, R. (1990) LISREL modelling. Genetic and environmental influences on IQ revisited. *Intelligence*, **14**, 11–21.
4. Jinks, J.L. and Fulker, D.W. (1970) A comparison of the biometrical genetical, MAVA and classical approaches to the analysis of human behaviour. *Psychol. Bull.*, **73**, 311–49.
5. Mather, K. and Jinks, J.L. (1982) *Biometrical Genetics*, 3rd edn, Chapman & Hall, London.
6. Eaves, L.J., Eysenck, H.J. and Martin, N.G. (1989). *Genes, Culture and Personality. An Empirical Approach*. Academic Press, San Diego.
7. Cypher, L.H., Fulker, D.W., Plomin, R. and DeFries, J.C. (1989) Cognitive abilities in the early school years. No effects of shared environment between parents and offspring. *Intelligence*, **13**, 369–86.
8. Haley, C.S., Jinks, J.L. and Last, K.L. (1981) The monozygotic twin half-sib method for analysing maternal effects and sex linkage in humans. *Heredity*, **46**, 227–38.
9. Jinks, J.L. (1954) The analysis of continuous variation in a diallel-cross of *Nicotiana rustica* varieties. *Genetics*, **39**, 767–88.

The consequences of linkage

10

So far in this book we have assumed that the individual polygenes are independent, both in their action (i.e. no epistasis) and distribution (i.e. unlinked). However, this is rarely true and interaction and linkage between genes are very common. In this chapter we will derive models for linkage between polygenes, explore the effects of linkage on selection and inbreeding, and examine the associated phenomenon of linkage disequilibrium.

10.1 Genetic variation with linkage

As earlier, we will start from a cross between two inbred lines. It was shown in Chapter 3 that with a single gene in an F_2 population we have the following genotypes and genetic values,

Genotype	A^+A^+	A^+A^-	A^-A^-	Mean
Frequency	$\frac{1}{4}$	$\frac{1}{2}$	$\frac{1}{4}$	
Genetic value	$m + a_A$	$m + d_A$	$m - a_A$	$m + \frac{1}{2}d_A$

Ignoring the constant term, m, the genetic variance of the F_2 can be calculated as:

$$\sigma^2_{gF2} = \tfrac{1}{4}a_A^2 + \tfrac{1}{2}d_A^2 + \tfrac{1}{4}(-a_A)^2 - (\tfrac{1}{2}d_A)^2$$
$$= \tfrac{1}{2}a_A^2 + \tfrac{1}{4}d_A^2$$

For many genes, this becomes,

$$= \frac{1}{2}\sum a_i^2 + \frac{1}{4}\sum d_i^2$$

$$\text{or} \quad = V_A^* + V_D^*$$

where $V_A^* = \frac{1}{2}\sum a_i^2$ and $V_D^* = \frac{1}{4}\sum d_i^2$, as defined in Chapter 3.

However, this is true for many genes only if there is no linkage, so let us now consider what happens if the genes A and B are linked. We will start with two F_1 hybrids, one in coupling and the other in repulsion.

$$F_1 \text{ (in coupling)} \qquad F_1 \text{ (in repulsion)}$$

$$\frac{A^+ \quad B^+}{A^- \quad B^-} \qquad \frac{A^+ \quad B^-}{A^- \quad B^+}$$

Recombination in these hybrids results in gametes with the following frequencies, where R represents the recombination frequency between them, i.e. $R = 0$ for completely linked genes and $R = 0.5$ when there is no linkage, as explained in Chapter 6.

Gamete	F_1 (in coupling)	F_1 (in repulsion)	Frequency
A^+B^+	$\frac{1}{2}(1-R)$	$\frac{1}{2}R$	$\frac{1}{2}p$
A^-B^-	$\frac{1}{2}(1-R)$	$\frac{1}{2}R$	$\frac{1}{2}p$
A^+B^-	$\frac{1}{2}R$	$\frac{1}{2}(1-R)$	$\frac{1}{2}q$
A^-B^+	$\frac{1}{2}R$	$\frac{1}{2}(1-R)$	$\frac{1}{2}q$

These F_1 gametes will combine at random to produce the 16 combinations of genotypes shown, together with their frequencies, in Table 10.1(i). There are in fact just nine genotypes (10 if we distinguish the coupling and the repulsion double heterozygotes) and these have been extracted and summarized in Table 10.1(ii). The means and variances can be derived from these nine genotypes in the usual way as illustrated in Table 10.2, from which we see the following.

1. The F_2 mean is the same as that derived in Chapter 2 with free recombination, i.e.

$$m + \tfrac{1}{2}d_A + \tfrac{1}{2}d_B.$$

Clearly, the mean depends on the heterozygote frequency at each gene which is not affected by linkage, when the additive–dominance model is adequate.

2. The variance is affected [1]. Whereas with no linkage,

$$\sigma_{gF2}^2 = \tfrac{1}{2}a_A^2 + \tfrac{1}{2}a_B^2 + \tfrac{1}{4}d_A^2 + \tfrac{1}{4}d_B^2$$

with linkage it becomes,

$$\sigma_{gF2}^2 = \tfrac{1}{2}a_A^2 + \tfrac{1}{2}a_B^2 + (p-q)a_Aa_B + \tfrac{1}{4}d_A^2 + \tfrac{1}{4}d_B^2 + \tfrac{1}{2}(p-q)^2 d_Ad_B.$$

The important feature to note is that there are additional, cross-product terms involving a_Aa_B and d_Ad_B in the F_2 variance, both of

Table 10.1 Derivation of genotypes in an F_2. (i) Gametic and zygotic frequencies; (ii) zygotic frequencies of (i) re-arranged together with genetic values which are given as deviations from m. Symbols: $++\ ++$ represent $A^+A^+B^+B^+$, etc.

(i)

			Female			
			A^+B^+	A^+B^-	A^-B^+	A^-B^-
	Gametes	Frequency	$\frac{1}{2}p$	$\frac{1}{2}q$	$\frac{1}{2}q$	$\frac{1}{2}p$
Male	A^+B^+	$\frac{1}{2}p$	$++\ ++$ $\frac{1}{4}p^2$	$++\ +-$ $\frac{1}{4}pq$	$+-\ ++$ $\frac{1}{4}pq$	$+-\ +-$ $\frac{1}{4}p^2$
	A^+B^-	$\frac{1}{2}q$	$++\ +-$ $\frac{1}{4}pq$	$++\ --$ $\frac{1}{4}q^2$	$+-\ +-$ $\frac{1}{4}q^2$	$+-\ --$ $\frac{1}{4}pq$
	A^-B^+	$\frac{1}{2}q$	$+-\ ++$ $\frac{1}{4}pq$	$+-\ +-$ $\frac{1}{4}q^2$	$--\ ++$ $\frac{1}{4}q^2$	$--\ +-$ $\frac{1}{4}pq$
	A^-A^-	$\frac{1}{2}p$	$+-\ +-$ $\frac{1}{4}p^2$	$+-\ --$ $\frac{1}{4}pq$	$--\ +-$ $\frac{1}{4}pq$	$--\ --$ $\frac{1}{4}p^2$

(ii)

		A^+A^+	A^+A^-	A^-A^-
B^+B^+	Frequency	$\frac{1}{4}p^2$	$\frac{1}{2}pq$	$\frac{1}{4}q^2$
	Genetic value	$a_A + a_B$	$d_A + a_B$	$-a_A + a_B$
B^+B^-	Frequency	$\frac{1}{2}pq$	$\frac{1}{2}(p^2 + q^2)$	$\frac{1}{2}pq$
	Genetic value	$a_A + d_B$	$d_A + d_B$	$-a_A + d_B$
B^-B^-	Frequency	$\frac{1}{4}q^2$	$\frac{1}{2}pq$	$\frac{1}{4}p^2$
	Genetic value	$a_A - a_B$	$d_A - a_B$	$-a_A - a_B$

which depend on $(p - q)$. Now, with linkage in coupling, $p = 1 - R$ and $q = R$, and therefore, $p - q = (1 - 2R)$. With linkage in repulsion, on the other hand, $p = R$ and $q = (1 - R)$, so $p - q = -(1 - 2R)$.

In general,

$$p - q = \delta(1 - 2R)$$

where $\delta = +1$ for coupling and $\delta = -1$ for repulsion.

Thus, for two unlinked genes,

$$V_A^* = \tfrac{1}{2}a_A^2 + \tfrac{1}{2}a_B^2 \quad \text{and} \quad V_D^* = \tfrac{1}{4}d_A^2 + \tfrac{1}{4}d_B^2$$

while with linkage, these expectations are modified to:

$$V_A^\dagger = \tfrac{1}{2}a_A^2 + \tfrac{1}{2}a_B^2 + \delta_{AB}(1 - 2R_{AB})a_A a_B \qquad \text{[Eqn 10.1]}$$

and

$$V_D^\dagger = \tfrac{1}{4}d_A^2 + \tfrac{1}{4}d_B^2 + \tfrac{1}{2}(1 - 2R_{AB})^2 d_A d_B. \qquad \text{[Eqn 10.2]}$$

Table 10.2 Derivation of means and variances for an F_2 with linkage

Using the genetic values in Table 10.1(ii):

$$\text{Mean of } F_2 = m + \tfrac{1}{4}p^2(a_A + a_B) + \tfrac{1}{2}pq(d_A + a_B) + \ldots + \tfrac{1}{4}p^2(-a_A - a_B)$$

$$= m + \tfrac{1}{2}d_A + \tfrac{1}{2}d_B$$

$$\text{Variance of } F_2 = \{\tfrac{1}{4}p^2(a_A + a_B)^2 + \tfrac{1}{2}pq(d_A + a_B)^2 + \ldots + \tfrac{1}{4}p^2(-a_A - a_B)^2\}$$

$$- \{(\tfrac{1}{2}d_A + \tfrac{1}{2}d_B)^2\}$$

$$= \tfrac{1}{2}a_A^2 + \tfrac{1}{2}a_B^2 + (p - q)a_A a_B + \tfrac{1}{4}d_A^2 + \tfrac{1}{4}d_B^2 + \tfrac{1}{2}(p - q)^2 d_A d_B$$

Now, with coupling linkage, $p = 1 - R$ and $q = R$,
therefore $p - q = +(1 - 2R)$
With linkage in repulsion, $p = R$ and $q = 1 - R$,
therefore $p - q = -(1 - 2R)$

Variance of
$$F_2 = \tfrac{1}{2}a_A^2 + \tfrac{1}{2}a_B^2 + \delta(1 - 2R)a_A a_B + \tfrac{1}{4}d_A^2 + \tfrac{1}{4}d_B^2 + \tfrac{1}{2}(1 - 2R)^2 d_A d_B$$
where $\delta = +1$ or -1 for coupling or repulsion, respectively

Some general consequences of changing R are apparent from these equations. Clearly, with no linkage (i.e. R = 0.5) $V_A^\dagger = V_A^*$ and $V_D^\dagger = V_D^*$. With linkage, however, estimates of the additive and dominance variance are biased. V_A^* is inflated by coupling $(V_A^\dagger > V_A^*)$ and deflated by repulsion linkage $(V_A^\dagger < V_A^*)$. On the other hand, V_D^* is inflated by both types of linkage, relative to its value with free recombination (R = 0.5) unless dominance is ambi-directional, in which case $d_A d_B$ will be negative. These biases have profound effects on the apparent amount of dominance as measured by the dominance ratio, as shown in Figure 10.1. With repulsion linkage and directional dominance, dominance is always inflated and the effect is particularly marked with tight linkage. There is also a slight underestimation of the amount of dominance with inter-mediate levels of coupling linkage.

10.2 Extension to more than two genes

Two genes represent a very special case as there can only be complete association or complete dispersion. With three or more genes it is still possible to have complete association and thus all pairs of genes can also display coupling linkage. However, dispersion does not auto-matically lead to a preponderance of repulsion linkages because

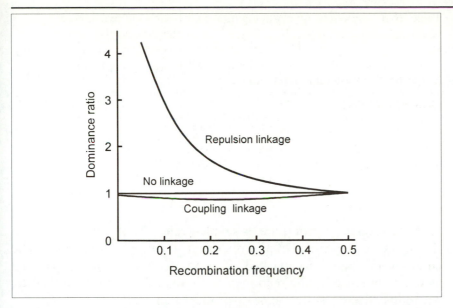

Figure 10.1 The effect of linkage on the dominance ratio. Two genes in repulsion and coupling, all gene effects are equal and there is complete dominance $(a_A = a_B = d_A = d_B = 1)$. The true dominance ratio is 1.

some gene pairs will be in repulsion while others will be in coupling. For instance, with three genes, the third gene has to be in coupling with one of the other two, whenever there is gene dispersion. For example, A^+C^+ are in coupling in $A^+B^-C^+$ and B^-C^- are in coupling in $A^+B^-C^-$.

The formulae for $V_A^\dagger$ and $V_D^\dagger$ with three genes become:

$$2V_A^\dagger = a_A^2 + a_B^2 + a_C^2 + 2\delta_{AB}(1 - 2R_{AB})a_A a_B + 2\delta_{AC}(1 - 2R_{AC})a_A a_C$$
$$+ 2\delta_{BC}(1 - 2R_{BC})a_B a_C$$

$$4V_D^\dagger = d_A^2 + d_B^2 + d_C^2 + 2(1 - 2R_{AB})^2 d_A d_B + 2(1 - 2R_{AC})^2 d_A d_C$$
$$+ 2(1 - 2R_{BC})^2 d_B d_C.$$

Although we have added only one a or d term, we have added two cross-product terms. For k loci we obtain:

$$2V_A^\dagger = \sum_{i=1}^{k} a_i^2 + 2\left\{ \sum_{i=1}^{k-1} \sum_{j=i+1}^{k} \delta_{ij}(1 - 2R_{ij})a_i a_j \right\} \qquad \text{[Eqn 10.3]}$$

$$4V_D^\dagger = \sum_{i=1}^{k} d_i^2 + 2\left\{ \sum_{i=1}^{k-1} \sum_{j=i+1}^{k} (1 - 2R_{ij})^2 d_i d_j \right\}. \qquad \text{[Eqn 10.4]}$$

It is not easy to visualize the meaning of these expressions even if we start with all the genes in complete association, while with dispersion there is such a wide range of possible gene combinations that the outcome can be very variable. Intuition might suggest that with many genes in dispersion, the coupling pairs would tend to cancel out those

Table 10.3 The effect of changing the distribution of five linked genes on estimates of $V_A^*(V_A^\dagger)$, $V_D^*(V_D^\dagger)$ and the dominance ratio, $(4V_D^*/2V_A^*)$. All gene effects are assumed to be equal, i.e. $a_A = a_B = d_A = d_B = 1.0$, and the sequence of alleles $(+, -$ etc.) shown is that for P_1, P_2 would be the converse

Gene distribution	$2V_A^\dagger$	$4V_D^\dagger$	$\sqrt{(4V_D^\dagger/2V_A^\dagger)}$
$+ + + + +$	11.13	7.45	0.82
$+ - + - +$	2.13	7.45	1.87
$+ + - - +$	3.63	7.45	1.43
$- + + + -$	4.13	7.45	1.34
$+ + + - -$	5.88	7.45	1.27

in repulsion but it is not quite as simple as that. Table 10.3 shows that the actual arrangement of the dispersed $+$ and $-$ alleles along the chromosome has a very marked affect on V_A^*, but not on V_D^*, as long as the dominance is unidirectional. With ambi-directional dominance, V_D^* too is affected, since the cross-product terms $d_i d_j$ can take sign.

The illustration involves five genes because it is not worth considering more than that number on a given chromosome. It is clear that the dominance ratio is reduced in the above case to 82% of its true value with complete association, a more marked effect than with two genes. The effect of dispersion is always to reduce V_A^* relative to V_D^* and hence cause the dominance ratio to be inflated. The extent to which this occurs depends highly on the actual gene distribution along the chromosome and it is most pronounced when the $+$ and $-$ alleles alternate in the sequence. In most breeding situations, the breeder is attempting to bring together dispersed sets of genes in the hope of producing a heterotic hybrid or transgressive segregants. Such dispersion of linked genes will obviously exaggerate the observed dominance and this effect has tended to reinforce researchers' belief in overdominance.

10.3 Linkage and random mating an F_2

We showed earlier that with two linked genes the genetic variance of an F_2 is,

$$\sigma_{gF2}^2 = V_A^\dagger + V_D^\dagger$$

where $V_A^\dagger$ and $V_D^\dagger$ are as defined in Equations 10.1 and 10.2. If we now mate the individuals of the F_2 at random, the derived population has the genetic variance,

$$\sigma_{gF2RM}^2 = V_A^\ddagger + V_D^\ddagger$$

where

$$V_A^{\ddagger} = \tfrac{1}{2}a_A^2 + \tfrac{1}{2}a_B^2 + \delta_{AB}(1 - 2R_{ab})(1 - R_{AB})a_A a_B \qquad \text{[Eqn 10.5]}$$

$$V_D^{\ddagger} = \tfrac{1}{4}d_A^2 + \tfrac{1}{4}d_B^2 + \tfrac{1}{2}[(1 - 2R_{AB})(1 - R_{AB})]^2 d_A d_B. \qquad \text{[Eqn 10.6]}$$

The contributions of the $a_A a_B$ and $d_A d_B$ terms to the genetic variance will decline with every cycle of random mating and asymptotically approach zero, as shown in Figure 10.2, because $0 \leq R \leq 0.5$ and $1 \geq (1 - R) \geq 0.5$. As the cross-product terms eventually get very close to zero, the population reaches a stage where there will no longer be any linkage bias. At this stage, the genes are said to be in '**linkage equilibrium**'. They are of course still linked as closely as before, but they have reached the state where the frequency of coupling heterozygotes ($\tfrac{1}{2}p^2$) equals the frequency of repulsion heterozygotes ($\tfrac{1}{2}q^2$). The tighter the linkage, the greater the initial bias and the longer the population will take to achieve complete linkage equilibrium.

In the case of wild or natural populations, there is no reason to assume that the coupling and repulsion heterozygotes would be present with equal frequency in the first place. Hence, linkage disequilibrium (LD) is a general problem in these populations and we shall deal with this later. Further, if we are starting with a population which by accident or design has two unlinked genes that happen to be in linkage disequilibrium (e.g. p^2 and q^2 are not equal), then the rate of decline of the linkage bias (under random mating) will be $(1 - R)$ per generation. Thus, even though the genes are unlinked (i.e. $R = 0.5$) the bias caused by the disequilibrium will only decline by 0.5 per generation, Figure 10.2.

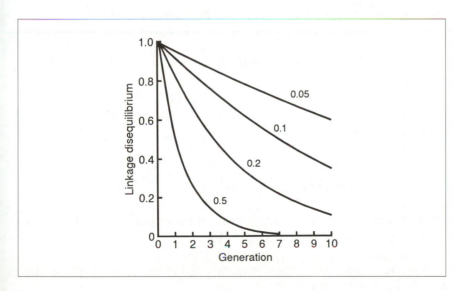

Figure 10.2 The effect of linkage on the approach to linkage equilibrium following various generations of random mating. Each line represents a different recombination frequency between the loci concerned. Even when the genes are unlinked, R = 0.5 (bottom line), LD reduces by only 0.5 per generation.

10.4 Linkage and inbreeding an F_2

The extent of bias to V_A^* and V_D^* due to linkage disequilibrium also declines with selfing, but the rapid approach to homozygosity due to recurrent selfing prevents linkage equilibrium ever being attained in the population. It was shown in Chapter 4 that if a population was inbred by SSD then, in the absence of linkage, the variance of the derived inbred line means would be $2V_A^*$. With two genes this would be $a_A^2 + a_B^2$, but if the genes were linked, it becomes,

$$a_A^2 + a_B^2 + 2\delta_{AB}\{(1 - 2R_{AB})/(1 + 2R_{AB})\}a_A a_B \qquad \text{[Eqn 10.7]}$$

i.e. the bias due to linkage disequilibrium is reduced from that in the F_2 (Equation 10.3) but only by a factor of $1/(1 + 2R_{AB})$. For many genes the variance of the inbred line means [2] is:

$$2V_A^\# = \sum_{i=1}^{k} a_i^2 + 2\left\{ \sum_{i=1}^{k-1} \sum_{j=i+1}^{k} \delta_{ij}[(1 - 2R_{ij})/(1 + 2R_{ij})]a_i a_j \right\}.$$

Therefore, this variance is also biased by linkage and hence a knowledge of $\sum a_i^2$, even if it were estimable, would not be relevant for predicting $V_A^\#$. Fortunately, $V_A^\#$ can be predicted with reasonable accuracy from the estimates of $V_A^\#$ obtained from the generation produced by random mating the F_2, Equation 10.5, because $(1 - R) \cong 1/(1 + 2R)$ [3]. This fact is very important for prediction purposes, as we shall show in Chapter 15.

10.5 Linkage disequilibrium in populations in general

If we are dealing with a population other than the one derived from an F_2, the problems of dealing with two loci become even more complex. Let us consider the four types of gamete which might occur in a given population with respect to two loci, A and B, and let their frequencies be c_1, c_2, c_3 and c_4 as follows:

Gamete	A^+B^+	A^+B^-	A^-B^+	A^-B^-
Frequency	c_1	c_2	c_3	c_4

Now, these frequencies can take any value within the restriction that they sum to unity. The actual values could reflect past selection, genetic drift or even linkage, and there is no reason why they should be equal. Providing that they unite at random to produce zygotes, the zygotic frequencies, by analogy to Table 10.1 will be as in Table 10.4. The mean and variance of this population can be derived by the same procedures as before, giving the expected values below.

$$\text{Mean} = m + a_A(p_A - q_A) + a_B(p_B - q_B) + 2p_Aq_Ad_A + 2p_Bq_Bd_B$$

$$\begin{aligned}
\text{Variance} = {}& 2p_Aq_Aa_A^2 + 2p_Bq_Ba_B^2 + 4(c_1c_4 - c_2c_3)a_Aa_B \\
& + 2p_Aq_A(1 - 2p_Aq_A)d_A^2 + 2p_Bq_B(1 - 2p_Bq_B)d_B^2 \\
& + 4\{(c_1c_4 + c_2c_3) - 2p_Aq_Ap_Bq_B\}d_Ad_B \\
& - 4p_Aq_A(p_A - q_A)a_Ad_A - 4p_Bq_B(p_B - q_B)a_Bd_B \\
& + (p_B - q_B)\{(p_B - q_B)(p_A - q_A) \\
& - (c_1 - c_2 - c_3 + c_4)\}a_Ad_B \\
& + (p_A - q_A)\{(p_B - q_B)(p_A - q_A) \\
& - (c_1 - c_2 - c_3 + c_4)\}a_Bd_A
\end{aligned}$$

where $p_A = c_1 + c_2$,

$p_B = c_1 + c_3$, etc.

As with the F_2, the mean is unchanged but the variance is biased, in this case by disequilibrium terms of the form, $(c_1c_4 - c_2c_3)$. The terms c_1c_4 and c_2c_3 are the frequencies of the coupling and repulsion heterozygotes. As such they are analogous to the term $(p - q)$ which caused the bias in the F_2. This term $(c_1c_4 - c_2c_3)$ is referred to by various names such as the coefficient of gametic phase disequilibrium, coefficient of linkage disequilibrium, or coefficient of epistatic disequilibrium. The former name describes the phenomenon while the other two imply the cause of the disequilibrium. We have already discussed linkage as a cause when the population is derived from an F_2, but selection which creates epistasis for fitness might also be a cause in populations. Whatever the cause, the disequilibrium will decline by a factor of $(1 - R)$ per generation following random mating as we showed for the F_2 earlier, Equations 10.5 and 10.6, so that even if the genes are unlinked, $R = 0.5$, the disequilibrium does not disappear immediately on random mating.

10.6 Tests of linkage

It is possible to construct simple tests of linkage between the genes underlying quantitative traits based on the assumption that if there

Table 10.4 Derivation of genotypes in a population in linkage disequilibrium. (i) Gametic and zygotic frequencies; (ii) zygotic frequencies of (i) rearranged together with genetic values. Symbols: $++$ $++$ represent $A^+A^+B^+B^+$, etc.

(i)

			Female			
			A^+B^+	A^+B^-	A^-B^+	A^-B^-
	Gametes	Frequency	c_1	c_2	c_3	c_4
Male	A^+B^+	c_1	$++$ $++$ c_1^2	$++$ $+-$ c_1c_2	$+-$ $++$ c_1c_3	$+-$ $+-$ c_1c_4
	A^+B^-	c_2	$++$ $+-$ c_2c_1	$++$ $--$ c_2^2	$+-$ $+-$ c_2c_3	$+-$ $--$ c_2c_4
	A^-B^+	c_3	$+-$ $++$ c_3c_1	$+-$ $+-$ c_3c_2	$--$ $++$ c_3^2	$--$ $+-$ c_3c_4
	A^-B^-	c_4	$+-$ $+-$ c_4c_1	$+-$ $--$ c_4c_2	$--$ $+-$ c_4c_3	$--$ $--$ c_4^2

(ii)

		A^+A^+	A^+A^-	A^-A^-
B^+B^+	Frequency	c_1^2	$2c_1c_3$	c_3^2
	Genetic value	$a_A + a_B$	$d_A + a_B$	$-a_A + a_B$
B^+B^-	Frequency	$2c_1c_2$	$2(c_1c_4 + c_2c_3)$	$2c_3c_4$
	Genetic value	$a_A + d_B$	$d_A + d_B$	$-a_A + d_B$
B^-B^-	Frequency	c_2^2	$2c_2c_4$	c_4^2
	Genetic value	$a_A - a_B$	$d_A - a_B$	$-a_A - a_B$

is no linkage, the frequencies of the various types of gamete will be the same from both the F_1 and the F_2, while with linkage they will be different. Such changes will result in different genotypic frequencies in the progeny and, therefore, any changes observed due to additional cycles of recombination will lead to significant differences in genotypic distributions if and only if the genes are linked. Thus, for example, the F_2 distribution will differ from that of the F_{2RM} but only when there is a reduction in the linkage disequilibrium in the latter and this will be true even under complex genetic conditions.

For example, consider the following generations derived from a cross between two inbred parents.

$$1(a)\ F_1 \times P_1 \quad 2(a)\ F_1 \times P_2 \quad 3(a)\ F_1 \times F_1$$
$$1(b)\ F_2 \times P_1 \quad 2(b)\ F_2 \times P_2 \quad 3(b)\ F_2 \times F_1$$

As we go from (a) to (b) in each case, there is a greater opportunity for recombination to occur. For instance, in 1(b) recombination may have occurred in the F_1 and in the F_2 gametes, while in 1(a) it can only have occurred in the F_1. Hence the effect of recombination on the resulting progenies increases from (a) to (b) and so if the total genetic variance differs between (a) and (b), then there must be linkage. Further, if the initial linkage phase had been predominantly repulsion, then the genetic variance would increase from (a) to (b) while the opposite would be observed if the genes were initially linked in coupling. In summary, with no linkage or linkage equilibrium:

$$\sigma^2(F_1 \times P_1) = \sigma^2(F_2 \times P_1),$$
$$\sigma^2(F_1 \times P_2) = \sigma^2(F_2 \times P_2),$$
and
$$\sigma^2(F_1 \times F_1) = \sigma^2(F_2 \times F_1);$$

with predominantly coupling linkage:

$$\sigma^2(F_1 \times P_1) > \sigma^2(F_2 \times P_1),$$
$$\sigma^2(F_1 \times P_2) > \sigma^2(F_2 \times P_2),$$
and
$$\sigma^2(F_1 \times F_1) > \sigma^2(F_2 \times F_1);$$

with predominantly repulsion linkage:

$$\sigma^2(F_1 \times P_1) < \sigma^2(F_2 \times P_1),$$
$$\sigma^2(F_1 \times P_2) < \sigma^2(F_2 \times P_2),$$
and
$$\sigma^2(F_1 \times F_1) < \sigma^2(F_2 \times F_1);$$

and any of these situations can be determined unambiguously by an F test [4,5].

10.7 Sex linkage

In most animals and dioecious plants, the sexes can be distinguished by whether they have one or two X chromosomes. In the majority of species the female has two X chromosomes while the male either has a single X or an X plus a Y chromosome. In some species this relationship is reversed. The sex with two Xs is referred to as the homogametic (like gametes) sex, because all its gametes contain an X chromosome, while the other sex has an X alone or an X and a Y chromosome, and is called the heterogametic sex, because half its gametes contain an X and half do not. Some common examples of

Table 10.5 Some examples of sex-determining mechanisms in plants and animals

Organism	Female XX	Male XX
Animals	Mammals	Fish
	Most insects	Birds
	Some amphibians	Lepidoptera
Plants	Hops	Wild strawberry
	Spinach	
	Cannabis	
	Campion	

these different patterns are shown in Table 10.5. However, for ease of reading and because it reflects the most common situation, we will refer to females when we mean the homogametic sex and to males when we mean the heterogametic sex.

Sex linkage specifically relates to the genes located on the X chromosomes, irrespective of whether they are closely linked among themselves or not. Sex-linked traits should not be confused with sex-limited traits, such as egg production, which are confined to one sex. The pattern of inheritance of the sex chromosomes, often called criss-cross inheritance, leads to predictable differences between reciprocal crosses which have to be accommodated in the biometrical genetical models and some of the most important effects will be described below. The Y chromosome is largely inert genetically and so we will not consider it further. The X chromosomes are referred to as the **sex-chromosomes** to distinguish them from the other chromosomes, called **autosomes.**

10.8 Basic generations of single crosses

It is necessary to define the genetic values of the homogametic and heterogametic sexes differently, although we will try to maintain the conventional symbolism. Thus,

$$\text{Females} \quad X^+X^+ = m + a_x$$

$$X^-X^- = m - a_x$$

$$\text{Males} \quad X^+Y = m + a_{x1}$$

$$X^-Y = m - a_{x1}$$

where a_{x1} and a_x represent the effects of one and two doses of alleles at gene A [1,6].

Using the standard terminology, we will classify the family with X^+X^+ females and X^+Y males as P_1 and that with X^-X^- females and X^-Y males as P_2. From here it follows that we can produce two types of cross from these homozygotes, i.e. the one between X^+X^+ females and X^-Y males ($= F_1$) and the other between X^-X^- females and X^+Y males ($= RF_1$). The phenotypes of these F_1 families are likely to be quite different. Of course, in both crosses the F_1 females will have the same genotype, X^+X^-, and consequently no difference is expected between their genetic values or phenotypes. The genotype of the males, on the other hand, will depend on the genotype of the mother; ♂ progeny of the cross $X^+X^+ \times X^-Y$ ($= F_1$) will have the genetic value $m + a_{x1}$ (being X^+Y), while in the reciprocal RF_1 it will be $m - a_{x1}$. The genetic values of these generations together with the reciprocal F_2s and Bcs are shown in Table 10.6. Dominance is confined to the females as the males have just one allele of any X-linked gene.

Extension of the one-gene model to many genes follows the same rules as for autosomal genes, i.e. replace,

$$a_x \text{ by } [a]_x,$$

$$a_{x1} \text{ by } [a]_{x1}, \text{ and}$$

$$d_x \text{ by } [d]_x.$$

The new parameters will be affected by gene dispersion and ambi-directional dominance in the same manner as $[a]$ and $[d]$ for autosomes described in Chapter 2. The model is further extendable to include epistatic parameters [6].

From Table 10.6 we can recognize several distinct patterns which, when observed in a set of generations, will reveal the presence of sex linkage and allow its effects to be measured and analysed. These are:

(i) an apparent difference between sexes in all the generations;
(ii) reciprocal differences between F_1 males but not the F_1 females;
(iii) reciprocal differences between F_2 females but not the F_2 males; and
(iv) reciprocal differences between both females and males in the $Bc_{1.1}$ and $Bc_{1.2}$ generations.

It follows from these patterns that the additive and dominance effects of the sex-linked genes can cause the simple additive dominance

Table 10.6 Expectations of the F_2 and the first backcross families for a QTL located on the X chromosome

Generation	Cross	Progeny		Expectation
$F_2(F_1$ sib-mated$)$	$X^+X^- \times X^+Y$	Females	$\frac{1}{2}X^+X^+ : \frac{1}{2}X^+X^-$	$m + \frac{1}{2}a_x + \frac{1}{2}d_x$
		Males	$\frac{1}{2}X^+Y : \frac{1}{2}X^-Y$	m
$F_2(F_1\female \times RF_1\male)$	$X^+X^- \times X^-Y$	Females	$\frac{1}{2}X^+X^- : \frac{1}{2}X^-X^-$	$m - \frac{1}{2}a_x + \frac{1}{2}d_x$
		Males	$\frac{1}{2}X^+Y : \frac{1}{2}X^-Y$	m
$RF_2(RF_1$ sib-mated$)$	$X^+X^- \times X^-Y$	Females	$\frac{1}{2}X^+X^- : \frac{1}{2}X^-X^-$	$m - \frac{1}{2}a_x + \frac{1}{2}d_x$
		Males	$\frac{1}{2}X^+Y : \frac{1}{2}X^-Y$	m
$RF_2(RF_1\female \times F_1\male)$	$X^+X^- \times X^+Y$	Females	$\frac{1}{2}X^+X^+ : \frac{1}{2}X^+X^-$	$m + \frac{1}{2}a_x + \frac{1}{2}d_x$
		Males	$\frac{1}{2}X^+Y : \frac{1}{2}X^-Y$	m
$Bc_{1.1}(F_1\female \times P_1\male)$	$X^+X^- \times X^+Y$	Females	$\frac{1}{2}X^+X^+ : \frac{1}{2}X^+X^-$	$m + \frac{1}{2}a_x + \frac{1}{2}d_x$
		Males	$\frac{1}{2}X^+Y : \frac{1}{2}X^-Y$	m
$Bc_{1.1}(RF_1\female \times P_1\male)$	$X^+X^- \times X^+Y$	Females	$\frac{1}{2}X^+X^+ : \frac{1}{2}X^+X^-$	$m + \frac{1}{2}a_x + \frac{1}{2}d_x$
		Males	$\frac{1}{2}X^+Y : \frac{1}{2}X^-Y$	m
$RBc_{1.1}(P_1\female \times F_1\male)$	$X^+X^+ \times X^+Y$	Females	All X^+X^+	$m + a_x$
		Males	All X^+Y	$m + a_{x1}$
$RBc_{1.1}(P_1\female \times RF_1\male)$	$X^+X^+ \times X^-Y$	Females	All X^+X^-	$m + d_x$
		Males	All X^+Y	$m + a_{x1}$
$Bc_{1.2}(F_1\female \times P_2\male)$	$X^+X^- \times X^-Y$	Females	$\frac{1}{2}X^+X^- : \frac{1}{2}X^-X^-$	$m - \frac{1}{2}a_x + \frac{1}{2}d_x$
		Males	$\frac{1}{2}X^+Y : \frac{1}{2}X^-Y$	m
$Bc_{1.2}(RF_1\female \times P_2\male)$	$X^+X^- \times X^-Y$	Females	$\frac{1}{2}X^+X^- : \frac{1}{2}X^-X^-$	$m - \frac{1}{2}a_x + \frac{1}{2}d_x$
		Males	$\frac{1}{2}X^+Y : \frac{1}{2}X^-Y$	m
$RBc_{1.2}(P_2\female \times F_1\male)$	$X^-X^- \times X^+Y$	Females	All X^+X^-	$m + d_x$
		Males	All X^-Y	$m - a_{x1}$
$RBc_{1.2}(P_2\female \times RF_1\male)$	$X^-X^- \times X^-Y$	Females	All X^-X^-	$m - a_x$
		Males	All X^-Y	$m - a_{x1}$

model to fail, particularly when not all the reciprocal families are included in the analysis. Averaging over all possible reciprocals of each generation, however, should combine the additive and dominance effects of the autosomal and the sex-linked genes in such a way that the standard scaling tests will again apply and allow the various parameters to be estimated by model fitting. However, $[a]$ and $[d]$ will not only represent the additive and dominance effects of the autosomal genes but also those on the X chromosome such that:

$$[a]' = [a] + [a]_x, \text{ and}$$

$$[d]' = [d] + [d]_x \text{ for the females and}$$

$$[a]'' = [a] + [a]_{x1}, \text{ and}$$

$$[d]'' = [d] \text{ for the males.}$$

Alternatively, the scaling tests may be applied to just those generations that have a common mother. For example,

$$P_1, F_1(P_1 \times P_2) \text{ and } Bc_{1.1}(P_1 \times F_1)$$

$$\text{or} \quad P_2, RF_1(P_2 \times P_1) \text{ and } Bc_{1.2}(P_2 \times F_1)$$

$$\text{or} \quad F_2(F_1 \times F_1), Bc_{1.1}(F_1 \times P_1) \text{ and } Bc_{1.2}(F_1 \times P_2).$$

Sex linkage has no effect on the variances of P_1, P_2 or F_1, which are purely environmental. Furthermore, these variances will not differ between the sexes, unless the males and females show different reactions to the same environmental variables. The variances among F_2 females, on the other hand, will differ between reciprocals; in those crosses with F_1 males, $\frac{1}{4}(a_x - d_x)^2$ is added to the F_2 variance while in those with RF_1 males the F_2 variance is increased by $\frac{1}{4}(a_x + d_x)^2$. So, the overall impact of these components will be to make the reciprocal F_2 variances different, provided that the genes are not completely dispersed along the X chromosome. Similarly, the variance of the female progeny is also expected to differ in the reciprocal backcrosses. In the $Bc_{1.1}$ generation, crosses with the P_1 mother do not have any genetic variance from the sex-linked loci while backcrosses to the F_1 and RF_1 mothers are supplemented by $\frac{1}{4}(a_x - d_x)^2$ for every locus that has an increasing allele in P_1 and by $\frac{1}{4}(a_x + d_x)^2$ for the remaining genes.

The impact of sex linkage on the variance of male progeny, on the other hand, will be consistent across the F_2, $Bc_{1.1}$ and $Bc_{1.2}$ generations. In any generation with an F_1 or RF_1 mother, the variance of the male progeny will be incremented by a_{x1}^2. The reciprocal families with a P_1 or P_2 mother, on the other hand, will not be affected by the sex-linked genes and consequently the former variances will always be larger than the latter in the presence of sex linkage.

It is important to remember that sex-linked effects will always be confounded with those of the autosomal genes and therefore may not be detected unless genes on the X chromosome control a considerable proportion of the total genetic variation for the trait.

10.9 Full-sib families

Sex-linked genes can have a marked effect on the variances of the male and the female progeny in FS families. With F_2 BIPs, for example, we obtain the genotypes shown in Table 10.7. Calculation of the between- and within-family variances from these expectations

Table 10.7 Male and female progeny genotypes of bi-parental families produced from an F_2 generation with sex linkage

$\female \times \male$ cross	Frequency	$\female$ Progeny (genotype)		$\male$ Progeny (genotype)	
$X^+X^+ \times X^+Y$	$\frac{1}{8}$	All X^+X^+	$m + a_x$	All X^+Y	$m + a_{x1}$
$X^+X^+ \times X^-Y$	$\frac{1}{8}$	All X^+X^-	$m + d_x$	All X^+Y	$m + a_{x1}$
$X^+X^- \times X^+Y$	$\frac{1}{4}$	$\begin{cases} \frac{1}{2}X^+X^+ \\ \frac{1}{2}X^+X^- \end{cases}$	$\begin{matrix} m + a_x \\ m + d_x \end{matrix}$	$\begin{matrix} \frac{1}{2}X^+Y \\ \frac{1}{2}X^-Y \end{matrix}$	$\begin{matrix} m + a_{x1} \\ m - a_{x1} \end{matrix}$
$X^+X^- \times X^-Y$	$\frac{1}{4}$	$\begin{cases} \frac{1}{2}X^+X^- \\ \frac{1}{2}X^-X^- \end{cases}$	$\begin{matrix} m + d_x \\ m - a_x \end{matrix}$	$\begin{matrix} \frac{1}{2}X^+Y \\ \frac{1}{2}X^-Y \end{matrix}$	$\begin{matrix} m + a_{x1} \\ m - a_{x1} \end{matrix}$
$X^-X^- \times X^+Y$	$\frac{1}{8}$	All X^+X^-	$m + d_x$	All X^-Y	$m - a_{x1}$
$X^-X^- \times X^-Y$	$\frac{1}{8}$	All X^-X^-	$m - a_x$	All X^-Y	$m - a_{x1}$

gives the following:

Statistic	Female progeny	Male progeny
(i) Progeny variance		
σ_B^2	$\frac{3}{8}a_x^2 + \frac{1}{8}d_x^2$	$\frac{1}{2}a_{x1}^2$
σ_W^2	$\frac{1}{8}a_x^2 + \frac{1}{8}d_x^2$	$\frac{1}{2}a_{x1}^2$
(ii) Covariance of progeny with parent		
Female parent	$\frac{1}{4}a_x^2$	$\frac{1}{4}a_x a_{x1}$
Male parent	$\frac{1}{4}d_x a_{x1}$	0

To represent the effects of any number of genes on the X chromosome we can replace $\frac{1}{2}a_x^2$ with V_{Ax} and $\frac{1}{2}a_x a_{x1}$ with $V_{Ax.x1}$.

Sex linkage therefore will affect both the between and the within family variances of the full-sib families. The intraclass correlation $t_{FS} = \sigma_B^2/(\sigma_B^2 + \sigma_W^2)$, will be larger for the females than the males. Similarly, we will expect the female progeny/female parent regression to be larger than the female progeny/male parent or the male progeny/female parent regressions while the male progeny/male parent regression will be the smallest of all.

10.10 Half-sib designs

Because the HS designs, such as the NCI and NCII, involve partitions of the variation between FS families, the within-family variance will remain the same as for the BIPs design. Thus, the contribution of a sex-linked gene to the σ_w^2 of the female progeny will be $\frac{1}{8}a_x^2 + \frac{1}{8}d_x^2$

and to that of the male progeny will be $\frac{1}{2}a_{x1}^2$. Further, the total genetic variance due to sex linkage will also stay the same, viz. $\frac{1}{2}a_x^2 + \frac{1}{4}d_x^2$ in the female sex and a_{x1}^2 in the males. Therefore, only σ_B^2 of the FS families is partitioned into further components as follows:

Component	Female progeny	Male progeny
(i) North Carolina I (males as common parent)		
σ_M^2	$\frac{1}{4}a_x^2$	0
$\sigma_{F/M}^2$	$\frac{1}{8}a_x^2 + \frac{1}{8}d_x^2$	$\frac{1}{2}a_{x1}^2$
(ii) North Carolina I (females as common parent)		
σ_F^2	$\frac{1}{8}a_x^2$	$\frac{1}{2}a_{x1}^2$
$\sigma_{M/F}^2$	$\frac{1}{4}a_x^2 + \frac{1}{8}d_x^2$	0
(iii) North Carolina II		
σ_M^2	$\frac{1}{4}a_x^2$	0
σ_F^2	$\frac{1}{4}a_x^2$	$\frac{1}{2}a_{x1}^2$
σ_{MF}^2	$\frac{1}{8}a_x^2 + \frac{1}{8}d_x^2$	0

As the genetic effects of the sex-linked and the autosomal genes will be compounded together, it will be difficult to separate and measure them, even when data are available from both sexes. Consequently, V_A^* and V_D^*, when estimated in the normal manner in the presence of sex linkage, will have the following expectations:

Estimate	Autosomal	Female progeny	Male progeny
(i) North Carolina I (males as common parent)			
$4 \times \sigma_M^2$	V_A^*	$V_A^* + a_x^2$	V_A^*
$4(\sigma_{F/M}^2 - \sigma_M^2)$	V_D^*	$V_D^* - \frac{1}{2}a_x^2 + \frac{1}{2}d_x^2$	$V_D^* + 2a_{x1}^2$
(ii) North Carolina I (females as common parent)			
$4 \times \sigma_F^2$	V_A^*	$V_A^* + \frac{1}{2}a_x^2$	$V_A^* + 2a_{x1}^2$
$4(\sigma_{M/F}^2 - \sigma_M^2)$	V_D^*	$V_D^* + \frac{1}{2}a_x^2 + \frac{1}{2}d_x^2$	$V_D^* - 2a_{x1}^2$
(iii) North Carolina II			
$4 \times \sigma_M^2$	V_A^*	$V_A^* + a_x^2$	V_A^*
$4 \times \sigma_F^2$	V_A^*	$V_A^* + a_x^2$	$V_A^* + 2a_{x1}^2$
$4 \times \sigma_{MF}^2$	V_D^*	$V_D^* + \frac{1}{2}a_x^2 + \frac{1}{2}d_x^2$	$V_D^* - 2a_{x1}^2$

The most appropriate estimates of the additive and dominance variances in the presence of sex linkage will be $V_A^* + \frac{1}{2}a_x^2$ and $V_D^* + \frac{1}{4}d_x^2$ for the female progeny and $V_A^* + \frac{1}{2}a_{x1}^2$ and V_D^* for the male progeny. Clearly, these expectations are obtained on only two occasions, once for the additive variance among the female progeny in the NCI (females as common parent) and secondly for dominance among the male progeny of the NCII design.

It is therefore possible that in many experiments the estimates of the additive and dominance variances will be biased in the presence of sex linkage. Further, the estimates from the same population may vary with the experimental design used to study them and also between the sexes, giving the impression of failure of one or more fundamental assumptions like random sampling or gametic selection, or both.

Specific studies of sex linkage are rare and it is therefore difficult to comment on the bias caused to the estimates of the additive and the dominance components of variance. However, these biases are not generally expected to be large because it is unlikely that a high proportion of the QTLs controlling a particular trait will be located on the X chromosome. The overall ratio of autosomes to X chromosome is often high, as in the case of man (22:1). On the other hand, these biases will be considerable in the case of *Drosophila* where the X chromosome represents approximately 20% of the genome.

Finally, we consider linkage between the loci that are located on the X chromosome. As with the autosomal genes, this linkage will affect the variances but not the means when the additive dominance model explains all the genetic variation between the individuals. Further, because recombination will occur only in the homogametic sex (XX) and not the heterogametic sex (XY), the linkage disequilibrium will linger longer on random mating and selfing than that between autosomal genes.

Summary

1. Providing there is no epistasis, linkage between autosomal genes does not affect the expected values of generation means.
2. Linkage between genes causes genetical covariances to occur between the linked genes which bias the estimates of additive and dominance variance.
3. Coupling linkages inflate and repulsion linkages deflate the additive genetic variance while dominance variance is generally inflated.
4. Biases due to linkage generally result in the exaggeration of the dominance ratio.
5. The biases to the additive genetic variance caused by linkage do not seriously mislead us in predicting the results of

continued

inbreeding or selection because similar biases often persist in the derived populations too.
6. Sex linkage does affect means and variances of reciprocal families in many experimental designs but its effects can be easily modelled and measured.

References

1. Mather, K. and Jinks, J.L. (1982) *Biometrical Genetics*, 3rd edn, Chapman & Hall, London.
2. Jinks, J.L. and Pooni, H.S. (1982) Predicting the properties of pure breeding lines extractable from a cross in the presence of linkage. *Heredity*, **49**, 265–70.
3. Kearsey, M.J. (1985) The effect of linkage on additive genetic variance with inbreeding an F_2. *Heredity*, **55**, 139–43.
4. Perkins, J.M. and Jinks, J.L. (1970) Detection and estimation of genotype-environmental, linkage and epistatic components of variation for a metrical trait. *Heredity*, **25**, 157–77.
5. Van der Veen, J.H. (1959) Tests of non-allelic interaction and linkage for quantitative characters in generations derived from two diploid pure lines. *Genetica*, **30**, 201–32.
6. Killick, R. (1971) Sex-linkage and sex limitation in quantitative inheritance. Random mating populations. *Heredity*, **27**, 175–88.

11 Epistasis

Epistasis is the interaction between alleles of different genes, i.e. non-allelic or intergenic interaction, as opposed to dominance, which is interaction between alleles of the same gene, called inter-allelic or intragenic interaction. Whenever an additive–dominance model is inadequate and complications like maternal effects, differential gametic viability and genotype by environment inter-actions are explicitly ruled out, then there must be epistasis.

In many cases, epistasis is caused by the scale of measurement and therefore can be removed by transforming the scale of the original data, e.g. by taking logarithms, square roots or angles. The concept of making adjustments to the scale of measurement frequently worries biologists. They feel suspicious that some decep-tion is being practised. There is no absolute scale, however, and those we normally use may be quite different from those used by the organism. Suppose, for example, that leaf length was controlled by a number of genes which behaved in an entirely additive manner. If we measure leaf length the genetic control would be additive but, were we to measure leaf area it would not because area is a function of length squared. A square root transformation of the area measurements would almost certainly remove the epistasis.

Nevertheless, there is no guarantee that any change of scale will lead to the adequacy of the additive–dominance model and the analysis of the same data on different scales generally alters the relative importance of gene effects. In other words, change of scale may actually lead to the failure of the additive–dominance model. As the traits are measured on scales that involve units (grams, centimetres, etc.) which are convenient and meaningful, it makes sense to try to understand the genetics of the trait on that scale even if this requires fitting and interpreting more complex models.

11.1 Definitions

The additive and dominance effects of genes which were developed in Chapter 2 are, in statistical terms, the main effects; interaction is implicated when these main effects do not account for all of the variation between a set of genotypes. For example, if we consider the four inbred lines that can be produced from continued selfing of an F_2, i.e. $A^+A^+B^+B^+$, $A^+A^+B^-B^-$, $A^-A^-B^+B^+$ and $A^-A^-B^-B^-$, all the differences between them must be accounted for by a_A and a_B, the additive genetic effects of the A and B genes, when there are no interactions. Under these circumstances, we shall find that

$$\tfrac{1}{4}(A^+A^+B^+B^+ + A^+A^+B^-B^- - A^-A^-B^+B^+ - A^-A^-B^-B^-) = a_A$$

$$\tfrac{1}{4}(A^+A^+B^+B^+ - A^+A^+B^-B^- + A^-A^-B^+B^+ - A^-A^-B^-B^-) = a_B$$

and

$$(A^+A^+B^+B^+ - A^+A^+B^-B^-) = (A^-A^-B^+B^+ - A^-A^-B^-B^-), \text{ or}$$

$$\tfrac{1}{4}(A^+A^+B^+B^+ - A^+A^+B^-B^- - A^-A^-B^+B^+ + A^-A^-B^-B^-) = 0.$$

In the presence of interaction, the last comparison will be equal to aa_{AB} and not zero. Because aa_{AB} represents the interaction between a_A and a_B, it is called the additive $\times$ additive interaction. In the presence of epistasis of this type, the genetic values of the four lines become:

Line	Genetic value
$A^+A^+B^+B^+$	$m + a_A + a_B + aa_{AB}$
$A^+A^+B^-B^-$	$m + a_A - a_B - aa_{AB}$
$A^-A^-B^+B^+$	$m - a_A + a_B - aa_{AB}$
$A^-A^-B^-B^-$	$m - a_A - a_B + aa_{AB}$

It is clear from these models that the external sign and coefficient of aa_{AB} is determined in precisely the same way as that of a statistical interaction, i.e. by the multiplication of the external signs and the coefficients of the main effects, a_A and a_B. For example, the contribution of aa_{AB} to the score of $A^+A^+B^-B^-$ is $(+1) \times (-1) = -1$, and so the expected genotype of $A^+A^+B^-B^-$ on a digenic interaction model is $m + a_A - a_B - aa_{AB}$.

Now consider the genotypes heterozygous for the B gene, $A^+A^+B^+B^-$ and $A^-A^-B^+B^-$. Half the difference between them should be equal to a_A on the additive–dominance model and when this difference is significantly greater or smaller than a_A, then $a_A \times d_B$ interaction is present. This interaction is represented by the symbol ad_{AB} and its coefficient and external sign are determined by

Table 11.1 Expectations of genotypes on the additive–dominance and digenic interaction model. Symbols $++\ ++$ represent genotype $A^+A^+B^+B^+$, etc.

Locus		Parameters								
A	B	m	a_A	a_B	d_A	d_B	aa_{AB}	ad_{AB}	ad_{BA}	dd_{AB}
$++$	$++$	1	1	1	0	0	1	0	0	0
$++$	$+-$	1	1	0	0	1	0	1	0	0
$++$	$--$	1	1	-1	0	0	-1	0	0	0
$+-$	$++$	1	0	1	1	0	0	0	1	0
$+-$	$+-$	1	0	0	1	1	0	0	0	1
$+-$	$--$	1	0	-1	1	0	0	0	-1	0
$--$	$++$	1	-1	1	0	0	-1	0	0	0
$--$	$+-$	1	-1	0	0	1	0	-1	0	0
$--$	$--$	1	-1	-1	0	0	1	0	0	0

the product of the coefficients and signs of a_A and d_B thus giving the genetic values:

$$A^+A^+B^+B^- = m + a_A + d_B + ad_{AB} \quad \text{and}$$

$$A^-A^-B^+B^- = m - a_A + d_B - ad_{AB}.$$

Similarly, the genetic values of $A^+A^-B^+B^+$ and $A^+A^-B^-B^-$ are modified to $m + d_A + a_B + ad_{BA}$ and $m + d_A - a_B - ad_{BA}$ when additive $\times$ dominance interaction is present. The ninth genotype is the double heterozygote, $A^+A^-B^+B^-$ and its genetic value is modified to $m + d_A + d_B + dd_{AB}$, where dd_{AB} is called the dominance $\times$ dominance interaction. The genetic values of the nine F_2 genotypes in terms of these parameters are presented in Table 11.1.

11.2 Relationship with classical epistasis

In classical genetics, non-allelic interaction is divided into two main types, complementary and duplicate epistasis. Both types reduce the number of phenotypes in an F_2 to two and give segregation ratios of $9:7$ and $15:1$ respectively. Marginally less extreme forms of these epistases are classical recessive and classical dominant epistasis, both of which give rise to three phenotypes with the frequencies of $9:3:4$ and $12:3:1$ respectively in an F_2 [1]. In terms of genotypes that have identical phenotypes, these types of epistasis have the relationships given in Table 11.2.

Table 11.2 Generalized phenotypic relationships in an F_2 for various types of epistasis

Genotype	Frequency	Types of epistasis			
		Complementary	Recessive	Duplicate	Dominant
$A^+A^+B^+B^+$	1/16				
$A^+A^+B^+B^-$	2/16				
$A^+A^-B^+B^+$	2/16	9/16	9/16		
$A^+A^-B^+B^-$	4/16			15/16	12/16
$A^+A^+B^-B^-$	1/16				
$A^+A^-B^-B^-$	2/16		3/16		
$A^-A^-B^+B^+$	1/16	7/16			
$A^-A^-B^+B^-$	2/16		4/16		3/16
$A^-A^-B^-B^-$	1/16			1/16	1/16

In quantitative genetics, the ratio of 9:7 is obtained under the following conditions:

(i)   equal gene effects ($a_A = a_B$);
(ii) complete dominance of A^+ over A^- and B^+ over B^- (i.e. $d_A = a_A$, $d_B = a_B$ and both d_A and d_B are *positive*); and
(iii) complete and *positive* epistasis ($aa_{AB} = a_A = a_B$; $ad_{AB} = a_A = d_B$; $ad_{BA} = d_A = a_B$ and $dd_{AB} = d_A = d_B$).

In other words, when dominance is for the higher score, a 9:7 ratio is obtained under the conditions where all the components have **positive** values and are of the same magnitude, that is:

$$a_A = a_B = d_A = d_B = aa_{AB} = ad_{AB} = ad_{BA} = dd_{AB} \quad (= \text{say } x \text{ units}).$$

Another difference between classical and quantitative genetics concerns the direction of dominance. In the latter case, dominance of A^- over A^+ is as likely as that of A^+ over A^-, and this situation is reflected by the internal sign of d_A which is positive when A^+ is dominant and negative when A^- is dominant. For negative dominance, the 9:7 ratio is obtained when the absolute values of the various components are still equal but the internal signs of d_A, d_B, aa_{AB} and dd_{AB} are negative while those of a_A, a_B, ad_{AB} and ad_{BA} are positive. In other words, $a_A = a_B = -d_A = -d_B = -aa_{AB} = ad_{AB} = ad_{BA} = -dd_{AB}$, when both internal and external signs are taken into account.

The 15:1 ratio of duplicate epistasis is obtained when all those genotypes that have at least one dominant allele (A^+ or B^+) take the dominant phenotype but $A^-A^-B^-B^-$ takes the recessive phenotype. This condition is fulfilled on the quantitative model when all components have the same magnitude but a_A, d_A, a_B and d_B are positive and aa_{AB}, ad_{AB}, ad_{BA} and dd_{AB} are all negative. That is, $a_A = d_A = a_B = d_B = -aa_{AB} = -ad_{AB} = -ad_{BA} = -dd_{AB}$, when both internal and external signs are considered. Similarly, the 15:1 ratio of duplicate epistasis is obtained with negative dominance when,

$$a_A = a_B = -d_A = -d_B = aa_{AB} = -ad_{AB} = -ad_{BA} = dd_{AB}.$$

Further, complementary epistasis is modified to classical recessive epistasis, giving a 9:3:4 ratio when we assume that the gene effects are unequal, i.e. $a_A = d_A = aa_{AB} = ad_{AB} = ad_{BA} = dd_{AB}$ (= say x units) but not equal to $a_B = d_B$, and the same conditions modify the duplicate ratio of 15:1 to the 12:3:1 of classical dominant epistasis [2].

11.3 What is parameter m?

The system that we have used to define the various parameters is known as the $\mathbf{F_\infty}$ **metric** [3]. It was proposed during the late 1950s and is one of several that have been used to define the expectations of the basic generations, the others being the $\mathbf{F_2}$ **metric** of Kempthorne [4] and the **mixed metric** of Hayman and Mather [5]. Because each of these metrics uses a different reference point for defining the expected generation means, the expectations look rather different. However, none has any specific statistical advantage over the others and the practical differences between them are marginal.

The F_∞ metric, however, is widely used, firstly because it is easy to follow and derive, and secondly because the associated parameters and models make better biological sense. On this metric, the reference point 'm' is defined as the overall mean of all the inbred lines that are possible from a cross, which for a two locus model is equal to

$$\tfrac{1}{4}(A^+A^+B^+B^+ + A^+A^+B^-B^- + A^-A^-B^+B^+ + A^-A^-B^-B^-).$$

Because this is also the expected mean of a random set of recombinant inbred lines (RILs) that can be produced from an F_1 by consecutive selfing ($F_1 \rightarrow F_2 \rightarrow F_3 \ldots \rightarrow F_\infty$), it is termed the '$F_\infty$ metric'.

In a biological sense, m represents all those effects that are common to every genotype, i.e. the effects of those genes that are monomorphic and fixed in the background, the common effects of the environment and the average effects of the genes showing polymorphism.

11.4 The effects of association and dispersion on epistasis

The distribution of alleles in the parents affects the definition of the epistatic terms in much the same way as it does the additive component of means, $[a]$. From the genetic values in Table 11.1 we see that the parental scores will be $m + a_A + a_B + aa_{AB}$ and $m - a_A - a_B + aa_{AB}$ when the alleles are in association, i.e. their genotypes are $A^+A^+B^+B^+$ and $A^-A^-B^-B^-$ respectively. The F_1 between these parents has the genotype $A^+A^-B^+B^-$ with genetic value $m + d_A + d_B + dd_{AB}$. Derivation of the expected means of the remaining generations is, however, complex because these generations are segregating and thus consist of more than one genotype. For example, the F_2 generation should contain all nine genotypes with $+ + + + : + + + - : \ldots . : - - - -$ in the expected ratios of $1:2:1:2:4:2:1:2:1$ respectively. Thus the mean of the F_2 generation is,

$$\bar{F}_2 = \{(m + a_A + a_B + aa_{AB}) + 2(m + a_A + d_B + ad_{AB}) + \cdots$$

$$+ (m - a_A - a_B + aa_{AB})\}/16$$

$$= m + \tfrac{1}{2}d_A + \tfrac{1}{2}d_B + \tfrac{1}{4}dd_{AB}, \text{ given in Table 11.3.}$$

The $Bc_{1.1}$ generation, a cross of F_1 ($A^+A^-B^+B^-$) and P_1 ($A^+A^+B^+B^+$), on the other hand, will consist of four genotypes $A^+A^+B^+B^+$, $A^+A^+B^+B^-$, $A^+A^-B^+B^+$ and $A^+A^-B^+B^-$ and they should be present in equal proportions in the absence of complications like linkage and gametic selection. Thus, the expected mean of $Bc_{1.1}$

Table 11.3 Expectations of generations derived from dispersion and association crosses

Family	\multicolumn{9}{c}{Parameters}								
	m	a_A	a_B	d_A	d_B	aa_{AB}	ad_{AB}	ad_{BA}	dd_{AB}
\multicolumn{10}{l}{Association cross $(A^+A^+B^+B^+ \times A^-A^-B^-B^-)$}									
P_1	1	1	1	0	0	1	0	0	0
P_2	1	-1	-1	0	0	1	0	0	0
F_1	1	0	0	1	1	0	0	0	1
F_2	1	0	0	$\frac{1}{2}$	$\frac{1}{2}$	0	0	0	$\frac{1}{4}$
$Bc_{1.1}$	1	$\frac{1}{2}$	$\frac{1}{2}$	$\frac{1}{2}$	$\frac{1}{2}$	$\frac{1}{4}$	$\frac{1}{4}$	$\frac{1}{4}$	$\frac{1}{4}$
$Bc_{1.2}$	1	$-\frac{1}{2}$	$-\frac{1}{2}$	$\frac{1}{2}$	$\frac{1}{2}$	$\frac{1}{4}$	$-\frac{1}{4}$	$-\frac{1}{4}$	$\frac{1}{4}$
\multicolumn{10}{l}{Dispersion cross $(A^+A^+B^-B^- \times A^-A^-B^+B^+)$}									
P_1	1	1	-1	0	0	-1	0	0	0
P_2	1	-1	1	0	0	-1	0	0	0
F_1	1	0	0	1	1	0	0	0	1
F_2	1	0	0	$\frac{1}{2}$	$\frac{1}{2}$	0	0	0	$\frac{1}{4}$
$Bc_{1.1}$	1	$\frac{1}{2}$	$-\frac{1}{2}$	$\frac{1}{2}$	$\frac{1}{2}$	$-\frac{1}{4}$	$\frac{1}{4}$	$-\frac{1}{4}$	$\frac{1}{4}$
$Bc_{1.2}$	1	$-\frac{1}{2}$	$\frac{1}{2}$	$\frac{1}{2}$	$\frac{1}{2}$	$-\frac{1}{4}$	$-\frac{1}{4}$	$\frac{1}{4}$	$\frac{1}{4}$

will be:

$$\tfrac{1}{4}(m + a_A + a_B + aa_{AB}) + \tfrac{1}{4}(m + a_A + d_B + ad_{AB})$$
$$+ \tfrac{1}{4}(m + d_A + a_B + ad_{BA}) + \tfrac{1}{4}(m + d_A + d_B + dd_{AB});$$

which simplifies to,

$$m + \tfrac{1}{2}a_A + \tfrac{1}{2}a_B + \tfrac{1}{2}d_A + \tfrac{1}{2}d_B + \tfrac{1}{4}aa_{AB} + \tfrac{1}{4}ad_{AB} + \tfrac{1}{4}ad_{BA} + \tfrac{1}{4}dd_{AB}.$$

Similarly, the $Bc_{1.2}$ mean is,

$$m - \tfrac{1}{2}a_A - \tfrac{1}{2}a_B + \tfrac{1}{2}d_A + \tfrac{1}{2}d_B + \tfrac{1}{4}aa_{AB} - \tfrac{1}{4}ad_{AB} - \tfrac{1}{4}ad_{BA} + \tfrac{1}{4}dd_{AB}.$$

Allele dispersion affects the expectations of four of the six basic generations. Firstly, the expectations of the parental lines are modified to $m + a_A - a_B - aa_{AB}$ and $m - a_A + a_B - aa_{AB}$ when P_1 and P_2 have $A^+A^+B^-B^-$ and $A^-A^-B^+B^+$ genotypes respectively. The other generations affected by gene dispersion are the $Bc_{1.1}$ and $Bc_{1.2}$ and their expectations are modified to $m + \tfrac{1}{2}a_A - \tfrac{1}{2}a_B + \tfrac{1}{2}d_A + \tfrac{1}{2}d_B - \tfrac{1}{4}aa_{AB} + \tfrac{1}{4}ad_{AB} - \tfrac{1}{4}ad_{BA} + \tfrac{1}{4}dd_{AB}$ and $m - \tfrac{1}{2}a_A + \tfrac{1}{2}a_B + \tfrac{1}{2}d_A + \tfrac{1}{2}d_B - \tfrac{1}{4}aa_{AB} - \tfrac{1}{4}ad_{AB} + \tfrac{1}{4}ad_{BA} + \tfrac{1}{4}dd_{AB}$ respectively. Allele association or dispersion does not affect the expectations of the F_1 and F_2 generations because heterozygosity is not determined by the distribution of alleles in the parental lines.

Extension of the two gene model to many $(= k)$ genes is comparatively easy providing the $+$ and $-$ alleles are completely associated in the parents, P_1 and P_2. In these circumstances, the various parameters of the digenic interaction model can be replaced by five new parameters, namely, $\sum a_i$, $\sum d_i$, $\sum aa_{ij}$, $\sum ad_{ij}$ and $\sum dd_{ij}$, which will represent the summed effects of the k genes or all possible pairs of genes as the case may be. However, complete association is extremely rare and dispersion is normal in most crosses, particularly those that form the source material for breeding programmes. In such circumstances, the actual magnitudes of the additive × additive and additive × dominance parameters may be obscured because, like the additive genetic component, they are also affected by gene dispersion.

Allele dispersion does not affect $\sum d_i$ and $\sum dd_{ij}$, but these parameters take their maximum values when dominance is unidirectional and all the interacting pairs display the same type of epistasis, i.e. all duplicate or all complementary. Because we cannot be sure about the direction of dominance and the type of epistasis prevailing at each pair of genes, it is therefore necessary to assume that these components are subject to internal cancellations, too. In general, therefore, we can detect and measure only the net effects of the additive, dominance and epistatic components which we represent by [a], [d], [aa], [ad] and $[dd]$ where:

$$[a] = r_a \sum a_i \text{ (where } r_a \text{ varies from 0 to 1)}$$

$$[d] = \sum d_i$$

$$[aa] = r_{aa} \sum aa_{ij} \text{ (where } r_{aa} \text{ varies from } -1 \text{ to 1)}$$

$$[ad] = r_{ad} \sum ad_{ij} \text{ (where } r_{ad} \text{ varies from } -1 \text{ to 1)}$$

and $$[dd] = \sum dd_{ij}.$$

Here, r_{aa} and r_{ad} stand for the coefficients of association–dispersion of the additive × additive and additive × dominance interaction components and are defined in exactly the same manner as r_a, that is:

$$r_{aa} = \{(\text{sum of } aa \text{ over all } \tfrac{1}{2}k(k-1) \text{ pairs of genes}) -$$
$$2 \times (\text{sum of } aa \text{ over } k'(k-k') \text{ pairs of dispersed genes})\}/$$
$$(\text{sum of } aa \text{ over all } \tfrac{1}{2}k(k-1) \text{ pairs of genes})$$

and

$$r_{ad} = \{(\text{sum of } ad \text{ over all } k(k-1) \text{ pairs of genes}) -$$
$$2 \times (\text{sum of } ad \text{ over } k'(k-1) \text{ pairs of dispersed genes})\}/$$
$$(\text{sum of } ad \text{ over all } k(k-1) \text{ pairs of genes}).$$

11.5 Deriving the expectations of generation means

There are several ways of deriving the expectations of generation means for a multigene, digenic interaction model. The simplest method is to obtain the expectations for an additive–dominance model and then extend them to the digenic interaction model using the following relationships:

The coefficient of $[aa]$ is the square of the coefficient of $[a]$.

The coefficient of $[ad]$ is the product of the coefficients of $[a]$ and $[d]$.

The coefficient of $[dd]$ is the square of the coefficient of $[d]$.

For example, on a multigene additive dominance model the genetic value of the $Bc_{1.2}$ generation is $m - \tfrac{1}{2}[a] + \tfrac{1}{2}[d]$ and on the digenic interaction model it becomes:

$$m - \tfrac{1}{2}[a] + \tfrac{1}{2}[d] + (-\tfrac{1}{2})^2[aa] + (-\tfrac{1}{2})(+\tfrac{1}{2})[ad] + (+\tfrac{1}{2})^2[dd],$$

Table 11.4 Generalized expectations of the generation means of a cross which is segregating at an unknown number of loci ($k \geq 2$)

Generation	m	$[a]$	$[d]$	$[aa]$	$[ad]$	$[dd]$
P_1	1	1	0	1	0	0
P_2	1	-1	0	1	0	0
F_1	1	0	1	0	0	1
F_2	1	0	$\frac{1}{2}$	0	0	$\frac{1}{4}$
$Bc_{1.1}$	1	$\frac{1}{2}$	$\frac{1}{2}$	$\frac{1}{4}$	$\frac{1}{4}$	$\frac{1}{4}$
$Bc_{1.2}$	1	$-\frac{1}{2}$	$\frac{1}{2}$	$\frac{1}{4}$	$-\frac{1}{4}$	$\frac{1}{4}$

which simplifies to

$$m - \tfrac{1}{2}[a] + \tfrac{1}{2}[d] + \tfrac{1}{4}[aa] - \tfrac{1}{4}[ad] + \tfrac{1}{4}[dd],$$

as given in Table 11.4.

11.6 Estimates and tests of significance

There are six parameters in a digenic interaction model, m, $[a]$, $[d]$, $[aa]$, $[ad]$ and $[dd]$, and therefore a minimum of six generation means will be required to estimate them all and to test their significance. Further, apart from the six basic generations, not every set of six generations will provide estimates of every parameter, i.e. we cannot estimate $[ad]$ from the P_1, P_2, F_1, F_2, F_3 and F_4 generation means. While the weighted least squares procedure described in Chapter 2 can be employed to obtain an estimate of each parameter, the result will be identical to the solutions provided by the following equations when estimates are obtained from the six basic generations:

$$m = \tfrac{1}{2}\bar{P}_1 + \tfrac{1}{2}\bar{P}_2 + 4\bar{F}_2 - 2\bar{B}c_{1.1} - 2\bar{B}c_{1.2}$$

$$[a] = \tfrac{1}{2}\bar{P}_1 - \tfrac{1}{2}\bar{P}_2$$

$$[d] = 6\bar{B}c_{1.1} + 6\bar{B}c_{1.2} - \bar{F}_1 - 8\bar{F}_2 - 1\tfrac{1}{2}\bar{P}_1 - 1\tfrac{1}{2}\bar{P}_2$$

$$[aa] = 2\bar{B}c_{1.1} + 2\bar{B}c_{1.2} - 4\bar{F}_2$$

$$[ad] = 2\bar{B}c_{1.1} - 2\bar{B}c_{1.2} - \bar{P}_1 + \bar{P}_2$$

$$[dd] = \bar{P}_1 + \bar{P}_2 + 2\bar{F}_1 + 4\bar{F}_2 - 4\bar{B}c_{1.1} - 4\bar{B}c_{1.2}$$

[Eqns 11.1]

(N.B. the mid parent $= \tfrac{1}{2}\bar{P}_1 + \tfrac{1}{2}\bar{P}_2 = m + [aa]$, not m).

The significance of each parameter may be tested by a t test in the usual manner and whenever a parameter proves not to be significant the remaining parameters can be estimated by weighted least squares.

We will describe the application and interpretation of the preceding analyses using simulated data in order to compare the estimates with the true situation. Duplicate epistasis was introduced into the simulation program such that $\sum aa_{ij} = 0$ and $-\sum ad_{ij} = -\sum dd_{ij} = \frac{1}{4}\sum a_i$ while keeping the other relationships as: $\sum d_i = \sum a_i = 28.28$ (i.e. complete dominance), $m = 65.00$ and $r_a = 1$ (i.e. complete association of alleles in the parents). Perfect fit estimates from the data in Table 11.5 gave the following results which indicate that all components are significant except [aa].

$$m = 62.32 \pm 3.67^{**}$$

$$[a] = 28.28 \pm 0.77^{**}$$

$$[d] = 35.15 \pm 8.84^{**}$$

$$[aa] = 2.80 \pm 3.59 \text{ n.s.}$$

$$[ad] = -8.31 \pm 2.55^{**}$$

$$[dd] = -11.11 \pm 5.47^{*}$$

Further, components [ad] and [dd] take negative sign while m, [a] and [d] are positive, as expected, even when some of these parameters are highly correlated. Exclusion of [aa] from the model and re-estimation of the remaining parameters by weighted least squares from the six means also provides a test of the adequacy of the digenic interaction model giving a non-significant $\chi^2_{(1)}$ value of 0.61 (Table 11.5).

Table 11.5 Digenic interaction components of the simulated data

Data				Model	
Family	$\bar{x}$	n	$s^2_{\bar{x}}$	Component	Estimate
P_1	93.39	40	1.2100	m	$65.12 \pm 0.77^{***}$
P_2	36.84	40	1.1881	[a]	$28.28 \pm 0.77^{***}$
F_1	86.35	80	0.5625	[d]	$28.65 \pm 2.96^{***}$
F_2	77.11	160	0.5476	[aa]	n.s.
$Bc_{1.1}$	89.87	160	0.3249	[ad]	$-7.98 \pm 2.52^{**}$
$Bc_{1.2}$	65.75	160	0.7056	[dd]	$-7.41 \pm 2.73^{**}$
				$\chi^2_{(1)}$	$= 0.61$ n.s.

11.7 *Determining the type of epistasis for a multigene case*

Based on the definitions given earlier, the magnitudes and directions of the various parameters can only be interpreted in relative terms and we can say little about their maximum values because they may be affected by allele dispersion or ambidirectionality of the intra- and the inter-genic effects. Allele dispersion is particularly awkward for interpreting the type of epistasis because it can affect both the magnitudes and the signs of [aa] and [ad]. Further, it is generally not possible to determine whether the effects of the various genes are equal or not. Consequently, it is rarely possible to distinguish complementary from recessive epistasis or duplicate from dominant epistasis. Thus, with polygenic traits, epistasis can only be classified as either predominantly duplicate or predominantly complementary, the distinction being based solely on the relative signs of the dominance [d] and the dominance × dominance [dd] components as follows:

Direction of [d] [dd]	Type of dominance and epistasis
+ +	Complementary epistasis between dominant increasers
− −	Complementary epistasis between dominant decreasers
+ −	Duplicate epistasis between dominant increasers
− +	Duplicate epistasis between dominant decreasers

11.8 *Epistasis and scaling tests*

In Chapter 2 we assumed that failure of the scaling tests proved the presence of non-allelic interactions. Now we can determine which test is affected by which particular epistatic component. For example, by substituting the genetic values of $\bar{B}c_{1.1}$, $\bar{F}_1$ and $\bar{P}_1$ in the equation for the **A** scaling test, $\mathbf{A} = 2\bar{B}c_{1.1} - \bar{F}_1 - \bar{P}_1$, we find that **A** is equal to $\frac{1}{2}(-[aa] - [dd] + [ad])$ in the presence of epistasis [6]. Similarly,

$$\mathbf{B} = \tfrac{1}{2}(-[aa] - [dd] - [ad]),$$

$$\text{and } \mathbf{C} = -2[aa] - [dd].$$

These expectations show clearly that each test represents a different combination of the [aa], [ad] and [dd] parameters and, therefore, the parameters may cancel in some tests. For example, if

$[ad] = [aa] + [dd]$, **A** will be zero while **B** will be large. This in turn will be reflected in the significance of each test. Thus all three tests will rarely detect epistasis simultaneously even when epistasis is a major source of heritable variation. Which parameters are important in a given situation can, however, sometimes be inferred from the pattern of results. For example, differences in the significance of the **A** and **B** scaling tests in the above example point to the presence of $[ad]$ type epistasis while the non-significance of the **C** scaling test reveals that $[aa]$ cannot be very large. Using the data in Table 11.4, the scaling tests take the following values.

$$\mathbf{A} = 0.00 \pm 1.75, \qquad c = 0 \text{ n.s.}$$

$$\mathbf{B} = 8.31 \pm 2.14, \qquad c = 3.88^{**}$$

$$\mathbf{C} = 5.51 \pm 3.66, \qquad c = 1.51^{*}$$

11.9 Higher order interactions

When a digenic interaction model is fitted to more than six generations using weighted least squares and the model fails (i.e. χ^2 is significant), this could be due to one of three factors:

1. Causes other than epistasis, such as maternal effects or sex linkage which are considered in Chapters 10 and 13.
2. Higher order interactions, such as trigenic interactions of the kind $a_A \times a_B \times a_C$, $a_A \times a_B \times d_C$, $a_A \times d_B \times d_C$ and $d_A \times d_B \times d_C$, and these can be defined, parameterized, tested and estimated in exactly the same manner as the digenic interactions [7].
3. Linkage between the genes displaying digenic epistasis. Linkage affects the relative frequencies of different combinations of genotypes in the progeny of a cross or self and because epistasis depends on these gene combinations, the nature and extent of the epistasis changes. For example, the F_2 mean has the expectation of $m + \frac{1}{2}(d_A + d_B) + \frac{1}{2}\delta(1 - 2R)aa_{AB} + \frac{1}{2}(p^2 + q^2)dd_{AB}$ when genes A and B are linked, instead of $m + \frac{1}{2}(d_A + d_B) + \frac{1}{4}dd_{AB}$ otherwise [3].

At this stage, one of the major problems is to identify the real cause of model failure, i.e. linkage or higher order interactions. This problem is difficult to resolve by weighted least squares estimation, firstly because there may not be enough generations to fit a full trigenic interaction model and secondly the number of parameters in a linkage model varies considerably with the particular set of generations in the experiment. However, this is easily resolved by applying scaling tests because they have a clear advantage over the weighted least squares in providing unbiased answers to specific questions. The following scaling tests determine unambiguously the presence of trigenic interactions (X, Y and Z) and linkage of interacting genes (A1, A2 and A3) and therefore can be used to identify the true cause or causes of the failure of a digenic interaction model, prior to fitting a suitable model by weighted least squares.

$$X = \tfrac{1}{2}(\bar{P}_1 - \bar{P}_2) - (\bar{Bc}_{2.1} + [\overline{Bc_{1.1} \times P_2}]) + ([\overline{Bc_{1.2} \times P_1}] + \bar{Bc}_{2.2})$$
$$= 0 \tag{3}$$

$$Y = \bar{F}_1 - \tfrac{1}{2}(\bar{P}_1 + \bar{P}_2) + (\bar{Bc}_{2.1} - [\overline{Bc_{1.1} \times P_2}]) + ([\overline{Bc_{1.2} \times P_1}] - \bar{Bc}_{2.2})$$
$$= 0 \tag{3}$$

$$Z = (\bar{Bc}_{1.1} - \bar{Bc}_{1.2}) - 2([\overline{Bc_{1.1} \times F_1}] - [\overline{Bc_{1.2} \times F_1}]) = 0 \tag{8}$$

$$A1 = \bar{Bc}_{1.1} - \bar{L}_1 \tag{9}$$

$$A2 = \bar{Bc}_{1.2} - \bar{L}_2 \tag{9}$$

$$A3 = \bar{F}_2 - \bar{L}_3 \tag{9}$$

11.10 Epistasis and variances

Epistasis also affects the magnitude of the variances, but only those of the segregating generations. We can demonstrate this by calculating the F_2 variance for a digenic situation from the nine genotypes in Table 11.1 as:

$$\{(m + a_A + a_B + aa_{AB})^2 + 2(m + a_A + d_B + ad_{AB})^2 + \ldots$$

$$+ (m - a_A - a_B + aa_{AB})^2\}/16 - (m + \tfrac{1}{2}d_A + \tfrac{1}{2}d_B + \tfrac{1}{4}dd_{AB})^2$$

which simplifies to:

$$\tfrac{1}{2}(a_A + \tfrac{1}{2}ad_{AB})^2 + \tfrac{1}{2}(a_B + \tfrac{1}{2}ad_{BA})^2 + \tfrac{1}{4}(d_A + \tfrac{1}{2}dd_{AB})^2 + \tfrac{1}{4}(d_B + \tfrac{1}{2}dd_{AB})^2$$

$$+ \tfrac{1}{4}aa_{AB}^2 + \tfrac{1}{8}(ad_{AB}^2 + ad_{BA}^2) + \tfrac{1}{16}dd_{AB}^2. \tag{Eqn 11.2}$$

In the absence of epistasis, the F_2 variance is $\tfrac{1}{2}(a_A^2 + a_B^2) + \tfrac{1}{4}(d_A^2 + d_B^2)$. It is therefore readily seen that the F_2 variance (Equation 11.2) will be inflated in the presence of complementary epistasis (i.e. d and dd have the same sign) and reduced when epistasis is of the duplicate type (i.e. d and dd have opposite signs). We can further surmise that the extent

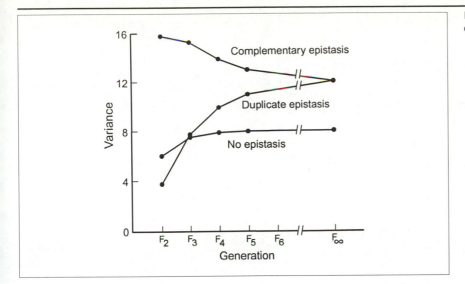

Figure 11.1 Effect of epistasis on variances with inbreeding.

of deflation by duplicate epistasis is likely to be smaller than the inflation by complementary epistasis because the aa_{AB}^2, ad_{AB}^2, ad_{BA}^2 and dd_{AB}^2 components will all be positive, irrespective of the type of epistasis prevailing in the material.

Similar effects should also be observed in the variances of other generations derived from the selfing series. While expectations of their variances are even more complex and beyond the scope of this book, such derivations have shown that epistasis upsets the standard relationships between the rank variances within generations. For example, in the absence of epistasis we expect σ_B^2 of the F_3 to be approximately equal to $2(\sigma_W^2 - V_E)$, but this will no longer be true when non-allelic interactions are present.

Further, alterations to the coefficients of aa, ad and dd and their cross-products with a and d modify the relationships between the total genetic variances of the various generations. This is better seen from the variances of the F_2, F_3, F_4, F_5 and F_∞ generations in Figure 11.1 which are derived for a two-gene model assuming complete dominance and complete epistasis. The first thing that becomes apparent is that the F_∞ variance is always inflated in the presence of epistasis while the F_2 variance receives maximum bias. In fact, in the presence of complementary epistasis the F_2 variance can be even larger than the F_∞ variance and this can have serious consequences for the prediction of the latter from the former.

The extent of bias to the variances, however, depends exclusively on the extent of epistasis prevailing in the experimental material. It is highly unlikely that there can ever be complete epistasis when the cross shows segregation at many genes because that will imply that

Table 11.6 Expectations of the $\bar{L}_{1i}$, $\bar{L}_{2i}$ and $\bar{L}_{3i}$ families and of the $\bar{L}_{1i} + \bar{L}_{2i} - 2\bar{L}_{3i}$ comparison on a digenic interaction model. For simplicity, the constant parameter m has been excluded from the $\bar{L}_{1i}$, $\bar{L}_{2i}$ and $\bar{L}_{3i}$ families

F_2	Frequency	$\bar{L}_{1i}\ (=\times\,A^+A^+B^+B^+)$								$\bar{L}_{2i}\ (=\times\,A^-A^-B^-B^-)$							
		a_A	a_B	d_A	d_B	aa_{AB}	ad_{AB}	ad_{BA}	dd_{AB}	a_A	a_B	d_A	d_B	aa_{AB}	ad_{AB}	ad_{BA}	dd_{AB}
$A^+A^+B^+B^+$	1/16	1	1	0	0	1	0	0	0	0	0	1	1	0	0	0	1
$A^+A^+B^+B^-$	2/16	1	$-\frac12$	0	$-\frac12$	$-\frac12$	$-\frac12$	0	0	0	$-\frac12$	1	$-\frac12$	0	0	$-\frac12$	$-\frac12$
$A^+A^+B^-B^-$	1/16	1	0	0	1	0	1	0	0	0	-1	1	0	0	0	-1	0
$A^+A^-B^+B^+$	2/16	$-\frac12$	1	$-\frac12$	0	$-\frac12$	0	$-\frac12$	0	$-\frac12$	0	$-\frac12$	1	0	$-\frac12$	0	$-\frac12$
$A^+A^-B^+B^-$	4/16	$-\frac12$	$-\frac12$	$-\frac12$	$-\frac12$	$-\frac14$	$-\frac14$	$-\frac14$	$-\frac14$	$-\frac12$	$-\frac12$	$-\frac12$	$-\frac12$	$-\frac14$	$-\frac14$	$-\frac14$	$-\frac14$
$A^+A^-B^-B^-$	2/16	$-\frac12$	0	$-\frac12$	1	0	$-\frac12$	0	$-\frac12$	$-\frac12$	-1	$-\frac12$	0	$-\frac12$	0	0	0
$A^-A^-B^+B^+$	1/16	0	1	1	0	0	0	1	0	-1	0	0	1	0	-1	0	0
$A^-A^-B^+B^-$	2/16	0	$-\frac12$	1	$-\frac12$	0	0	$-\frac12$	$-\frac12$	-1	$-\frac12$	0	$-\frac12$	$-\frac12$	$-\frac12$	0	0
$A^-A^-B^-B^-$	1/16	0	0	1	1	0	0	0	1	-1	-1	0	0	1	0	0	0

F_2	Frequency	$\bar{L}_{3i}\ (=\times\,A^+A^+B^+B^+)$								$\bar{L}_{1i} + \bar{L}_{2i} - 2\bar{L}_{3i}$			
		a_A	a_B	d_A	d_B	aa_{AB}	ad_{AB}	ad_{BA}	dd_{AB}	aa_{AB}	ad_{AB}	ad_{BA}	dd_{AB}
$A^+A^+B^+B^+$	1/16	$-\frac12$	$-\frac12$	$-\frac12$	$-\frac12$	$-\frac14$	$-\frac14$	$-\frac14$	$-\frac14$	$-\frac12$	$-\frac12$	$-\frac12$	$\frac12$
$A^+A^+B^+B^-$	2/16	$-\frac12$	0	$-\frac12$	$-\frac12$	0	$-\frac14$	0	$-\frac14$	$-\frac12$	0	$-\frac12$	0
$A^+A^+B^-B^-$	1/16	$-\frac12$	$-\frac12$	$-\frac12$	$-\frac12$	$-\frac14$	$-\frac14$	$-\frac14$	$-\frac14$	$-\frac12$	$-\frac12$	$-\frac12$	$-\frac12$
$A^+A^-B^+B^+$	2/16	0	$-\frac12$	$-\frac12$	$-\frac12$	0	0	$-\frac14$	$-\frac14$	$-\frac12$	$-\frac12$	0	0
$A^+A^-B^+B^-$	4/16	0	0	$-\frac12$	$-\frac12$	0	0	0	$-\frac14$	$-\frac12$	0	0	0
$A^+A^-B^-B^-$	2/16	0	$-\frac12$	$-\frac12$	$-\frac12$	0	0	$-\frac14$	$-\frac14$	$-\frac12$	$-\frac12$	0	0
$A^-A^-B^+B^+$	1/16	$-\frac12$	$-\frac12$	$-\frac12$	$-\frac12$	$-\frac14$	$-\frac14$	$-\frac14$	$-\frac14$	$-\frac12$	$-\frac12$	$-\frac12$	$-\frac12$
$A^-A^-B^+B^-$	2/16	$-\frac12$	0	$-\frac12$	$-\frac12$	0	$-\frac14$	0	$-\frac14$	$-\frac12$	0	$-\frac12$	0
$A^-A^-B^-B^-$	1/16	$-\frac12$	$-\frac12$	$-\frac12$	$-\frac12$	$-\frac14$	$-\frac14$	$-\frac14$	$-\frac14$	$-\frac12$	$-\frac12$	$-\frac12$	$\frac12$

the variance due to epistasis is many times larger than that due to the additive and dominance effects of the genes. Even with just two genes, we are unlikely to find complete epistasis because it will yield as much variance as the additive and dominance effects combined. Another problem that we face concerns the detection of epistasis. It is difficult to detect in variances because there are not that many tests available.

One of the most commonly used tests of epistasis comes from the Triple Test Cross (TTC), (see Chapter 5) [10]. The TTC involves crossing each of a number (n) of F_2 individuals to P_1, P_2 and F_1 to produce 3n families. It was shown in Chapter 5 that, for every F_2 individual, $\bar{L}_1 + \bar{L}_2 - 2\bar{L}_3 = 0$ when the additive–dominance model is adequate, but not otherwise. The relevant ANOVA was given in Table 5.12. Table 11.6 gives the expectations of $\bar{L}_1$, $\bar{L}_2$ and $\bar{L}_3$ for the F_2 assuming a digenic interaction model, together with the expected values of $\bar{L}_1 + \bar{L}_2 - 2\bar{L}_3$. It follows from these expectations that the k_i^2 and σ_e^2 in Table 5.13(ii) take the form:

$$\text{Correction factor } (k_i^2) = \tfrac{1}{144} a a_{AB}^2$$

$$\text{Variance } (\sigma_e^2) = \tfrac{1}{288}(ad_{AB}^2 + ad_{BA}^2) + \tfrac{1}{576} dd_{AB}^2.$$

This test has shown that in many plant and animal species the variance due to epistasis is often very small compared with its directional effect on the means. In other words, epistasis is less important with variances, either because its magnitude is small or, being an interaction of the main effects, it is not a major source of genetic variation [11].

Summary

1. Epistasis, like dominance, is a function of the scale of measurement and can be removed in some cases by modifying the scale.
2. In most cases, however, non-allelic interactions need to be accommodated in the model either because a suitable scale cannot be found or, if such a scale can be found, it is not very meaningful.
3. It is possible to build suitable models to analyse and study epistasis and interpret its effects on both the generation means and variances.

continued

4. The impact of epistasis is often more profound on the means compared with variances and the latter generally show less bias.
5. Epistatic effects of the generation means are more amenable to analysis than those of the variances.

References

1. Klug, W.S. and Cummings, M.R. (1983) *Concepts of Genetics*, Charles E. Merrill Publishing Co., Columbus, Ohio.
2. Mather, K. (1967) Complementary and duplicate interactions in biometrical genetics. *Heredity*, **22**, 97–103.
3. Van der Veen, J.H. (1959) Tests of non-allelic interaction and linkage for quantitative characters in generations derived from two diploid pure lines. *Genetica*, **30**, 201–32.
4. Kempthorne, O. (1957) *An Introduction to Genetic Statistics*. John Wiley and Sons, NY.
5. Hayman, B.I. and Mather, K. (1955) The description of genic interactions in continuous variation. *Biometrics*, **11**, 69–82.
6. Mather, K. and Jinks, J.L. (1982) *Biometrical Genetics*, 3rd edn, Chapman & Hall, London.
7. Jinks, J.L. and Perkins, J.M. (1969) The detection of linked epistatic genes for a metrical trait. *Heredity*, **24**, 465–75.
8. Hill, J. (1966) Recurrent backcrossing in the study of quantitative inheritance. *Heredity*, **21**, 85–120.
9. Jinks, J.L. (1978) Unambiguous test for linkage of genes displaying nonallelic interactions for a metrical trait. *Heredity*, **40**, 171–3.
10. Kearsey, M.J. and Jinks, J.L. (1968) A general method of detecting additive, dominance and epistatic variation for metrical traits. I Theory. *Heredity*, **23**, 403–9.
11. Pooni, H.S. and Jinks, J.L. (1979) Sources and biases of the predictors of the properties of recombinant inbreds produced by single seed descent. *Heredity*, **42**, 41–8.

Genotype by environment interaction 12

Genotype by environment interaction (G × E) is a major problem in the study of quantitative traits because it complicates the interpretation of genetical experiments and makes predictions difficult. It is particularly a problem in plant and animal breeding where genotypes have to be selected in one environment and used in another. For example, early generations derived from a cross in cereals may be used to make decisions about choices of crosses for further development. In these early generations, seed quantities are small, and selection takes place in relatively small plot trials at one location in one year, while the final cultivars are grown in multi-location trials in subsequent years. In this chapter we will attempt to explain the nature and genetical analysis of G × E for quantitative traits and draw together some of the more important conclusions that have emerged from G × E analyses.

12.1 The nature and causes of G × E

In theory, G × E may be recognized in at least three forms.

Form 1. In a given environment, the environmental variation in phenotype for a particular trait among genetically identical individuals may vary with the genotype,

$$\text{e.g. } s_{P1}^2 \neq s_{P2}^2 \neq s_{F1}^2.$$

Form 2. The environmental variance exhibited by a particular genotype may alter with the environment,

$$\text{e.g. } s_{P1}^2 (\text{in environment A}) \neq s_{P1}^2 (\text{in environment B}).$$

Form 3. The genetical variance among a collection of genotypes may alter with the environment, i.e. the effects of given allele substitutions may be quite different in one environment than in another.

e.g. V_G(in environment A) $\neq V_G$(in environment B).

All three forms can, in principle, be related in a single theoretical model. Consider a situation where a number of different genotypes (homozygotes or heterozygotes but each with many identical replicates) are raised and scored in a single environment which is completely uniform except for a single factor, E'. If we were to plot the phenotype, P, of two genotypes, G_1 and G_2, against E' we might obtain a relationship similar to that in Figure 12.1(i). The two genotypes have different response curves and this implies G × E interaction.

Now consider G × E of form 1. In practice, the environment is not completely controlled, even for the factor E'. For example, if plants are raised in a growth cabinet at 25°C, the temperature will vary throughout the cabinet within some tolerance. Thus, although one attempts to provide environment A (25°C), the plants are in fact exposed to $\bar{A} \pm \delta E'A$ where $\delta E'A$ is the margin of error in the distribution of factor A. Clearly at $\bar{A}$ on our graph, G_1 is responding more to change in E' than is G_2. Given the range of $\delta E'A$, therefore, G_1 will appear to be more variable than G_2 in that environment, which explains point 1, i.e. why $s_{G1}^2 \neq s_{G2}^2$, Figure 12.1(ii).

Figure 12.1 G × E interaction. Two genotypes, G_1 and G_2 show different responses to the environment, E'. This results in the lines having different additive genetic effects (i) and environmental variances (ii) in environments A and B.

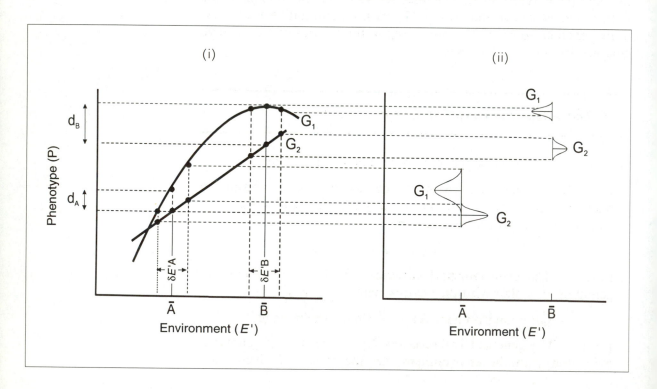

In any experiment one tries to keep $\delta E'$ as small as possible. However, there are limitations on the extent to which it can be reduced, such as lack of knowledge of the important micro-environmental variables and the costs which will be incurred in controlling them. Thus, one can never hope to eliminate the micro-environmental variability nor the developmental errors entirely and experiments must be adequately randomized over this heterogeneity. Generally, these environmental elements are referred to as **uncontrolled**.

Figure 12.1 also assumes that all the G_1 individuals, or G_2 individuals are potentially identical. In that they are not, either because of environmental or developmental heterogeneity, our single curve simply represents the mean performance at different E's. There should in fact be several approximately parallel lines representing the true variants within G_1 and G_2, as is shown for G_1 in Figure 12.2.

If we turn to G × E of form 2, it is now clear from Figure 12.1(ii) that the combination of differences in response curves of P on E, together, result in

$$s^2_{G1A} > s^2_{G1B}$$

while

$$s^2_{G2A} = s^2_{G2B}$$

Finally, $\bar{G}_1 - \bar{G}_2$ (at B) $> \bar{G}_1 - \bar{G}_2$ (at A) or, put another way, the variance of G_1 over environments (A, B) is different (greater) from that of G_2 (i.e. G × E of form 3).

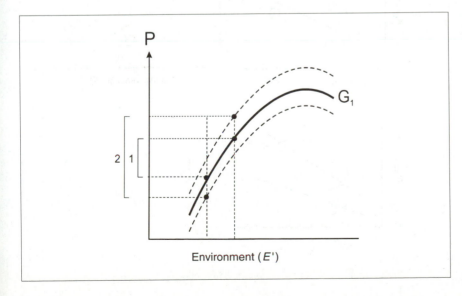

Figure 12.2 The contribution of environment (1) and development (2–1) to variation in phenotype.

In practice, there will seldom be just one environmental factor, but many acting simultaneously. Furthermore, it is quite likely that the factors contributing to differences in phenotype between environments A and B may not be the same factors as those which influence variation within environments A and B. Consider a field experiment in which environments A and B represent two sowing dates. A and B will differ in day length, rainfall and temperature (= X factors) while within A and B, differences in soil nutrients (= Y factors) may be all important, while the impact of climatic factors will be fairly constant over all the genotypes. We may find that, if factors X and Y could be analysed, the relationship shown in Figure 12.3 might exist. Normally, our knowledge of what factors are important is limited and so we must concentrate on those that we know or can control.

12.2 Tests of G × E

There are various statistical tests that are appropriate for detecting the three basic types of G × E described at the start of this chapter.

Figure 12.3 The effects of two different types of G × E interaction concerning environments X and Y occurring alone (top) or together (bottom).

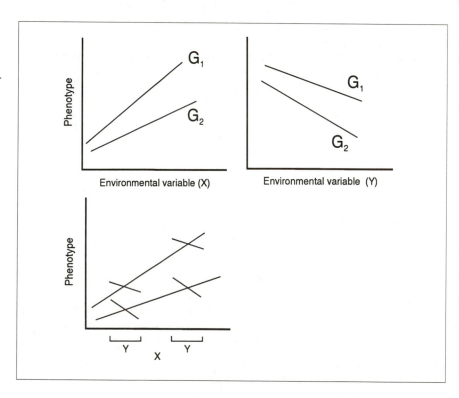

Heterogeneity of variances of non-segregating generations within an environment can be tested by means of variance ratio (F), Bartlett's or, preferably, Levene's test. If there are just two or three variances to be compared, the ratio of the largest to the smallest (s_L^2/s_S^2) can be tested as a variance ratio, but because in this case F will always be greater than 1.0, the resulting probability must be corrected, as explained in Chapter 3. Bartlett's test is particularly susceptible to non-normality in the data and hence one cannot reliably assume that significance is due to heterogeneous variances. Heterogeneous variances could be due to large $\delta E'$, somatic variation, scoring errors (not mistakes) or scalar effects. Whatever the cause, they will create problems in estimating components of variance and in predicting response to selection because one will always be in doubt as to what value of V_E to use. It will be a particular problem in the case of generations derived from two inbred lines and usually, if variances of non-segregating generations are unequal, then they should be weighted in the F_2 as:

$$V_E = \tfrac{1}{4}s_{P1}^2 + \tfrac{1}{2}s_{F1}^2 + \tfrac{1}{4}s_{P2}^2.$$

When estimating h_n^2 by standard direct methods from a random mating or natural population, $G \times E$ may not be too much of a problem, because the environmental components will probably be weighted correctly. One can try to change the scale in order to eliminate non-linearity in the data, but this often creates other problems, including modification of the gene effects, and possibly the creation of epistasis (Chapter 11).

The third type of $G \times E$ in which the genetical variance changes with the environment can also be detected by means of tests for heterogeneous variances or by standard ANOVA as will be illustrated later.

12.3 Macro-environmental variables

12.3.1 Controlled macro-environments

Where macro-environmental variables are controllable and measurable, one can explore the situation graphically by plotting the phenotypic score against the level of the variable, as in Figure 12.1. A two-way ANOVA of genotypes, macro-environments and $G \times E$ will indicate the relative magnitudes of these effects, while the graph will reveal the nature of the interaction. A typical situation in which such an analysis is applied is when varieties of a crop are grown at different fertility levels. If there is no interaction, then all lines respond to the fertilizer to the same extent and hence the best genotype at one level is also the best at any other. However, this may also be true with G × E but only a direct comparison of the means as provided by a graph will indicate this.

There are various criteria that a breeder may require of a new variety, such as (i) good performance under average management and input levels, (ii) superior performance at high input levels, or (iii) superior performance and high stability. From the ANOVA and the graphical representation, such genotypes can be identified. Further, in the case of performance under high inputs the projected increase in performance must adequately cover the cost of the additional inputs, such as protein in animal feed or fertilizer in a plant trial, which are necessary in order to exploit the full potential of such a variety.

12.3.2 Uncontrolled macro-environments

Such environments are commonly encountered and pose many problems for the analysis and exploitation of G × E. Typical examples of such environments are locations and seasons. As the major components of these environments, like temperature, light, rainfall, etc. are largely undefined, breeders often attempt to produce varieties that show minimal response to such variables. The handling and analysis of data from such environments is complex and we describe it by analysing a well-documented set of plant height data recorded on two lines of *Nicotiana rustica*, P_1 and P_2, and their F_1 during a 16-year period (t = 16 environments) [1,2]. These data are given in Table 12.1(i) and presented graphically in Figure 12.4. The major macro-environmental variables are not known exactly but they include season and location effects (London and Birmingham), in addition to the continual improvement in field fertility and change in plant-handling techniques. Clearly, the physical components of these environments are difficult to measure and partition because relevant meteorological and soil quality data are not available and there is no way of quantifying changes in plant management practice. It is clear from Figure 12.4 that although P_1 is generally the taller, the difference in height between the two varieties changes considerably and erratically over the 16 years, while P_2 is actually the taller in some years. An ANOVA of the parental data shown in Table 12.1(ii) confirms that G × E is, indeed, highly significant. So, how do we proceed to disentangle the genetic and G × E effects in such a situation?

Because we have no physical measure of the climatic and edaphic factors, one approach is to let the plants themselves assess the

Table 12.1 Plant height (cm) of two inbred lines and their F_1 in 16 environments (years). (i) Data; (ii) ANOVA

(i)

Year	$\bar{P}_2$	$\bar{P}_1$	$\bar{F}_1$
1	99.97	124.71	126.34
2	100.58	127.36	129.21
3	100.28	122.33	124.84
4	94.31	104.57	110.06
5	102.11	109.07	115.98
6	97.38	94.92	108.28
7	96.88	102.97	105.56
8	106.43	113.49	126.44
9	107.06	107.95	119.66
10	91.03	83.46	95.05
11	112.90	120.47	133.10
12	93.35	113.33	118.75
13	123.70	154.25	161.62
14	120.17	146.53	152.68
15	114.88	135.86	140.49
16	117.75	152.48	151.16
Mean	104.92	119.61	126.20
	112.27		

(ii)

Source	df	SS	MS	F	P
Genotypes	1	1725.33	1725.33	199.69	***
Environments	15	6459.76	430.65	49.84	***
G × E	15	1200.20	80.01	9.26	***
Error	∞	–	8.64		

environment, i.e. instead of relating phenotype (height) of each line to some physical measure of the environment such as temperature, the mean height of the two lines in each environment is used as a measure of environmental quality, the **environmental index**. One advantage of defining the environment in this way is that the phenotypes of the two parents in any environment, j, can be written in the form of a genetical model as follows, where e_j and ge_{aj} are the environmental and additive G × E effects respectively:

$$\bar{P}_{1j} = m + [a] + e_j + ge_{aj}$$
$$\bar{P}_{2j} = m - [a] + e_j - ge_{aj}$$

Figure 12.4 G × E interaction
illustrated by two inbred lines and
their F_1 in 16 environments. The
character is final plant height in
Nicotiana rustica.

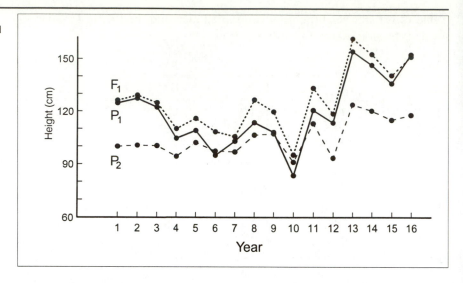

where

$$m = \sum_{j=1}^{t} (\bar{P}_{1j} + \bar{P}_{2j})/2t$$

$$\equiv (\bar{P}_1 + \bar{P}_2)/2 \text{ averaged over } t(= 16) \text{ environments}$$

$$[a] = \sum_{j=1}^{t} (\bar{P}_{1j} - \bar{P}_{2j})/2t \equiv (\bar{P}_1 - \bar{P}_2)/2 \text{ averaged over } t \text{ environments.}$$

In the jth environment,

$$\tfrac{1}{2}(\bar{P}_{1j} + \bar{P}_{2j}) = m + e_j$$

$$\tfrac{1}{2}(\bar{P}_{1j} - \bar{P}_{2j}) = [a] + ge_{aj}$$

thus

$$\sum_{j=1}^{t} e_j = \sum_{j=1}^{t} ge_{aj} = 0.$$

Since m and $[a]$ can be calculated, the values of every environmental effect, e_j, and additive G × E effect, ge_{aj}, can be obtained using the formulae above as shown in Table 12.2(i). This operation, however, does not provide a summary of the data but merely rearranges the information into sensible parameters, e.g., m, $[a]$, $t(e_j)$ and $t(ge_{aj})$ values instead of the 2t means of the P_1 and P_2 genotypes. In order to simplify the situation further and enable predictions to be made across novel environments it is necessary to look for simple relationships between the e_j and ge_{aj} values. If ge_{aj} is regressed onto e_j using

Table 12.2 Environmental and G × E effects estimated from the data in Table 12.1(i). (i) Parameter estimates; (ii) regression of ge_{aj} on e_j

(i)

Environment	e_j	ge_{aj}	ge_{dj}
1	0.0734	5.0272	0.0653
2	1.7034	6.0472	1.3053
3	−0.9616	3.6822	−0.3997
4	−12.8266	−2.2128	−3.3147
5	−6.6766	−3.8628	−3.5447
6	−16.1166	−8.5728	−1.8047
7	−12.3416	−4.4978	−8.2997
8	−2.3066	−3.8128	2.5453
9	−4.7616	−6.8978	−1.7797
10	−25.0216	−11.1278	−6.1297
11	4.4184	−3.5578	2.4803
12	−8.9266	2.6472	1.4753
13	26.7084	7.9322	8.7103
14	21.0834	5.8372	5.3953
15	13.1034	3.1472	1.1853
16	22.8484	10.0222	2.1103
Mean	0.0004	0.0002	−0.0002

(ii)

Source	df	SS	MS	F	P
Regression	1	401.20	401.20	92.87	***
Residual	14	198.90	14.21	3.29	***
G × E	15	600.10	40.01	–	–
Error	∞	–	4.32	–	–
$\beta_a = 0.35$					

the values in Table 12.2(i), the ANOVA in Table 12.2(ii) is obtained which shows that there is a highly significant linear regression with slope, $\beta_a = 0.35$. Hence we can write:

$$ge_{aj} = \beta_a.e_j$$
$$= 0.35e_j.$$

As a result, the models for the expected generation means can be written as,

$$\bar{P}_{1j} = m + [a] + e_j + \beta_a.e_j$$
$$= (m + [a]) + (1 + \beta_a).e_j$$

Figure 12.5 Linear relationship between height and environmental index (e_j) for the two parents and their F_1 after analysis of data in Figure 12.4.

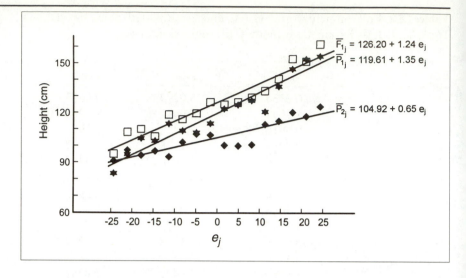

and

$$\bar{P}_{2j} = m - [a] + e_j - \beta_a.e_j$$
$$= (m - [a]) + (1 - \beta_a).e_j.$$

Because the F_1 was also raised with the parents in the same environments, we can write its mean as:

$$\bar{F}_{1j} = m + [d] + e_j + ge_{dj}$$

where m and e_j are the same as for P_1 and P_2, and

$$[d] = \sum_{j=1}^{t} \bar{F}_{1j}/t - m(= \text{mean of } F_1 \text{ over all the t environments} - m).$$

Therefore, t values of ge_{dj} can be calculated as shown in Table 12.2(i) and also regressed onto e_j. These, too, show a significant regression indicating that ge_{dj} is also a linear function β_d of e_j, where $\beta_d = 0.24$, Table 12.3. Thus we can write,

$$\bar{F}_{1j} = (m + [d]) + (1 + \beta_d)e_j \quad \text{(Figure 12.5)}.$$

The linear regression analysis carried out on the parental data is, in effect, a partition of the original $G \times E$ SS in Table 12.1(ii), the difference of a factor of two arising from the fact that the regression analysis was performed on means across environments. It is therefore possible to compare the residual variation against the original error to see to what extent the linear regression approach has been successful in explaining the $G \times E$. Although the residual is still significant, the linear regression, β_a, has explained 67% of the $G \times E$ SS.

Table 12.3 Regression ANOVA of the F_1 data

Source	df	SS	MS	F	P
Regression	1	180.81	180.81	26.43	***
Residual	14	83.11	5.94	<1.0	n.s.
Error	∞	–	6.84	–	–
$\beta_d = 0.24$					

12.4 G × E with many lines

The analytical approach with many lines is essentially the same as with two [3–6]. Initially, G × E interaction is detected by a two-way ANOVA followed by a regression analysis of each line separately. With s lines raised in t environments, the variation in the s × t table of line means can be partitioned into the 'Between lines SS', 'Between environments SS' and 'Lines × Environments SS' with $(s - 1)$, $(t - 1)$ and $(s - 1)(t - 1)$ df respectively. When the 'Lines × Environments' component is significant, a regression analysis can be carried out to determine if any of the G × E components is linearly dependent on e_j. As before, the environmental index, e_j, is calculated as a deviation of the environmental mean $(= \mu + e_j)$ from the overall mean $(= \mu)$.

This is best illustrated by an example. The data consist of five recombinant inbred lines raised in seven environments. These lines were derived by SSD from the cross between the same two *Nicotiana rustica* lines described above. In each year, 10 plants of each line were raised in a completely randomized block giving the means presented in Table 12.4(i) and the ANOVA in Table 12.4(ii). The latter shows that there is significant G × E interaction. The statistical model which applies to these data is:

$$\bar{y}_{ij} = \mu + a_i + e_j + ge_{aij} + \epsilon_{ijk} \qquad \text{[Eqn 12.1]}$$

where

$\bar{y}_{ij}$ is the mean of line *i* in environment *j*,

μ = overall mean of s × t means,

a_i = mean of the *i*th line $-\mu$,

e_j = mean of *j*th environment $-\mu$,

$ge_{aij} = \bar{y}_{ij} - a_i - e_j - \mu$,

and ϵ_{ijk} represents the random error of the line means.

If ge_{aij} is linearly dependent on e_j, then it can be replaced by $\beta_{a_i}.e_j$, or, alternatively, by $\beta_{a_i}.e_j + \delta_{ij}$ if there is some non-linearity remaining, and Equation 12.1 then becomes:

$$\bar{y}_{ij} = (\mu + a_i) + (1 + \beta_{a_i})e_j + \delta_{ij} + \epsilon_{ijk}.$$

In order to test for the relation between ge_{aij} and e_j, the values of each line in each year are regressed onto the environmental index which is the mean of all lines for each environment. For example, the values of the means of line 1 in each environment, $\bar{y}_{1j}$, and the corresponding environmental means, $e_j + \mu$, from Table 12.4 are,

$$\bar{y}_{1j}, \qquad 126.0, 126.0, 105.0, 75.0, 121.0, 106.0, 124.0$$

$$e_j + \mu, \quad 130.2, 124.8, 108.4, 81.8, 120.6, 103.6, 118.4$$

The overall mean of line 1 ($\bar{y}_{1j}$) is 111.86 ($= \mu + a_1$), the regression slope of $\bar{y}_{1j}$ onto $e_j + \mu$ is 1.111 ($= 1 + \beta_{a_1}$), while the regression SS is 1979.8026. The corresponding statistics for this genotype and the other four lines are given in Table 12.4(iii).

This model allows tests of the lines, environments and $G \times E$ components as well as the linear and the remainder parts of $G \times E$ in the form of the so-called 'joint regression analysis' which is presented in Table 12.5. This analysis has the following characteristics.

1. The combined or joint regression of all the lines takes a value of exactly 1.00 and consequently its SS and df are identical to those for 'Environments'.
2. The sum of β_{a_i} is always equal to 0.00.
3. The 'Heterogeneity of regressions', calculated as total regression SS − joint regression SS, tests for differences between β_{a_i} and consequently has s − 1 df.
4. The 'Remainder MS' has $(s - 1)(t - 2)$ df and not $s(t - 2)$ which one normally obtains by pooling the remainder SS (with t − 2 df) of all the s lines. The final result of the analysis is presented in the form of equations given in Table 12.6.

12.5 Interpretation of $G \times E$ analysis

This approach allows the variation in the line means to be partitioned into two components; the constant part which is due to the true genetic potential of the genotype, $(\mu + a_i)$, and the variable part which depends on the environment, $(1 + \beta_{a_i})e_j$. These components are meaningful because they are directly related to the mean performance and the environmental sensitivity of the line. Thus, line 2 has a high mean

Table 12.4 Average heights (cm) of five inbred lines of *Nicotiana rustica* in seven environments (years). (i) Data; (ii) ANOVA; (iii) SS and SP for the regression ANOVA

(i)

Line	Environments (years)							Mean
	1	2	3	4	5	6	7	
1	126	126	105	75	121	106	124	111.86
2	156	151	135	99	149	127	140	136.71
3	88	91	78	67	90	74	85	81.86
4	148	130	108	81	122	108	119	116.57
5	133	126	116	87	121	103	124	115.71
Mean	130.2	124.8	108.4	81.8	120.6	103.6	118.4	112.54

(ii)

Source	df	SS	MS	F	P
Lines	4	10868.40	2727.10	239.81	**
Environments	6	8017.49	1336.25	117.94	**
$G \times E$	24	908.80	37.87	3.34	**
Error	315	3568.95	11.33	–	–

(iii)

Line	SSy	SPxy	Reg.SS	Rem.SS
1	2070.8571	1781.7429	1979.8026	91.0545
2	2257.4286	1887.2857	2221.2995	36.1291
3	494.8571	847.5429	447.9765	46.8806
4	2615.7143	1983.2286	2452.8861	162.8282
5	1487.4286	1517.6857	1436.4665	50.9621
Total	–	8017.4858	8538.4312	387.8545

performance but is very sensitive to the environment, while line 3 is a low performing stable line. Overall there is a positive correlation between the mean and sensitivity among these five lines.

The basic model is readily extendable to accommodate any other relationship between the $G \times E$ and the environmental index e_j. For instance, the above linear equation can be easily modified to:

$$\bar{y}_{ij} = (\mu + a_i) + (1 + \beta_{a1_i})e_j + \beta_{a2_i}.e_j^2 + \ldots + \delta_{ij} + \epsilon_{ijk}$$

where β_{a1_i} and β_{a2_i} represent the linear and quadratic relationships between ge_{aij} and e_j. Similarly, the simple additive–dominance

Table 12.5 Joint regression analysis of the *Nicotiana rustica* data in Table 12.4

Source	df	SS	MS	F	P
Genotypes	4	10868.40	2717.10	239.81	***
Environments	6	8017.49	1336.25	117.94	***
$G \times E$	24	908.8	37.87	3.34	**
Joint regression	6	8017.49	1336.25	117.94	***
Heterogeneity between regressions	4	520.94	130.24	11.49	***
Remainder	20	387.85	19.40	1.71	*
Between plants, within environments	315	3568.95	11.33	–	–

model can be extended to accommodate the epistatic components, $[aa]$, $[ad]$ and $[dd]$, and their interaction with the environment.

Even when simple relationships cannot be identified between the $G \times E$ and the environmental index, the sensitivity of any genotype can be measured in the form of its variance over the environments in which it is raised. This enables one to identify the best-performing and the most stable (lowest variance) genotypes for a given set of environments. It does not, however, enable us to analyse the situation genetically nor, perhaps more importantly, to allow predictions to be made.

There is no theoretical reason why $G \times E$ should be simply related to the environmental index and in many cases it does not reveal such a relationship. In those cases where it does, and particularly where the relationship is linear, it can be exploited for genetical and breeding purposes.

12.6 Predictions in the presence of $G \times E$

When the linear component is either the only or the major source of $G \times E$ among the parental lines and their F_1, then, based solely on

Table 12.6 Regression parameters for the five *Nicotiana rustica* lines

Line	Mean ($= \mu + a_i$)	Regression ($= 1 + \beta_{a_i}$)
1	111.86	1.111
2	136.71	1.177
3	81.86	0.529
4	116.57	1.237
5	115.71	0.946

the performance of the parents and F_1, we can predict the performance of any other generation derived from that cross in any environment. For example, in the presence of G × E the F_2 mean in the jth environment is:

$$\bar{F}_{2j} = m + \tfrac{1}{2}[d] + e_j + \tfrac{1}{2}ge_{dj}$$

if

$$ge_{dj} = \beta_d.e_j$$

then

$$\bar{F}_{2j} = m + \tfrac{1}{2}[d] + (1 + \tfrac{1}{2}\beta_d)e_j$$

or

$$\bar{B}c_{1.1j} = m + \tfrac{1}{2}[a] + \tfrac{1}{2}[d] + e_j + \tfrac{1}{2}ge_{aj} + \tfrac{1}{2}ge_{dj}$$
$$= m + \tfrac{1}{2}[a] + \tfrac{1}{2}[d] + (1 + \tfrac{1}{2}\beta_a + \tfrac{1}{2}\beta_d)e_j.$$

One can also use these equations to predict the range of environments in which, for example, the F_1 will perform better than the best parent. Such predictions have been found to be in good agreement with the observed generation means in practice [7].

12.7 Conclusions from the genetic analysis of G × E

Several important conclusions concerning the nature of G × E for quantitative traits have been obtained from studies similar to those above, as well as from selection experiments in diverse environments. Some of these are listed below.

1. Comparison of β_a and β_d indicates the relative stability of homozygotes and heterozygotes. Generally $\beta_d < \beta_a$ and heterozygotes are found to show an intermediate response to the lower and higher scoring homozygotes, i.e.

$$(1 - \beta_a) < (1 + \beta_d) < (1 + \beta_a).$$

2. The generally greater stability of heterozygotes described in (1) above affects heterosis. Positive heterosis is present when

$$\bar{F}_1 - \bar{P}_1 > 0.$$

In the jth environment,

$$\bar{F}_{1j} = m + [d] + (1 + \beta_d)e_j$$
$$\bar{P}_{1j} = m + [a] + (1 + \beta_a)e_j$$

i.e.

$$\bar{F}_{1j} - \bar{P}_{1j} = ([d] - [a]) + e_j(\beta_d - \beta_a).$$

Clearly, $\bar{F}_1 - \bar{P}_1$ will vary over environments only if β_d is not equal to β_a. If, as is generally the case, $\beta_d < \beta_a$, then, $\beta_d - \beta_a$ is negative. Thus, in 'poor' environments, i.e. when e_j is negative, the heterosis will be greater than in the 'good' environments, where e_j is positive. Similarly, the potence ratio in the jth environment is given by the formula $([d] + ge_{dj})/([a] + ge_{aj})$, or $([d] + \beta_d.e_j)/([a] + \beta_a.e_j)$ when G × E is a linear function of the environmental index. Therefore, when $\beta_d < \beta_a$, potence is also expected to decline as the environment improves.

3. The nature and extent of G × E is a function of the trait and not of the individual, i.e. a particular set of genotypes would rank differently for G × E for different traits. For example, Perkins and Jinks [8] studied two inbred lines of *Nicotiana rustica* in a range of environments varying in sowing dates, fertilizer levels and seasons. They found that line V_{42} was the most sensitive to the environment for final height, while line V_2 was the most sensitive for flowering time.

4. If we consider two different environmental variables the response of the trait may or may not be the same. Returning to the example in (3), although the same β_a could be fitted to all the environmental variables for final height, different β_as were required for flowering time across seasons and fertilizers. These conclusions also apply to the micro-environmental stability because, as shown previously, the various types of G × E are simply different manifestations of the same phenomena.

5. It is generally found that the means and variances of random collections of lines are correlated, no matter whether measured over micro- or macro-environments. This phenomenon occurred in the data in Table 12.4. Such correlations, however, can be broken by selection to obtain genotypes which have a high mean with a low variance and others with low mean and high variance. Thus, these correlations are, at least in part, due to linkage.

The fact that the mean performance and environmental sensitivity are due to different sets of genes was well demonstrated in *Drosophila melanogaster* [9]. In this study, chromosome assay techniques (Chapter 8) were used to show that different chromosomes controlled the variation in the mean and the variance for two traits, bristle number and yield of progeny. The main genetic determinants for the differences in the mean bristle number were on chromosome II, while those responsible for the variance were located on chromosome III. Similarly, for yield of progeny, the genes controlling the mean performance were on chromosome II while those responsible for the variance were on chromosomes II and III.

There are, however, limits to the extent to which the correlation between mean and environmental sensitivity can be separated, as part of the correlation is a function of scale. Low scoring genotypes are less able to vary around their mean than high scoring ones.

6. Epistasis is more pronounced in extreme environments, although this has only received limited attention. This can be shown in several ways, e.g. take two inbred lines;

$$\bar{P}_1 = m + [a] + [aa]$$

$$\bar{P}_2 = m - [a] + [aa]$$

where m = mean of all possible inbreds derived from the $P_1 \times P_2$ cross. Therefore, if a random sample of inbred lines is produced by SSD from the F_2 of a cross between these two parents, and these SSD lines together with P_1 and P_2 are raised in diverse environments, m and $[aa]$ can be estimated in each environment. It has been shown that in such a situation, the magnitude of $[aa]$ increases as the environment becomes more extreme [10].

7. It is possible in some situations to select for environmental sensitivity in a single environment [11]. For example, it has been found that the environmental sensitivity shown by the *Nicotiana rustica* plants for their final height at the end of the season is highly correlated with the proportion of the height they achieve during the first 6 weeks after transplanting in the field. The slow-growing lines are usually the more sensitive genotypes while those that achieve in excess of 70% of their overall height during these 6 weeks are highly stable (see Figure 12.6).

12.8 Selection in heterogeneous environments

There are not many published experiments where selection has been carried out over several generations in diverse environments, because of the time and effort required. However there is one well-documented example using the Basidiomycete fungus, *Schizophyllum commune*, as a model [12].

This fungus is haploid, can be raised on nutrient media under very controlled environmental conditions and several generations of

Figure 12.6 Growth curves of two tall and two short inbred lines of *Nicotiana rustica*. D3 and D34 are highly sensitive, while D50 and D10 are comparatively stable lines. It is apparent from the growth curves that the ratio of height at 6 weeks to height at the end of the season is comparatively low for the former and high for the latter pair of lines. This ratio can be used to select for sensitivity when material is raised in a single environment.

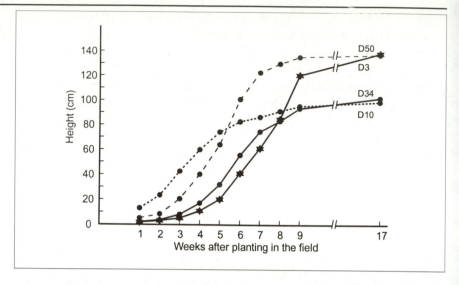

selection are possible in a short period of time. It can be reproduced vegetatively as well as sexually. The trait scored was growth rate, measured as the distance the hyphae grew down a narrow tube in a given time. The environments were two different temperatures, 20°C and 30°C, the latter being the temperature at which the fungus would grow fastest. Selection was practised for both high (H) and low (L) growth rate.

Genotypes were selected on the basis of their growth rate under three different conditions, (i) at 20°C, (ii) at 30°C and (iii) at both temperatures. After several generations of selection the growth rates of the three L lines and three H lines were tested at 20°C and 30°C. The results in Table 12.7 show two very important features.

First, in every case the best-performing line in any given environment is the line selected in that environment. For example, the slowest growing line at 20°C is the L line selected at 20°C, while the fastest

Table 12.7 Results of selecting for growth rate in *Schizophyllum commune* at different temperatures

Direction of selection	Temperature of selection (°C)	Mean growth rate			Variance over temperatures
		20°C	30°C	20 and 30°C	
L	20	24.25	48.75	32.36	168.6
L	30	32.00	36.25	33.67	61.7
L	20 and 30	28.50	46.50	32.17	86.7
H	20	60.55	93.75	68.83	494.6
H	30	54.50	104.25	68.25	674.8
H	20 and 30	58.92	100.75	73.64	611.5

growing line at both temperatures is the H line selected for fast growth at both temperatures.

Second, the most stable lines, as measured by the variance in growth rates over temperature, are the lines selected in the poorest environments, while the least stable are those selected in the best environments. For example, when selecting for high growth rate, the poor environment is 20°C and the best environment 30°C, and we see that the H line selected at 20°C has the lowest variance when raised at both temperatures, i.e. 494.60. Conversely, the H line selected at 30°C has the largest variance, 674.84. The same effect can be seen in reverse when selecting for slow growth rate, where the 'best' environment will be 20°C. Here the most stable L line is that selected at 30°C, the 'poor' environment, with a variance of 61.7, while the least stable is that selected at 20°C with a variance of 168.60. Those lines selected for their performance at both temperatures were intermediate.

These results have strong implications for plant and animal breeding. For example, if a breeder wishes to produce a strain or variety that performs well in a particular environment, then selection should be carried out in that environment. On the other hand, in order to produce a stable variety, selection should be carried out in a poor environment. What happens in practice is quite the reverse. Breeding is carried out under the 'best' environments, with good management and high inputs, and this is necessary, at least from the plant breeder's point of view, in order to have their varieties approved. However, they should not be surprised if these varieties are unstable and perform relatively badly under poor management regimes. Such a policy will be least appropriate for many developing countries where there can be major differences between the breeder's environments and those which prevail on farmer's fields.

12.9 Other methods of analysis

One serious criticism of the regression analysis concerns the use of the environmental mean as an independent measure of the environmental value. This is a valid criticism because each genotype contributes to the environmental mean and thus the line values and the environmental indices cannot be considered completely independent. Consequently, the regressions are biased upwards and the regression coefficients of up to ± 0.2 are often observed even when there are no linear relationships between ge_{aij} and e_j. Several suggestions have been made to resolve this problem and some of the more important ones are listed below.

1. Use the performance of a standard variety as a measure of the environmental index.
2. Raise additional samples of the same lines and use their data to measure the environmental index.
3. Regress each line on to the environmental indices calculated from the remaining s − 1 lines.
4. Increase the number of genotypes in the experiment in order to make 1/s very small.

While these suggestions solve the problem of independence of e_j, at least partially, they create some other difficulties. Suggestion (1) makes the regression analysis even less reliable, (2) and (4) are expensive on resources, while (3) increases the complexity of the analysis without offering much improvement. Furthermore, none of these solutions has been observed to provide a markedly different ranking of the lines to that of the original joint regression.

An alternative approach to the analysis of G × E is called the **Additive Main Effects and Multiplicative Interaction Effects Model,** or AMMI in short [13]. It utilizes the standard two way ANOVA and the Principal Components Analysis (PCA) to identify any patterns in the data. The basic model is as follows:

$$\bar{y}_{ij} = \mu + a_i + e_j + \sum_{n=1}^{N} \alpha_n \Gamma_{in} Z_{jn} + \delta_{ij} + \epsilon_{ijk}$$

where, in addition to other parameters,

N = number of IPCA (Interaction Principal Component Axes)

α_n = the singular value for IPCA axis n

Γ_{in} = genotypic eigenvector for IPCA axis n

Z_{jn} = environmental eigenvector for IPCA axis n, and

δ_{ij} = matrix of the residuals.

The results of AMMI are generally presented both in the form of an ANOVA and a bi-plot which allows one to visualize any relationships between the eigenvalues for PCA_1 and the means of genotypes and environments [14]. The analysis of *Nicotiana rustica* data from Table 12.4 (in Table 12.8) shows that a much larger proportion of G × E SS is accounted for by the two PCA components of the AMMI model and the remainder is non-significant. Figure 12.7 further reveals that environments 2, 3, 4 and 5 and genotypes 1 and

Table 12.8 AMMI analysis of *Nicotiana rustica* data in Table 2.4. (i) Analysis; (ii) principal components

(i)

Source	df	SS	MS	F	P
Genotypes	4	10868.40	2717.10	239.81	***
Environments	6	8017.49	1336.25	117.94	***
G × E	24	908.80	37.87	3.34	***
PCA 1	9	614.68	68.30	6.03	***
PCA 2	7	171.54	24.51	2.16	***
Remainder	8	122.57	15.32	1.35	n.s.
Between plants within environments	315	3568.95	11.33	–	–

(ii)

Source	Mean	Eigenvector
e_1	81.80	3.57
e_2	103.60	0.03
e_3	118.40	−0.02
e_4	108.40	0.55
e_5	120.60	0.01
e_6	124.80	−0.80
e_7	130.20	−3.33
g_1	111.86	−0.53
g_2	136.71	−1.23
g_3	81.86	3.94
g_4	116.57	−2.68
g_5	115.71	0.50

5 are the least interactive while the highest G × E is shown by genotype 3 in environment 1 and by genotypes 2 and 4 in environment 7.

Another problem with the regression analysis is its inability to handle more than one trait. Multiple trait analysis is necessary if one wishes to study the totality of G × E in the experimental material and understand its biological causes. Virtually all types of multivariate analyses have been applied to analyse G × E. For instance, principal components (PCA) and cluster (CA) analyses have been used to group genotypes and environments into cohesive subsets [15,16] while pattern (PA) [17] and factor (FA) [18] analyses classified environments on the basis of G × E. Clearly, a large number of procedures and their variants have been developed for the analysis of G × E and it is not possible to discuss each and every one of them in detail in this text. The reader is, therefore, encouraged to consult the considerable literature that is available on this topic.

Figure 12.7 Biplot of eigenvalues for PCA$_1$ obtained from the AMMI analysis and the overall means of the genotypes and the environments. The plot indicates that the highest G × E is shown by genotype 3 in environment 1 and by genotypes 2 and 4 in environment 7.

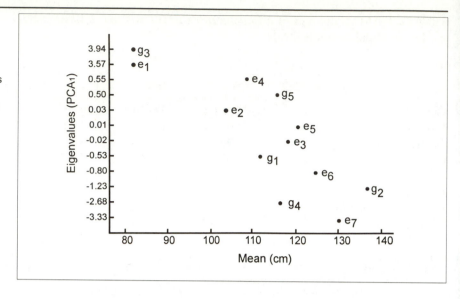

12.10 G × E – stability versus flexibility

It is very commonly found in the literature that terms such as stability and flexibility are used to describe G × E interactions. However, such statements involve views on the fitness value of the interaction. A given genotype, when exposed to diverse environments can either maintain a constant phenotype or alter its phenotype. In theory, one can attempt to interpret both responses in terms of an adaptive or a non-adaptive phenomenon. Thus, a genotype may have a constant phenotype because it is advantageous for it to do so or it may maintain that phenotype because it is incapable of altering it even though an alteration would be desirable. Similarly a genotype may be forced to alter its phenotype even though it would be better off not doing so. Whether or not a particular response is valuable can only be decided by relating it to the fitness of the individual, as shown in Table 12.9 [19].

Table 12.9 Relationships between G × E and fitness

Fitness	Genotype determines same phenotype in different environments	Genotype determines different phenotypes in different environments	Regulation
High	Developmentally stable	Developmentally flexible	Well regulated
Low	Developmentally inflexible	Developmentally unstable	Poorly regulated

Summary

1. G × E is a complex phenomenon which undermines the repeatability of experimental results and consequently reduces the efficiency of selection.
2. G × E occurs at both micro- and macro-environmental levels and due to the confounding of its effects with those of the genetic sources, it is usually cumbersome to analyse and interpret.
3. Under well-defined situations, G × E is as analysable and predictable as any other genetic effect. Empirical studies have shown that it can be measured in the form of environmental sensitivity parameters which provide reasonable predictions of genotypic performance of existing and new populations across environments and generations.
4. A wide range of methods, ranging from a straightforward comparison of the phenotypic variances of varieties in a set of environments to the most sophisticated multivariate analysis, have been developed for the analysis and interpretation of G × E, and the experimenter therefore can choose the most suitable for his or her needs.
5. Selection experiments have revealed that the selection environment determines the sensitivity of the selected material. Selections frequently show their best performance in the same environment in which they were initially selected. Further, selection for (a) high performance in good environment(s) or (b) low performance in poor environment(s) leads to improved performance in the desired direction but also increases sensitivity. Selection of genotypes on the basis of their performance across a range of environments, therefore, provides the best means of manipulating and exploiting G × E for breeding purposes.

References

1. Bucio Alanis, L. (1966) Environmental and genotype environmental components of variability. I. Inbred lines. *Heredity*, **21**, 387–97.
2. Bucio Alanis, L. and Hill, J. (1966) Environmental and genotype environmental components of variability. I. Heterozygotes. *Heredity*, **21**, 399–405.

3. Eberhart, S.A. and Russell, W.A. (1966) Stability parameters for comparing varieties. *Crop Sci.*, **6**, 36–40.
4. Finlay, K.W. and Wilkinson, G.N. (1963) The analysis of adaptation in a plant breeding programme. *Aust. J. Agric. Res.*, **14**, 742–54.
5. Perkins, J. M. and Jinks, J.L. (1968) Environmental and genotype environmental components of variability. III. Multiple lines and crosses. *Heredity*, **23**, 339–56.
6. Yates, F. and Cochran, W.G. (1938) The analysis of groups of experiments. *J. Agric. Sci.*, **28**, 556–80.
7. Bucio Alanis, L., Perkins, J.M. and Jinks, J.L. (1969) Environmental and genotype environmental components of variability. V. Segregating generations. *Heredity*, **24**, 115–27.
8. Perkins, J. M. and Jinks, J.L. (1973) The assessment and specificity of environmental and genotype environmental components of variability. *Heredity*, **30**, 111–26.
9. Caligari, P.D.F. and Mather, K. (1975) Genotype–environment interaction. III. Interactions in *Drosophila melanogaster*. *Proc. R. Soc. Lond. B.*, **191**, 387–411.
10. Jinks, J.L., Perkins, J.M. and Pooni, H.S. (1973) Incidence of epistasis in average and extreme environments. *Heredity*, **31**, 263–9.
11. Boughey, H. and Jinks, J.L. (1978) Joint selection for both extremes of mean performance and of environmental sensitivity to a macro-environmental variable. III. The determinants of sensitivity. *Heredity*, **40**, 363–9.
12. Jinks, J.L. and Connolly, V. (1973) Selection for specific and general response to environmental differences. *Heredity*, **30**, 33–40.
13. Gauch, H.G. (1988) Model selection and validation for yield trials with interaction. *Biometrics*, **44**, 705–15.
14. Kempton, R.A. (1984) The use of bi-plots in interpreting variety by environment interactions. *J. Agric. Sci.*, **103**, 123–35.
15. Perkins, J.M. (1972) The principal component analysis of genotype environmental interactions and physical measures of the environment. *Heredity*, **29**, 51–70.
16. Abou-El-Fittouh, H.A., Rawlings, J.O. and Miller, P.A. (1969) Classification of environments to control genotype by environment interactions with an application to cotton. *Crop Sci.*, **9**, 135–40.
17. Mungomery, V.E., Shorter, R. and Byth, D.E. (1974) Genotype × environment interactions and environmental adaptation. I. Pattern analysis-application to soya bean populations. *Aust. J. Agric. Res.*, **25**, 59–72.
18. Peterson, C.J. and Pfieffer, W.H. (1989) International winter wheat evaluation: relationships among test sites based on cultivar performance. *Crop Sci.*, **29**, 276–82.
19. Westerman, J. and Lawrence, M.J. (1970) Genotype-environment interaction and developmental regulation in *Arabidopsis thaliana*. *Heredity*, **25**, 609–27.

Further reading

Kang, M.S. (ed.) (1990) *Genotype by Environment Interaction and Plant Breeding*. Louisiana State University, Baton Rouge, Louisiana, USA.

13 Maternal effects and non-diploids

13.1 Maternal effects

Up to now we have assumed that the phenotype of an individual is simply the product of its own genotype and the environment. However, different families have different mothers and fathers, and these parents can also affect the phenotype of their progeny either directly or indirectly. With some exceptions, it is normally the female parent who is capable of having the greatest effect on the progeny, and hence we will concentrate exclusively on maternal effects.

It is important to be aware of these effects because they can bias the means and variances of families and mislead us in our attempts to understand the genetics of a given quantitative trait. A knowledge of their effects can at least help us to avoid obvious pitfalls and sometimes make an important contribution to the study of QTL.

13.1.1 Types of maternal effect

There are essentially three routes by which the mother can influence her progeny. The first is through cytoplasmic factors, such as mitochondria or chloroplasts which contain their own DNA and genes. These organelles are almost entirely inherited through the female line and hence give rise to permanent differences between lines over many generations of selfing or outcrossing.

The second involves the effects of the mother's own genes on her progeny. Thus a particular female plant may have a genotype that results in large seeds and the plants raised from these seeds may have an initial advantage over plants from a different mother. Because these effects are genetic in origin they will be inherited in a Mendelian fashion and thus can be handled by similar models to those discussed in earlier chapters, except that the genotype is one generation removed.

Third, the mother's environment may affect the phenotype of her progeny, just as her genes do. Thus a plant which was grown free of disease and given optimum fertilizer might also produce larger seeds, the effect on the offspring being the same as if the larger seeds had arisen from the mother's genotype. Unlike the genetic effects, however, these environmental influences will not be transmitted to the subsequent generations in a Mendelian fashion.

13.2 Models for generation means

These three types of maternal effect can easily be added to the models of generation means and their consequences to genetic analysis explored. We will pursue this through the six basic generations and their selfed and backcrossed progeny.

Because cytoplasmic effects are consistent over generations, they cause permanent differences between reciprocal crosses. Such differences can be recognized in the generations of a cross in the form of:

$$\bar{F}_1 - \overline{RF}_1 = \bar{F}_2 - \overline{RF}_2 = \bar{F}_3 - \overline{RF}_3 = \ldots = \bar{F}_\infty - \overline{RF}_\infty$$

where F_1 and RF_1, etc. descend from P_1 and P_2 mothers, respectively. Similar relationships will also hold between the reciprocal families of other generations such as the recurrent backcrosses provided they, too, descend from different mothers. These cytoplasmic effects can easily be handled by adding an additional parameter, $[c]$, which will be preceded by a coefficient of $+1$ or -1 depending on whether the original mother was P_1 or P_2 respectively.

The effects of the mother's genes or of her environment (e.g. health) will tend to be greatest on juvenile traits and decline as the offspring get older. This is well illustrated by the effects of seed weight in the cross between two varieties (V2, V12) of *Nicotiana rustica* (Table 13.1). The initial weight of 50 F_1 seeds produced from the cross V2 × V12 is more than twice that of the reciprocal. This results in the F_1 having larger cotyledons and leaves than its reciprocal for up to 6 weeks after sowing, although the relative size of the effect continually declines.

These delayed effects of the mother's genotype are the genetic effects of its alleles on the progeny phenotype and can be modelled as such. For example, the maternal contributions of P_1, P_2 and F_1 to their progeny (selfs and crosses) can simply be represented by the parameters $+[a]_m$, $-[a]_m$ and $+[d]_m$ where $[a]_m$ and $[d]_m$ account for the additive and dominance effects of all those genes in P_1 and P_2 that directly affect the phenotype of the progeny. Thus, for a trait

Table 13.1 The waning effect of seed size on the F_1 performance of a cross between varieties V2 and V12 of *Nicotiana rustica*. R in the last column represents the ratio $(\bar{F}_1 - \overline{RF}_1)/(\bar{F}_1 + \overline{RF}_1)$

Character	Age in weeks	$\bar{F}_1$	$\overline{RF}_1$	$\bar{F}_1 - \overline{RF}_1$	Probability	R
50-seed weight (μg)	0	14.20	6.20	8.00	**	0.78
Cotyledon spread (mm)	2	15.60	10.50	5.10	**	0.39
Spread of first pair of real leaves (mm)	3	35.45	24.53	10.92	**	0.36
Leaf length (mm)	4	37.30	28.06	9.24	**	0.28
Leaf length (mm)	5	64.55	53.83	10.72	**	0.18
Leaf length (mm)	6	81.45	76.00	5.45	*	0.07
Height (mm)	7	30.60	29.67	0.93	n.s.	0.02
Height (cm)	10	21.50	21.40	0.10	n.s.	0.01
Height (cm)	12	88.80	87.75	1.05	n.s.	0.01
Leaf length (cm)	14	30.60	28.95	1.65	n.s.	0.06
Leaf width (cm)	14	26.70	26.80	−0.10	n.s.	−0.004
Flowering time (days)	9–15	88.60	88.50	0.10	n.s.	0.001
Height at flowering (cm)	9–15	119.40	119.25	0.15	n.s.	0.001
Final height (cm)	19	169.75	173.80	−4.05	n.s.	−0.024

completely determined by the mother's genotype, the genetic model for the reciprocal F_1s and F_2s would be,

$$\bar{F}_1 = m + [a]_m$$

$$\overline{RF}_1 = m - [a]_m$$

$$\bar{F}_2 = m + [d]_m$$

$$\overline{RF}_2 = m + [d]_m.$$

In addition, the normal genetical models accounting for the progeny's own genotype must be added. As with the genetic models described in Chapter 2, these models can, if necessary, be extended to incorporate interactions, both between the genes exerting maternal influence and between those controlling the maternal and the progeny genotype. Adapting Mather and Jinks [1] terminology, the following symbols can represent these interactions.

Type of interaction	Symbol
$[a]_m \times [a]_m$	$[aa]_m$
$[a]_m \times [d]_m$	$[ad]_m$
$[d]_m \times [d]_m$	$[dd]_m$
$[a] \times [a]_m$	$a.a_m$
$[d] \times [a]_m$	$d.a_m$
$[a] \times [d]_m$	$a.d_m$
$[d] \times [d]_m$	$d.d_m$

Table 13.2 The genetic, maternal and interaction effects model for generation means

Family	Progeny genotype			Maternal genotype			Genotype/maternal interaction			
	m	$[a]$	$[d]$	$[a]_m$	$[d]_m$	$[c]$	$a.a_m$	$a.d_m$	$d.a_m$	$d.d_m$
P_1	1	1	0	1	0	1	1	0	0	0
P_2	1	−1	0	−1	0	−1	1	0	0	0
F_1 (1 × 2)	1	0	1	1	0	1	0	0	1	0
RF_1 (2 × 1)	1	0	1	−1	0	−1	0	0	−1	0
F_2 ($F_1 × F_1$)	1	0	0.5	0	1	1	0	0	0	0.5
F_2 ($F_1 × RF_1$)	1	0	0.5	0	1	1	0	0	0	0.5
RF_2 ($RF_1 × F_1$)	1	0	0.5	0	1	−1	0	0	0	0.5
RF_2 ($RF_1 × RF_1$)	1	0	0.5	0	1	−1	0	0	0	0.5
F_3 (F_2 selfed)	1	0	0.25	0	0.5	1	0	0	0	0.125
RF_3 (RF_2 selfed)	1	0	0.25	0	0.5	−1	0	0	0	0.125
$Bc_{1.1}$ (1 × F_1)	1	0.5	0.5	1	0	1	0.5	0	0.5	0
$Bc_{1.1}$ (1 × RF_1)	1	0.5	0.5	1	0	1	0.5	0	0.5	0
$RBc_{1.1}$ (F_1 × 1)	1	0.5	0.5	0	1	1	0	0.5	0	0.5
$RBc_{1.1}$ (RF_1 × 1)	1	0.5	0.5	0	1	−1	0	0.5	0	0.5
$Bc_{1.2}$ (2 × F_1)	1	−0.5	0.5	−1	0	−1	0.5	0	−0.5	0
$Bc_{1.2}$ (2 × RF_1)	1	−0.5	0.5	−1	0	−1	0.5	0	−0.5	0
$RBc_{1.2}$ (F_1 × 2)	1	−0.5	0.5	0	1	1	0	−0.5	0	0.5
$RBc_{1.2}$ (RF_1 × 2)		−0.5	0.5	0	1	−1	0	−0.5	0	0.5
F_∞	1	0	0	0	0	1	0	0	0	0
RF_∞	1	0	0	0	0	−1	0	0	0	0

It is, of course, possible to include cytoplasmic effects, maternal genotype effects and the progenies' own genetic effects to these models, as well as their interactions (Table 13.2). However, as more parameters are added the model becomes more difficult to test and more prone to systematic errors that the experimenter may have failed to take into account during the replication process and material production.

13.3 Maternal effects and scaling tests

In the presence of maternal effects the standard scaling tests for epistasis will be inappropriate and may well fail even when there is no epistasis. For example, assuming that the parameters m, $[a]$, $[d]$, $[a]_m$ and $[d]_m$ are significant and we apply the **C** scaling test to, say, the means of $P_1, P_2, F_1(1 \times 2)$ and $F_2(F_1$ self$)$ families, the expected value of C will not be zero but $4[d]_m - 2[a]_m$ which can lead to the conclusion that epistasis is present when in fact it is not. Use of $RF_1(2 \times 1)$ and $RF_2(RF_1$ self$)$ instead of $F_1(1 \times 2)$ and $F_2(F_1$ self$)$ families changes the expected value of C to $4[d]_m + 2[a]_m$, suggesting that the two versions of C may provide markedly different interpretations concerning the presence and the type of epistasis, even though there is no epistasis present. The A and B scaling tests, on the other hand, will still apply so long as the generations used in the test descend from a common mother and there are no interactions between the maternal and the progeny genotypes. Another test, $2\bar{F}_2 - \bar{B}c_{1.1} - \bar{B}c_{1.2}$, will also provide an unambiguous test of nonallelic interactions but, again, only when the generations have a common mother (F_1 or RF_1).

We illustrate the application of the maternal effects model by analysing the reciprocal families of the F_1, F_2, F_3 and F_∞ generations derived from the V2 × V12 cross of *Nicotiana rustica*. The trait is the spread of the second pair of real leaves measured at 4 weeks after sowing in the glasshouse, a character known to be highly influenced by the maternal genotype. The average scores of families (each based on a minimum of 30 plants/family), tests of the reciprocal effects and the results of model fitting are given in Table 13.3.

Tests of reciprocal differences in the various generations show that $[a]_m$ is significant but $[c]$ is not. The existence of $[d]_m$, however, cannot be determined from the reciprocal differences but can be inferred from the significance of $[a]_m$. Consequently, two of the following three scaling tests,

$$2\bar{F}_2 - \bar{F}_1 - \bar{F}_\infty,$$

$$2\overline{RF}_2 - \overline{RF}_1 - \bar{F}_\infty \quad \text{and}$$

$$2\bar{F}_2 + 2\overline{RF}_2 - \bar{F}_1 - \overline{RF}_1 - 2\bar{F}_\infty,$$

that are more appropriate in the present case, take significant t values ($9.4 \pm 3.6^{**}$; 2.3 ± 3.7 n.s. and $11.70 \pm 5.1^*$) suggesting the presence of $[dd]$ type of epistasis which is shown to be non-significant by the weighted least squares procedure (Table 13.3).

Maternal effects have two major consequences for the interpretation of generation means. The first concerns the classification of parents into the larger scoring P_1 and the lower scoring P_2. As $[a]$ and $[a]_m$ appear together in the parents and when $[a]_m$ is negative and greater in magnitude than $[a]$, it will make P_1 smaller than P_2 leading to the misclassification of parents which may have far-reaching consequences for the interpretation of components of means and of variances. Secondly, when $[d]$ and $[d]_m$ take opposing signs and

Table 13.3 Mean, variance and weight of various generations, t tests of reciprocal differences and estimates of components of means for the character leaf spread (mm). (i) Data; (ii) components

(i)

Generation	n	Mean	MS	df	s^2_{mean}	t (recips)	Weight
V12	30	32.8	42.2	29	1.407		0.7107
V2	30	29.9	44.3	29	1.477		0.6772
F_1 (12 × 2)	30	32.0	63.2	29	2.108 $\}$ 2.86**		0.4744
RF_1 (2 × 12)	30	38.4	85.2	29	2.840		0.3521
F_2 ($F_1 \times F_1$)	40	37.4	68.8	39	1.720 $\}$ 0.37 n.s.		
RF_2 ($RF_1 \times RF_1$)	40	38.1	71.6	39	1.791		
F_2 pooled	**80**	**37.7**	**70.0**	**79**	**0.875**		**1.1429**
F_3	99	35.3	203.2	9	2.053 $\}$ 0.55 n.s.		
RF_3	100	34.2	207.2	9	2.072		
F_3 pooled(Between)	**199**	**34.7**	**197.7**	**19**	**0.994**		**1.0066**
(Within)				68.4	179		
F_∞	100	33.4	344.9	9	3.797 $\}$ 0.78 n.s.		
RF_∞	100	35.5	273.5	9	3.676		
F_∞ pooled(Between)	**200**	**34.4**	**366.1**	**19**	**1.830**		**0.5464**
(Within)				67.1	180		

(ii)

Parameter	Estimate	Significance
m	32.19 ± 0.68	**
$[a]$	4.60 ± 1.40	**
$[d]$	3.00 ± 1.31	*
$[a]_m$	-3.19 ± 1.11	**
$[d]_m$	3.95 ± 1.11	**
$\chi^2_{(2df)}$	3.72	n.s.

$[d]_m > \frac{1}{2}[d]$, then the F_2 mean will fall outside the F_1, F_3 range and this can be misconstrued as an indication of gametic selection.

However, neither of the above conditions apply to our example and the results show that V12 $(= P_1)$ has proportionately more plus alleles than V2 $(= P_2)$ that are responsible for the increased size of young leaves. But, V12 is an extremely poor maternal parent, reducing the progeny mean by approximately 70% of $[a]$ in the form of a negative maternal effect. Further, the alleles controlling the progeny genotype generally show dominance for increased score while the opposite is true for the maternal effect.

13.4 Generation variances

Maternal effects will make the genetic variances of segregating generations differ. Variances of the parental and F_1 families of course are not affected and in each case the variance provides an estimate of V_E in the absence of genotype $\times$ micro-environmental interactions. Maternal effects, however, make the progeny of different mothers less alike and sibs within the same families more alike. Thus, on a purely maternal model where the mother controls all the variation in her progeny the variances of the various generations are as follows:

Generation	Expectation
P_1, P_2 and F_1	V_E
F_2, $Bc_{1.1}$ and $Bc_{1.2}$	V_E
F_3 (σ_B^2)	$V_{Am} + V_{Dm}$
F_3 (σ_W^2)	V_E
F_∞ (σ_B^2)	$2V_{Am}$
F_∞ (σ_W^2)	V_E

where $V_{Am} = \frac{1}{2}\sum a_m^2$ and $V_{Dm} = \frac{1}{4}\sum d_m^2$.

However, the effects of the maternal and the progeny genotypes often overlap each other and consequently their effects on the variances are expressed simultaneously. In other words, the F_2 and the within F_3 families mean square are rarely just equal to V_E as above, even when maternal control is very strong. Therefore, in general:

$$F_2 \text{ variance} = V_A^* + V_D^* + V_E;$$

$$Bc_{1.1} \text{ variance} = \frac{1}{2}V_A^* + V_D^* - \frac{1}{2}V_{AD} + V_E;$$

$$Bc_{1.2} \text{ variance} = \frac{1}{2}V_A^* + V_D^* + \frac{1}{2}V_{AD} + V_E;$$

$$F_3 \; (\sigma_B^2) = (V_A^* + \frac{1}{4}V_D^*) + (V_{Am} + V_{Dm}); \text{ and}$$

$$F_3 \; (\sigma_W^2) = \frac{1}{2}V_A^* + \frac{1}{2}V_D^* + V_E;$$

$$F_\infty \; (\sigma_B^2) = 2V_A^* + 2V_{Am}; \text{ and}$$

$$F_\infty \; (\sigma_W^2) = V_E.$$

Thus, when maternal effects are not allowed for in the model, they will inflate the 'between families' variance and subsequently lead to an inflated estimate of the genetic variance and a corresponding underestimate of the environmental component V_E. From the F_2 and the within F_3 variances in Table 13.3 we can calculate the approximate value of $2V_A^*$ as:

$$\text{Pooled } V_E \text{ (from parents, } F_1s \text{ and } F_\infty s) = 63.82$$

$$2V_A^* = 1.33\{F_2 \text{ variance} + F_3(\sigma_W^2) - 2V_E\}$$

$$= 1.33\{(70.0 + 68.4) - 2 \times 63.8\} = 14.40$$

and this is smaller than the σ_B^2 of F_∞ $\{(366.1 - 67.1)/10\} = 29.90$ which represents $2V_A^* + 2V_{Am}$. From these estimates, $2V_{Am} = 29.90 - 14.40 = 15.50$ which indicates that both the genetic and the maternal variances are equally important in the V2 × V12 cross.

13.5 Maternal effects in FS and HS designs

In the presence of maternal effects, the total variation of a randomly mating population is no longer $V_A^* + V_D^* + V_E$ but $(V_A^* + V_D^*) + (V_{Am} + V_{Dm}) + V_E$ and all of the maternal variation is confounded with the genotypic differences between the families. Thus, expectations of the various components of each design are modified as follows:

Design	Component	Expectation
All designs	σ_W^2	$= \frac{1}{2}V_A^* + \frac{3}{4}V_D^* + V_E$
BIPs	σ_B^2	$= \{\frac{1}{2}V_A^* + \frac{1}{4}V_D^*\} + \{V_{Am} + V_{Dm}\}$
NCI (female	σ_F^2	$= \frac{1}{4}V_A^* + \{V_{Am} + V_{Dm}\}$
common)	$\sigma_{M/F}^2$	$= \frac{1}{4}V_A^* + \frac{1}{4}V_D^*$
NCI (male	σ_M^2	$= \frac{1}{4}V_A^*$
common	$\sigma_{F/M}^2$	$= \frac{1}{4}V_A^* + \frac{1}{4}V_D^* + \{V_{Am} + V_{Dm}\}$
NCII	σ_M^2	$= \frac{1}{4}V_A^*$
	σ_F^2	$= \frac{1}{4}V_A^* + \{V_{Am} + V_{Dm}\}$
	σ_{MF}^2	$= \frac{1}{4}V_D^*$

Providing that an NCII is based on an equal number of males and females, maternal effects can be tested as a variance ratio of 'Between females MS'/'Between males MS' although such a test is not very powerful because it will be based on few df. In other designs, however, direct tests are not possible, except in the case of Augmented BIPs [1], a design which we have not described in this text. In this design, the parents are crossed reciprocally and also selfed, so reciprocal differences are tested as a part of the design. In the case of the NCI, an indirect test of maternal effects is possible by model fitting, but only when a measure of the parent–offspring covariance is available.

When a full set of reciprocal crosses is produced between inbred lines in a diallel cross, a thorough analysis of the maternal as well as the genotypic effects is possible. For further elaborations of the design and analysis of diallels, the reader should consult Mather and Jinks [1].

13.6 Haploids and polyploids

So far we have concentrated on procedures that can be used to study the nature of metrical variation displayed by diploid organisms but now we shall consider briefly how similar procedures can be employed to investigate material that show either haploid or polyploid inheritance.

13.7 Basic generations with haploids

Allelic effects in haploid organisms can only be of an additive type because heterozygosity cannot exist among the haploid progeny. To illustrate the basics of such analysis, we consider a cross between two isolates of a fungal species such as *Schyzophyllum commune* or *Aspergillus nidulans* and develop the relevant theory. For a single gene difference between the isolates, their genotypes will be A^+ and A^- where the allelic effects are as defined in Chapter 2. A cross between these isolates will result in a transitory F_1 diploid which will enter meiosis and produce haploid segregants. Thus, the so-called F_1 generation consists of haploids which will show segregation analogous to the doubled haploids discussed in Chapter 4, as will the $Bc_{1.1}$ and $Bc_{1.2}$ families. Other generations of the selfing series, on the other hand, are difficult to obtain because it is not possible to self the monokaryons. Pair-wise crosses can be made between the haploid monokaryons which will give rise to progenies that are equivalent to FS families or BIPs. The expected means and variances of these generations on a single gene model are given below:

Generation	Genotype	Genetic value	Total variance
P_1	A^+	$m + a_A$	V_E
P_2	A^-	$m - a_A$	V_E

F_1	$\frac{1}{2}A^+ : \frac{1}{2}A^-$	m	$a_A^2 + V_E$
F_1 BIPs	$\frac{1}{2}A^+ : \frac{1}{2}A^-$	m	$a_A^2 + V_E$
$Bc_{1.1}$	$\frac{3}{4}A^+ : \frac{1}{4}A^-$	$m + \frac{1}{2}a_A$	$\frac{3}{4}a_A^2 + V_E$
$Bc_{1.2}$	$\frac{3}{4}A^- : \frac{1}{4}A^+$	$m - \frac{1}{2}a_A$	$\frac{3}{4}a_A^2 + V_E$

When many genes are segregating in the cross, their effects can be accommodated in the model by replacing a_A and a_A^2 with $[a]$ and $2V_A^*$ respectively. Further, replication of individual isolates of the F_1 BIPs, $Bc_{1.1}$ and $Bc_{1.2}$ generations will allow us to separate not only the genetic and environmental parts but also facilitate the partitioning of the former into its components, between crosses (σ_B^2) and within crosses (σ_W^2). For example, the total genetic variance of the $Bc_{1.1}$ generation, $\frac{3}{4}a_A^2$, will be partitioned into that between the means of the $A^+ \times A^+$ and $A^+ \times A^-$ types of matings which will be present with equal frequency ($= \frac{1}{4}a_A^2$) and the remainder ($= \frac{3}{4}a_A^2 - \frac{1}{4}a_A^2$) which will represent the variation within matings. These components take the following expectations in the absence of epistasis.

Generation	Component	Expectation	
		Single gene	Many genes
F_1 BIPs	σ_B^2	$\frac{1}{2}a_A^2$	V_A^*
	σ_W^2	$\frac{1}{2}a_A^2$	V_A^*
$Bc_{1.1}$ ($= Bc_{1.2}$)	σ_B^2	$\frac{1}{4}a_A^2$	$\frac{1}{2}V_A^*$
	σ_W^2	$\frac{1}{2}a_A^2$	V_A^*

The above expectations show that we require only P_1, P_2 and F_1 families in order to test and estimate all the components of a simple additive genetic/additive environmental model. For example, m and $[a]$ can be estimated as the mid-parental value and the average difference between P_1 and P_2, and epistasis tested as:

$$2\bar{F}_1 - \bar{P}_1 - \bar{P}_2 = 0.$$

The additive genetic variance $2V_A^*$ and the environmental component V_E, on the other hand, can also be estimated from the one-way analysis of variance of a random sample of F_1 monokaryons when they are replicated in the experiment. Here σ_B^2 and σ_W^2 can be directly equated with $2V_A^*$ and V_E respectively and the significance of $2V_A^*$ will be determined by the variance ratio of the 'between monokaryons MS' and 'within monokaryons MS'. Alternatively, V_E can be estimated as an average variance of the parental monokaryons and then $2V_A^* = (F_1 \text{ variance} - V_E)$ can be obtained. Once again, the significance of $2V_A^*$ can be determined from the F ratio between the F_1 variance and the pooled parental variance. Furthermore, a heterogeneity test of the parental variances will confirm the presence or absence of genotype × micro-environmental interaction in the cross.

Data from generations like $Bc_{1.1}$, $Bc_{1.2}$ and F_1 BIPs, when available, will provide additional tests of epistasis and linkage equilibrium, and even allow estimates of their components from both the means and variances. There is only one kind of epistasis which can exist in haploids and that is the additive × additive type, i.e. aa_{AB}, and it will affect means of the P_1, P_2, $Bc_{1.1}$ and $Bc_{1.2}$ generations but not of the F_1 or the F_1 BIPs. The magnitude and direction of this component will depend on the degree of allele dispersion between the parental lines and it will take its maximum value when interactions between the various pairs of genes have the same sign and there is a high degree of gene association in the parents. The expectations of the various generations in the presence of epistasis will be as follows, where $[a] = r_a \sum a_i$ and $[aa] = r_{aa} \sum aa_{ij}$.

Generation	Expectation
$\bar{P}_1$	$m + [a] + [aa]$
$\bar{P}_2$	$m - [a] + [aa]$
$\bar{F}_1$	m
$\bar{F}_1$ BIPs	m
$\bar{B}c_{1.1}$	$m + \frac{1}{2}[a] + \frac{1}{4}[aa]$
$\bar{B}c_{1.2}$	$m - \frac{1}{2}[a] + \frac{1}{4}[aa]$

It is now apparent that the scaling test described earlier not only detects the presence of epistasis but also provides an estimate of $[aa]$. Similar tests can also be devised from other generations and all of them will provide estimates of $[aa]$. For example,

$$\bar{B}c_{1.1} + \bar{B}c_{1.2} - 2\bar{F}_1 = \tfrac{1}{2}[aa],$$

$$4\bar{B}c_{1.1} - 3\bar{P}_1 - \bar{P}_2 = -3[aa] \quad \text{and} \quad 4\bar{B}c_{1.2} - 3\bar{P}_2 - \bar{P}_1 = -3[aa].$$

Alternatively, all the components, m, $[a]$ and $[aa]$, can be estimated by weighted least squares.

The effect of epistasis can also be identified in variances. For example, the genetic component of the F_1 variance will now have the expectation of $2V_A^* + V_{AA}$ where $V_{AA} = \sum aa_{ij}^2$, the additive × additive interaction variance. Similarly, the heritable part of the combined backcross variances ($Bc_{1.1}$ variance + $Bc_{1.2}$ variance) takes the expectation of $3V_A^* + 1\frac{7}{8}V_{AA}$. Thus, one can estimate V_A^*, V_{AA}

and V_E from the P_1, P_2, F_1, $Bc_{1.1}$ and $Bc_{1.2}$ variances as:

$$V_E = \tfrac{1}{2}P_1 \text{ variance} + \tfrac{1}{2}P_2 \text{ variance}$$

$$2V_A^* = 5F_1 \text{ variance} - 2\tfrac{2}{3}(Bc_{1.1} \text{ variance} + Bc_{1.2} \text{ variance}) + \tfrac{1}{3}V_E$$

$$V_{AA} = 2\tfrac{2}{3}(Bc_{1.1} \text{ variance} + Bc_{1.2} \text{ variance}) - 4F_1 \text{ variance} - 1\tfrac{1}{3}V_E.$$

V_A^* and V_{AA} can also be estimated by simultaneous equations from the σ_B^2s of the F_1, $Bc_{1.1}$ and $Bc_{1.2}$ generations when replicated random samples of their monokaryons are assessed in a single experiment. Now $\sigma_{B(F1)}^2 = 2V_A^* + V_{AA}$ and $\sigma_{B(Bc1.1)}^2 + \sigma_{B(Bc1.2)}^2 = V_A^* + \tfrac{3}{8}V_{AA}$. Alternatively, information from the within variances can be utilized to obtain the weighted least squares estimates of V_A^*, V_{AA} and V_E and to test the adequacy of the three-parameter model. The model fitting procedure is already explained in Chapter 3 and the expectations of the various MS in terms of V_A^*, V_{AA} and V_E are given below.

	Source	df	EMS	V_A^*	V_{AA}	V_E
(F_1)	Between isolates	$n_1 - 1$	$\sigma_{W1}^2 + r'\sigma_{B1}^2$	$2r'$	r'	1
	Replicates	$n_1(r'-1)$	σ_{W1}^2	0	0	1
$(Bc_{1.1} + Bc_{1.2})$	Between isolates	$2(n_2-1)$	$2\sigma_{W2}^2 + 2r\sigma_{B2}^2$	$2(1+r)$	$1\tfrac{1}{2}(1+\tfrac{1}{2}r)$	2
	Replicates	$2n_2(r-1)$	$2\sigma_{W2}^2$	2	$1\tfrac{1}{2}$	2

The test of linkage disequilibrium, on the other hand, will be comparatively simple. Linkage will make the additive genetic variance in the F_1 and F_1 BIPs differ irrespective of epistasis. The significance of this difference can be tested by comparing the total variances of these generations or their 'between isolates' MS using variance ratio. A significant difference between the means of these generations will also indicate the presence of linkage but only when the linked genes display epistasis.

13.8 Multiple mating designs with haploids

Multiple mating designs such as BIPs, NCI and NCII are more suitable for studying the quantitative genetics of fungi because incompatibility barriers often restrict crossing. Further, vegetative replication of each monokaryon also generates an additional statistic which allows the various components to be tested and estimated with reasonable efficiency. The σ^2 representing variability between replicates within monokaryons estimates V_E and providing the replicates are completely randomized, all the remaining σ^2s should provide some measure of the genetic variation in the population. These σ^2s have the following expectations for the various designs.

All designs	σ_R^2 (replicates)	$= V_E$
	σ_W^2 (within families)	$= V_A^*\{+\frac{3}{4}V_{AA}\}$
BIPs	σ_B^2 (between families)	$= V_A^*\{+\frac{1}{4}V_{AA}\}$
NCI	σ_C^2 (common parents)	$= \frac{1}{2}V_A^*\{+\frac{1}{16}V_{AA}\}$
	σ_{NC}^2 (non-common parents)	$= \frac{1}{2}V_A^*\{+\frac{3}{16}V_{AA}\}$
NCII	σ_M^2 (male parents)	$= \frac{1}{2}V_A^*\{+\frac{1}{16}V_{AA}\}$
	σ_F^2 (female parents)	$= \frac{1}{2}V_A^*\{+\frac{1}{16}V_{AA}\}$
	σ_{MF}^2 (m × f interaction)	$= 0\{+\frac{1}{8}V_{AA}\}$

All components (V_A^*, V_{AA} and V_E) can be estimated and the adequacy of the model tested both in NCI and NCII. A similar test can also be applied to the BIPs provided that an estimate of the parent–offspring covariance can be obtained; this covariance is $V_A^*\{+\frac{1}{4}V_{AA}\}$. Further, the NCII allows an independent test of epistatic variance V_{AA} in the form of 'male × female' interaction MS. To find out more about the application of these procedures to the study of quantitative variation in fungi, the reader is referred to the works of Jinks and his colleagues [2–4].

13.9 Basic generations with polyploids

Providing a polyploid species behaves like a diploid at meiosis, i.e. it shows normal bivalent pairing and disomic inheritance, as in wheat or tobacco, then it can be treated as a diploid and the normal procedures of analysis apply. Otherwise, the complex mode of inheritance of triploids, tetraploids and species of higher ploidy makes genetic modelling very difficult and whenever models are proposed they are subjected to highly restrictive assumptions which can rarely be tested and shown to hold in practice. Nevertheless, many workers have attempted to develop models for polyploids [5–9].

The quantitative analysis of triploid material, as found in the endosperm of seed, is well developed because the inheritance is systematic as the female parent contributes two identical gametes and the male parent, just one. If only one gene is segregating with two alleles (A^+ and A^-), four genotypes can be recognized in the endosperm and their genetic values on an additive–dominance model are as follows.

Endosperm genotype	Genetic value
$A^+A^+A^+$ $(= P_1)$	$m + a_A$
$A^+A^+A^-$ $\{= F_1(1 \times 2)\}$	$m + \frac{1}{3}a_A + d_{A1}$
$A^+A^-A^-$ $\{= RF_1(2 \times 1)\}$	$m - \frac{1}{3}a_A + d_{A2}$
$A^-A^-A^-$ $\{= P_2\}$	$m - a_A$

Here, m is the mid parental value and a_A is the average difference between the two homozygotes $A^+A^+A^+$ and $A^-A^-A^-$, as defined before. There are, however, two dominance components, d_{A1} and d_{A2}, which represent the unique dominance relationships of having two A^+ alleles and one A^- allele or one A^+ allele to two A^- alleles respectively [10].

Extension to many (k) genes is simple but only when P_1 and P_2 display complete association and $r_a = 1$. In these circumstances, a_A is replaced by $\sum a_i$, d_{A1} by $\sum d_{i1}$ and d_{A2} by $\sum d_{i2}$. However, when P_1 has 'decreasing' alleles fixed at k' out of the k genes and 'increasing' alleles at the remaining $k - k'$ genes, the above extension is inappropriate and the effects of 'gene dispersion' have to be accommodated in all the parameters. Now, the definitions of the various parameters are:

$$m = \text{mid-parent} = \text{mean of } F_\infty \text{ generation}$$

$$[a] = \sum_{i=1}^{k-k'} a_i - \sum_{i=1}^{k'} a_i$$

$$[d]_1 = \sum_{i=1}^{k-k'} d_{i1} + \sum_{i=1}^{k'} d_{i2}$$

$$[d]_2 = \sum_{i=1}^{k-k'} d_{i2} + \sum_{i=1}^{k'} d_{i1}.$$

This shows clearly that we will rarely be able to test and estimate both the d_{A1} and d_{A2} types of dominance, unless we are sure that the $+$ and $-$ alleles are completely associated in the parents. Table 13.4 shows the contributions of these and other parameters to the means of various generations and, using these as the basis for defining the additive–dominance relationships between the family means, we can test for the presence of epistasis using the following

Table 13.4 The additive–dominance model for endosperm characters

Generation	Parameters of means				Parameters of variances							
	m	$[a]$	$[d]_1$	$[d]_2$	V_A^*	V_{D1}	V_{D2}	V_{D12}	V_{AD}	V'_{AD}	V''_{AD}	V_E
P_1	1	1	0	0	0	0	0	0	0	0	0	1
P_2	1	-1	0	0	0	0	0	0	0	0	0	1
F_1 (1×2)	1	$\frac{1}{3}$	1	0	0	0	0	0	0	0	0	1
RF_1 (2×1)	1	$-\frac{1}{3}$	0	1	0	0	0	0	0	0	0	1
F_2 (all)	1	0	$\frac{1}{4}$	$\frac{1}{4}$	$\frac{10}{9}$	$\frac{3}{4}$	$\frac{3}{4}$	$-\frac{1}{2}$	$\frac{1}{3}$	0	0	1
F_3 (all) Between	1	0	$\frac{1}{8}$	$\frac{1}{8}$	$\frac{10}{9}$	$\frac{1}{16}$	$\frac{1}{16}$	$\frac{1}{8}$	0	0	0	0
Within					$\frac{5}{9}$	$\frac{3}{8}$	$\frac{3}{8}$	$-\frac{1}{4}$	$\frac{1}{6}$	0	0	1
F_∞ (all) Between	1	0	0	0	2	0	0	0	0	0	0	0
Within					0	0	0	0	0	0	0	1
$Bc_{1.1}$ $(1 \times F_1)$	1	$\frac{2}{3}$	$\frac{1}{2}$	0	$\frac{2}{9}$	1	0	0	0	$-\frac{2}{3}$	0	1
$Bc_{1.1}$ $(1 \times RF_1)$	1	$\frac{2}{3}$	$\frac{1}{2}$	0	$\frac{2}{9}$	1	0	0	0	$-\frac{2}{3}$	0	1
$RBc_{1.1}$ $(F_1 \times 1)$	1	$\frac{1}{3}$	0	$\frac{1}{2}$	$\frac{8}{9}$	0	1	0	0	0	$-\frac{4}{3}$	1
$RBc_{1.1}$ $(RF_1 \times 1)$	1	$\frac{1}{3}$	0	$\frac{1}{2}$	$\frac{8}{9}$	0	1	0	0	0	$-\frac{4}{3}$	1
$Bc_{1.2}$ $(2 \times F_1)$	1	$-\frac{2}{3}$	0	$\frac{1}{2}$	$\frac{2}{9}$	0	1	0	0	0	$\frac{2}{3}$	1
$Bc_{1.2}$ $(2 \times RF_1)$	1	$-\frac{2}{3}$	0	$\frac{1}{2}$	$\frac{2}{9}$	0	1	0	0	0	$\frac{2}{3}$	1
$RBc_{1.2}$ $(F_1 \times 2)$	1	$-\frac{1}{3}$	$\frac{1}{2}$	0	$\frac{8}{9}$	1	0	0	0	$\frac{4}{3}$	0	1
$RBc_{1.2}$ $(RF_1 \times 2)$	1	$-\frac{1}{3}$	$\frac{1}{2}$	0	$\frac{8}{9}$	1	0	0	0	$\frac{4}{3}$	0	1

scaling tests.

$$2\bar{F}_2 + 2\overline{RF}_2 - \bar{F}_1 - \overline{RF}_1 - \bar{P}_1 - \bar{P}_2 = 0$$

$$\bar{Bc}_{1.1} + \bar{Bc}_{1.2} - \bar{F}_2 - \overline{RF}_2 = 0$$

$$\overline{RBc}_{1.1} + \overline{RBc}_{1.2} - \bar{F}_2 - \overline{RF}_2 = 0$$

We can illustrate the application of this model by the analysis of amylose content in rice which is known to be controlled by several genes of unequal effect. For simplicity, we consider only the parental, F_1 and F_2 generations of the cross IR8 × IR24, although data are also available from other generations. Table 13.5 shows generation means, variances and weights together with the results of the scaling test and model fitting. The scaling test indicates that the additive–dominance model should be adequate and model fitting confirms this, i.e. χ^2 is n.s. Further, the direction of dominance is predominantly for high amylose content and a single dose of the dominant allele is often sufficient for the fuller expression of dominance as well as to compensate for the reduction caused by an extra allele of decreasing effect.

Table 13.5 Data and estimates of components for amylose content in the IR8 × IR24 cross of rice

		Data		Model fitting	
Generation	n	Mean	Variance of mean	Parameter	Estimate
P_1	20	27.26	0.0405	m	21.25 ± 0.14**
P_2	20	15.11	0.0410	$[a]$	6.08 ± 0.15**
$F_1 \ (1 \times 2)$	20	25.24	0.1110	$[d]_1$	2.13 ± 0.36**
$RF_1 \ (2 \times 1)$	20	27.96	0.0450	$[d]_2$	8.80 ± 0.26**
F_2	396	24.20	0.0701	$\chi^2_{(1)}$	1.89 n.s.

Scaling test $4\bar{F}_2 - \bar{F}_1 - \overline{RF}_1 - \bar{P}_1 - \bar{P}_2 = 1.23 \pm 1.17$ n.s.

The impact of trisomic segregation on the components of variances is even more profound. Now separate components are needed for the variances of the selfing and backcrossing series. Altogether, the following seven components are needed and their contributions to the various variances are given in Table 13.4.

$$2V_A^* \text{ (additive genetic variance)} = \sum_{i=1}^{k} a_i^2$$

$$4V_{D1} \text{ (dominance variance)} = \sum_{i=1}^{k-k'} d_{i1}^2 + \sum_{i=1}^{k'} d_{i2}^2$$

$$4V_{D2} \text{ (dominance variance)} = \sum_{i=1}^{k-k'} d_{i2}^2 + \sum_{i=1}^{k'} d_{i1}^2$$

$$4V_{D12} \text{ (dominance covariance)} = \sum_{i=1}^{k} d_{i1} d_{i2}$$

$$2V_{AD} \text{ (additive/dominance covariance)} = \sum_{i=1}^{k} a_i(d_{i1} + d_{i2})$$

$$2V'_{AD} \text{ (additive/dominance covariance)} = \sum_{i=1}^{k-k'} a_i d_{i1} - \sum_{i=1}^{k'} a_i d_{i2}$$

$$2V''_{AD} \text{ (additive/dominance covariance)} = \sum_{i=1}^{k-k'} a_i d_{i2} - \sum_{i=1}^{k'} a_i d_{i1}.$$

Further details of how this model can be applied and the components measured and interpreted are beyond the scope of this text and can be found elsewhere [10,11–15].

13.10 Multiple mating designs with polyploids

As for multiple mating designs, pair-wise matings of the four geno-types and the assessment and analysis of the FS families provide the following expectations:

$$\sigma_B^2 = \tfrac{5}{18}a_A^2 + \tfrac{5}{64}(d_{A1}^2 + d_{A2}^2) + \tfrac{1}{12}(a_A d_{A1} - a_A d_{A2}) - \tfrac{3}{32}d_{A1}d_{A2}$$

$$\sigma_W^2 = \tfrac{5}{18}a_A^2 + \tfrac{7}{64}(d_{A1}^2 + d_{A2}^2) + \tfrac{1}{12}(a_A d_{A1} - a_A d_{A2}) - \tfrac{1}{32}d_{A1}d_{A2}.$$

When extended to many genes, these expectations become:

$$\sigma_B^2 = \tfrac{5}{9}V_A^* + \tfrac{5}{16}(V_{D1} + V_{D2}) + \tfrac{1}{6}(V_{AD}' - V_{AD}'') - \tfrac{3}{8}V_{D12}$$

$$\sigma_W^2 = \tfrac{5}{9}V_A^* + \tfrac{7}{16}(V_{D1} + V_{D2}) + \tfrac{1}{6}(V_{AD}' - V_{AD}'') - \tfrac{1}{8}V_{D12}.$$

Similarly, the expected components of NCI and NCII are as given in Table 13.6.

Apparently none of the designs provide a direct test and estimate of the additive genetic variance V_A^* or of the dominance variances V_{D1} or V_{D2}. These parameters therefore can only be tested and estimated by weighted least squares and even there several of the parameters will appear confounded. An estimate of V_A^*, however, can be obtained from the diallel and the Triple Test-Cross analyses and their application to experimental data are discussed elsewhere [14,15].

Table 13.6 Expectations of σ^2s for NCI and NCII for triploid endosperm traits

(i) NCII

$$\sigma_M^2 = \tfrac{1}{9}V_A^* + \tfrac{1}{8}(V_{D1} + V_{D2}) + \tfrac{1}{3}(-V_D' + V_{AD}'') - \tfrac{1}{4}V_{D12}$$

$$\sigma_F^2 = \tfrac{4}{9}V_A^* + \tfrac{1}{8}(V_{D1} + V_{D2}) + \tfrac{1}{3}(V_{AD}' - V_{AD}'') - \tfrac{1}{4}V_{D12}$$

$$\sigma_{MF}^2 = \tfrac{1}{16}(V_{D1} + V_{D2}) + \tfrac{1}{6}(V_{AD}' - V_{AD}'') + \tfrac{1}{8}V_{D12}$$

$$\sigma_W^2 = \sigma_W^2 \text{ of FS families (BIPs)}$$

(ii) NCI when males are the common parent

$$\sigma_C^2 = \sigma_M^2 \text{ of NCII}$$

$$\sigma_{NC}^2 = \sigma_F^2 + \sigma_{MF}^2 \text{ of NCII}$$

$$\sigma_W^2 = \sigma_W^2 \text{ of FS families}$$

(iii) NCI when females are the common parent

$$\sigma_C^2 = \sigma_F^2 \text{ of NCII}$$

$$\sigma_{NC}^2 = \sigma_M^2 + \sigma_{MF}^2 \text{ of NCII}$$

$$\sigma_W^2 = \sigma_W^2 \text{ of FS families}$$

Analysis of other polyploids, particularly of auto-tetraploids, is often tackled at the diploid level. Using the familiar techniques of anther/ovule culture, reduced gametes of crops like potato can be induced to grow as the so-called 'dihaploid' plants which can then be used in 2x.2x and 4x.2x matings [6–9].

Summary

1. Maternal effects are very common and can often be confused with complex gene interaction when not all the reciprocal generations are raised in the experiment.
2. They are easy to detect and model for both means and variances. The most commonly observed maternal effects are those that are attributable to the mother's own genotype while cytoplasmic effects are detected very infrequently.
3. Effects of the mother's own genotype on the progeny are usually more pronounced at the juvenile stages of growth and they generally wane as the progeny come to maturity and their own genotype takes full control.
4. The quantitative genetics of haploids is much simpler than that of diploids or polyploids. Haploid species allow tests of all types of genetic variability, including that due to non-allelic interactions and linkage.
5. Analyses of polyploids, on the other hand, are extremely complex and the models are frequently subject to untestable assumptions. Consequently, not many QTL studies have been conducted with polyploid species.

References

1. Mather, K. and Jinks, J.L. (1982) *Biometrical Genetics*, 3rd edn, Chapman & Hall, London.
2. Caten, C.E. (1979) Quantitative genetic variation in fungi, in *Quantitative Genetic Variation* (eds J.N. Thomson, Jr and J.M. Thoday), Academic Press, New York, pp. 35–60.
3. Connolly, V. and Jinks, J.L. (1975) The genetical architecture of general and specific environmental sensitivity. *Heredity*, **35**, 249–59.
4. Simchen, G. and Jinks, J.L. (1964) The determination of growth rate in the Basidiomycete, *Schizophyllum commune*. *Heredity*, **19**, 629–49.

5. Fisher, R.A. and Mather, K. (1936) The inheritance of style length in *Lythrum salicaria. Ann. Eugenics*, **12**, 1–23.

6. Kempthorne, O. (1957) *An Introduction to Genetic Statistics*. Wiley, New York.

7. Killick, R. (1971) The biometrical genetics of autotetraploids. I. Generations derived from a cross between two pure lines. *Heredity*, **27**, 331–46.

8. Tai, G.C.C. (1982) Estimation of double reduction and genetic parameters in autotetraploids based on 4x.2x and 2x.2x matings. *Heredity*, **49**, 331–5.

9. Tai, G.C.C. (1986) Biometrical genetical analysis of tetrasomic inheritance based on matings of diploid parents which produce 2x gametes. *Heredity*, **57**, 315–17.

10. Gale, M.D. (1976) High alpha amylase breeding and genetical aspects of the problem. *Cereal Res. Commun.*, **4**, 231–43.

11. Bogyo, T.P., Lance, R.C.M., Chevalier, P. and Nilan, R.A. (1988) Genetic models for quantitatively inherited endosperm characters. *Heredity*, **60**, 61–7.

12. Huidong, M.O. (1987) *Genetic Expression of Endosperm Traits*. Proceedings of the Second International Conference on Quantitative Genetics. Sinaur Associates, Massachusetts, pp. 478–87.

13. Pooni, H.S., Kumar, I. and Khush, G.S. (1992) A comprehensive model for disomically inherited metrical traits expressed in triploid tissues. *Heredity*, **69**, 166–74.

14. Pooni, H.S., Kumar, I. and Khush, G.S. (1993) Genetical control of amylose content in selected crosses of *indica* rice. *Heredity*, **70**, 269–80.

15. Pooni, H.S., Kumar, I. and Khush, G.S. (1993) Genetical control of amylose content in a diallel set of rice crosses. *Heredity*, **71**, 603–13.

Correlated and threshold characters

14

14.1 Correlations between characters

Until now we have considered the analysis of individual characters assuming that they were independently inherited, i.e. there are no correlations between them. However, this is rarely true and barring a few exceptions, most traits are found to show a range of mild to strong associations with others. For example, it is a matter of common observation that increased egg production in poultry frequently results in smaller eggs. Similarly, while increased seed size in sunflower might lead to higher seed yield, it is known to reduce the oil content of the kernel. There is clearly a lack of independence in the control of these characters. Some correlations are, however, conceptually trivial in that the traits concerned are structurally related such as height and weight in most organisms.

When breeders attempt to improve an animal or plant they are generally interested in upgrading several attributes of the phenotype simultaneously. Thus, they may wish not only to increase the egg production of hens or seed yield of sunflower but also the size of the eggs or the oil content of the seed. The extent to which these characters are correlated will, therefore, influence the breeder's success. From these illustrations we can see that a knowledge of the extent, direction and cause of correlation between characters is important and we devote part of this chapter to their analysis.

Suppose we grow a random sample of n plants from a population of unknown constitution and score each plant for two characters X and Y (e.g. height and flowering time). We can then calculate the correlation between the two traits as shown in Table 14.1. This would provide an estimate of the phenotypic correlation, r_P, because its underlying cause is not known. Further, r_{XY} could take any value between $+1$ and -1 and its significance could be tested as

Table 14.1 Correlation between the height at flowering (X) and flowering time (Y) of 20 *Nicotiana rustica* plants. (i) Data; (ii) statistics; (iii) calculations

(i)

Plant no.	X	Y	Plant no.	X	Y
1	118	84	11	84	81
2	122	90	12	112	87
3	108	86	13	78	83
4	115	87	14	102	86
5	131	90	15	122	83
6	125	86	16	92	83
7	114	85	17	99	81
8	111	84	18	98	80
9	111	87	19	113	86
10	120	84	20	109	86

(ii)

$$r_{XY} = SP_{XY}/\sqrt{(SS_X \times SS_Y)}$$

OR

$$r_{XY} = s_{(XY)}/\sqrt{(s_X^2 \times s_Y^2)}$$

where $s_{(XY)} = SP_{XY}/(n-1)$

$$s_X^2 = SS_X/(n-1)$$

$$s_Y^2 = SS_Y/(n-1)$$

(iii)

$SS(X) = 3915.25 \qquad SS(Y) = 138.95 \qquad SP(XY) = 452.25 \qquad n = 20$

$s_{(X)}^2 = 185.22 \qquad s_{(Y)}^2 = 7.31 \qquad s_{(XY)} = 23.80$

$r_P = 23.80/\sqrt{(185.22 \times 7.31)} = 0.65$

$sd(r_P) = \sqrt{\{(1 - 0.65^2)/(20 - 2)\}} = 0.18$

$t_{(18)} = r_P/sd = 0.65/0.18 = 3.61^{***}$

$t_{(n-2)} = r/\sqrt{\{(1 - r^2)/(n - 2)\}}$ [1]. A positive significant correlation, as in our example in Table 14.1, merely indicates that either genes or micro-environments (or both) that increase X also increase Y, but it does not reveal the extent to which the causation is genetic or environmental. Likewise, a zero r_P could arise from genetical and environmental effects on the two characters cancelling out, so again no real insight can be gained.

Both environmental and genetic correlations are important to a breeder because they collectively determine how effective the selection programme would be. While the breeder can do little about the environmental correlations, except note their direction and try to minimize them or exploit them as the situation may demand, the genetic correlations would have to be manipulated to improve the efficiency of selection. It is therefore important to examine the underlying causes of the phenotypic correlation and identify and separate its causal components.

14.2 Environmental correlations

Let us suppose that we now repeat the experiment but use replicates of the same genotype as is possible with an inbred line or F_1 hybrid. There can be no genetic variation among the replicates and thus there can be no genetic covariation either. So, the correlation between X and Y in such a situation would be entirely environmental in origin, i.e.

$$r_E = s_{E(XY)}/\sqrt{(s^2_{E(X)} \times s^2_{E(Y)})},$$

where $s_{E(XY)}$ stands for the environmental covariance of X and Y, and $s^2_{E(X)}$ and $s^2_{E(Y)}$ are the corresponding environmental variances, as defined in Chapter 3.

If we have several non-segregating families, we can calculate the environmental correlation for each family and, if they are homogeneous, we can combine the separate values of r into a single correlation [2]. The procedure that we will follow to achieve this involves transforming the correlation coefficients, r into z, because the latter is normally distributed whereas the r is not, particularly when the expected value of correlation is not zero. For every correlation $z = \frac{1}{2}\{\log_e(1+r) - \log_e(1-r)\}$ and $s^2_{(z)} = 1/(n-3)$ where n is the number of pairs of X and Y values [1,2]. We can better demonstrate the procedure using the example in Table 14.1 and two other correlations. The correlation in Table 14.1 was derived from the data obtained from a non-segregating, homozygous, *Nicotiana rustica* variety V12 and therefore is of purely environmental origin. Another variety, V2, and the F_1 hybrid (V2 × V12), were also raised in the same experiment with V12 and their correlations took the values of 0.39 and 0.64 respectively (Table 14.2). The test of homogeneity of these correlations is obtained as $F_{(2,51)}$ = variance of z_i/error variance of $z_i = 0.81$, which is not significant. The pooled

Table 14.2 Test of homogeneity and pooling of correlations

Genotype	r_i	$n_i - 3$	z_i	$z_i \times (n_i - 3)$	s_{zi}^2
V2	0.39	17	0.3884	6.6028	0.0588
V12	0.65	17	0.7753	13.1801	0.0588
$F_1(V2 \times V12)$	0.64	17	0.7582	12.8890	0.0588
Total		51		32.6719	0.1764

Variance $(z_i) = 0.0478$; Pooled $s_{zi}^2 = 0.0588$

Homogeneity $F_{(2,51)}$ = Variance (z_i)/Pooled s_{zi}^2

$$= 0.0478/0.0588$$

$$= 0.81 \text{ n.s.}$$

$$z = \frac{32.6719}{51} = 0.6406; \quad r_P = \frac{e^{(2 \times 0.6406)} - 1}{e^{(2 \times 0.6406)} + 1} = 0.5673$$

Pooled correlation $= 0.57$

estimate of r is given by the formula $r = (e^{2z'} - 1)/(e^{2z'} + 1) = 0.57$, where z' is the pooled value of z_i, as given in Table 14.2.

Homogeneity of the correlations in this example indicates, statistically, the additive relationship between plant height and flowering time in both the parents and their F_1. On the other hand, significant differences between them would have revealed differential responses of the various genotypes, so indicating the presence of G × E. The overall positive and significant correlation implies that any environmental effects that delayed flowering had also increased the height at flowering in this material, a finding which is common in plants.

14.3 Genetic correlations

Returning to a situation in which we have a number of pure breeding lines, we could design the following experiment. A number (n) of F_∞ inbred lines derived from a cross between two pure breeding lines are taken and $2n'$ individuals from every line raised in an experiment with complete, individual randomization. If half the $2n'$ individuals of every line are scored at random for X and the remaining half for Y, we can obtain the mean score of every line for X and for Y and proceed to calculate the covariance $s_{(XYmeans)}$ between the line means. Since for every line, X and Y are scored on different individuals, there can be no environmental reason for a relationship between X and Y. Hence the covariance obtained in this manner must be due to genetic causes alone. However, the variances of the line means for the characters X and Y will contain some environmental variation, albeit a small proportion because:

$$s_X^2 = (\sigma_{W(X)}^2)/n' + \sigma_{B(X)}^2$$

and

$$s_Y^2 = (\sigma_{W(Y)}^2)/n' + \sigma_{B(Y)}^2$$

where $\sigma_W^2 = V_E$ and $\sigma_B^2 = 2V_A^*$ for any given trait. Thus, the correlation based on these statistics, calculated as:

$$r_{G'} = s_{(XYmeans)}/\sqrt{(s_X^2 \times s_Y^2)}$$

will always be underestimated unless the family size is very large and the environmental variance is low, i.e. heritability is high.

The above expectations suggest that the genetic correlation can also be estimated from the components of variances as:

$$r_G = \frac{s_{(XYmeans)}}{\sigma_{B(X)} \times \sigma_{B(Y)}}$$

Clearly, this estimate of r_G will be unbiased and should, at least in theory, cover the whole range of values, from +1 to −1. However, as the σ_B^2 are obtained by subtraction they will be negatively correlated with the σ_W^2; consequently r_G will sometimes take values that fall outside the limits of +1 to −1. In other words, this estimate of r_G will be subjected to a larger standard error than that based on the variance and covariance of family means [3,4].

While taking independent measurements of X and Y has been useful in demonstrating that the covariance between the family means was purely genetic, it is very wasteful, both in terms of data and experimental resources. In fact, it would be more informative to score all individuals for X and Y because that alone would double the sample size for each family mean. We could then carry out ANOVA of X and Y separately and use the same data to perform an analysis of covariance also. In the analysis of covariance, the sums of products for the between- and within-families items are computed in an identical fashion to those of an ANOVA except that $\sum XY, \sum X$ and $\sum Y$ terms will be used in the calculations rather than $\sum X^2$ and $\sum X$ or $\sum Y^2$ and $\sum Y$. The ems of these analyses, given in Table 14.3, provide all the information needed to compute both the genetic and environmental correlations. Table 14.4 provides a numerical example of these analyses for the height and flowering time data collected on 30 recombinant inbred lines extracted from

Table 14.3 Generalized expectations of ANOVA and analysis of covariance for two characters, X and Y

Source	df	MS	ANOVA (X)	ANOVA (Y)	Analysis of covariance (XY)
Between lines	$n-1$	MS_B	$\sigma^2_{W(X)} + 2n'\sigma^2_{B(X)}$	$\sigma^2_{W(Y)} + 2n'\sigma^2_{B(Y)}$	$\sigma_{W(XY)} + 2n'\sigma_{B(XY)}$
Within lines	$n(2n'-1)$	MS_W	$\sigma^2_{W(X)}$	$\sigma^2_{W(Y)}$	$\sigma_{W(XY)}$

the V2 × V12 cross by the method of single seed descent. The data were collected on ten plants of each line that were raised in a trial with individual plant randomization. In the same way as the genetic and environmental variances of trait X are estimated from MS_B and MS_W of the one way ANOVA, we can estimate the corresponding genetic and environmental covariances from the analysis of covariance and calculate the genetic and environmental correlations as described in Table 14.4(ii).

The analysis therefore indicates that there is a positive environmental and a positive genetic correlation between height and flowering time. The genetic correlation further indicates that not all variation in

Table 14.4 ANOVA and analysis of covariance of height at flowering (X) and flowering time (Y) of 30 inbred lines of *Nicotiana rustica*. (i) Analyses; (ii) formulae and calculations

(i)

Source	df	ANOVA (X) MS	ANOVA (Y) MS	Analysis of covariance (XY) MSP
Lines	29	3970.09**	692.73**	809.88**
Plants/lines	270	177.83	23.21	34.95

(ii)

$\sigma^2_{W(X)} = 177.83 \qquad \sigma^2_{W(Y)} = 23.21 \qquad \sigma_{W(XY)} = 34.95$

$\sigma^2_{B(X)} = 379.23 \qquad \sigma^2_{B(Y)} = 66.95 \qquad \sigma_{B(XY)} = 77.49$

$r_G = \sigma_{B(XY)}/(\sigma_{B(X)} \times \sigma_{B(Y)}) = 77.49/\sqrt{(379.23 \times 66.95)} = 0.49$

$r_E = \sigma_{W(XY)}/(\sigma_{W(X)} \times \sigma_{W(Y)}) = 34.95/\sqrt{(177.83 \times 23.21)} = 0.54$

$sd(r_G) = \sqrt{\{(1-r^2)/df\}} = \sqrt{\{(1-0.49^2)/28\}} = 0.16$

$sd(r_E) = \sqrt{\{(1-r^2)/df\}} = \sqrt{\{(1-0.54^2)/270\}} = 0.05$

$t_{(28)} = r_G/sd = 0.49/0.16 = 3.06^{**}$

$t_{(270)} = r_E/sd = 0.54/0.05 = 10.80^{**}$

height is accounted for by the variation in flowering time and that selection for either character will show some response which will be independent of the other character. Only a proportion $r^2 (= 0.24)$ of the genetic variation in each character is completely correlated and the remainder is uncorrelated and therefore free to respond to selection.

14.4 Genetic covariation and design of experiment

Although we have illustrated the use of ANOVA and analysis of covariance in estimating genetic and environmental correlations of inbred lines, such analyses can be applied to any set of families, only the interpretation of the correlations would change. For example, had we raised FS families instead of inbred lines, the variance components would provide estimates of the following:

$$\sigma^2_{W(X)} = \tfrac{1}{2} V_{A(X)} + \tfrac{3}{4} V_{D(X)} + V_{E(X)}$$

$$\sigma^2_{W(Y)} = \tfrac{1}{2} V_{A(Y)} + \tfrac{3}{4} V_{D(Y)} + V_{E(Y)}$$

$$\sigma^2_{B(X)} = \tfrac{1}{2} V_{A(X)} + \tfrac{1}{4} V_{D(X)}$$

$$\sigma^2_{B(Y)} = \tfrac{1}{2} V_{A(Y)} + \tfrac{1}{4} V_{D(Y)}.$$

The corresponding covariance components would be:

$$\sigma_{W(XY)} = \tfrac{1}{2} W_{A(XY)} + W_{D(XY)} + W_{E(XY)}$$

$$\sigma_{B(XY)} = \tfrac{1}{2} W_{A(XY)} + \tfrac{1}{4} W_{D(XY)}.$$

It is clear that what we previously called the environmental correlation now contains genetic effects. The genetic correlation, r_G, however, maintains its integrity but it is altered in detail, i.e. it now involves both the additive and dominance effects of genes segregating in the families.

The principles of covariance partitioning also apply to the basic generations because the nature and extent of the covariation will differ between the segregating and non-segregating generations. In this case, the genetic covariance of the F_2, for example, will be estimated simply as the phenotypic covariance of the F_2 minus the environmental covariance obtained from the non segregating generations. The genetic correlation is then calculated as:

$$\frac{\text{Genetic covariance(XY)}}{\sqrt{\{\text{Genetic variance(X)} \times \text{Genetic variance(Y)}\}}}.$$

Data of other designs such as NCI, NCII, TTC and diallel, etc., are also amenable to covariance analysis and such designs would provide further partitioning of the genetic covariance into additive, dominance and epistatic components. However, the reliability of such analyses is much reduced when the sample sizes are small.

14.5 Causes of covariation

Genetic covariances between traits can occur for two reasons, i.e.

 (i) linkage/linkage disequilibrium; and
(ii) pleiotropy.

The effects of linkage on the variances of individual traits have already been discussed in Chapter 10. Table 10.1(ii) gives the expected values of the nine genotypes when two genes (A and B) are linked. We will now extend that model to two traits by assuming that gene A controls character X and gene B, character Y. Table 14.5 gives the genetic

Table 14.5 Genetic covariance among F_2 individuals

Genotype	Frequency	Genetic value for character X	Genetic value for character Y
$A^+A^+B^+B^+$	$\frac{1}{4}p^2$	$m + a_A$	$m' + a_B$
$A^+A^+B^+B^-$	$\frac{1}{2}pq$	$m + a_A$	$m' + d_B$
$A^+A^+B^-B^-$	$\frac{1}{4}q^2$	$m + a_A$	$m' - a_B$
$A^+A^-B^+B^+$	$\frac{1}{2}pq$	$m + d_A$	$m' + a_B$
$A^+A^-B^+B^-$	$\frac{1}{2}p^2 + \frac{1}{2}q^2$	$m + d_A$	$m' + d_B$
$A^+A^-B^-B^-$	$\frac{1}{2}pq$	$m + d_A$	$m' - a_B$
$A^-A^-B^+B^+$	$\frac{1}{4}q^2$	$m - a_A$	$m' + a_B$
$A^-A^-B^+B^-$	$\frac{1}{2}pq$	$m - a_A$	$m' + d_B$
$A^-A^-B^-B^-$	$\frac{1}{4}p^2$	$m - a_A$	$m' - a_B$
Mean	1	$m + \frac{1}{2}d_A$	$m' + \frac{1}{2}d_B$

Covariance $= \frac{1}{4}p^2(a_A \times a_B) + \frac{1}{2}pq(a_A \times d_B) + \ldots + \frac{1}{4}p^2(-a_A)(-a_B)$

$$- (\tfrac{1}{2}d_A)(\tfrac{1}{2}d_B)$$

$$= \tfrac{1}{2}(1 - 2R)a_A a_B + \tfrac{1}{4}(1 - 2R)^2 d_A d_B$$

where R = recombination frequency between A and B, and

$$1 - 2R = p - q$$

values of the genotypes of an F_2 generation for X and Y, from which their covariance takes the value,

$$\sigma_{G(XY)} = \tfrac{1}{2}\delta(1 - 2R)a_A a_B + \tfrac{1}{4}(1 - 2R)^2 d_A d_B$$

where $\delta = +1$ if $+$ alleles are linked in coupling and -1 otherwise.

When generalized by summing all the pair-wise covariances between the k_1 and k_2 genes that may be controlling the two characters:

$$\sigma_{G(XY)} = \sum_{i}^{k_1} \sum_{j}^{k_2} [\tfrac{1}{2}\delta(1 - 2R)a_i a_j + \tfrac{1}{4}(1 - 2R)^2 d_i d_j]$$

which like the F_2 genetic variance can be written as $\sigma_{A(XY)} + \sigma_{D(XY)}$ where:

$$\sigma_{A(XY)} = \sum_{i}^{k_1} \sum_{j}^{k_2} [\tfrac{1}{2}\delta(1 - 2R)a_i a_j]$$

and

$$\sigma_{D(XY)} = \sum_{i}^{k_1} \sum_{j}^{k_2} [\tfrac{1}{4}(1 - 2R)^2 d_i d_j].$$

These expectations reveal that the genetic covariance of F_2 would be positive and large when:

(i) there are coupling linkages;
(ii) linkages are tight; and
(iii) dominance at the linked genes is in the same direction.

Alternative arrangements, i.e. repulsion linkages between the k_1 and k_2 groups of genes and ambi-directional dominance, on the other hand, will give a negative covariance which will ultimately indicate that the characters are negatively correlated.

Pleiotropy, on the other hand, is a condition where the same gene is involved in the control of more than one character. Falconer (1989) argues that pleiotropy is the main cause of genetic covariance. However, this situation can also be obtained when genes A and B are tightly linked and there is no measurable recombination between them, i.e. $R = 0$. In other words, segregation of alleles at gene A also determines the genotype at gene B and there are in F_2 only three genotypes $A^+A^+(B^+B^+)$, $A^+A^-(B^+B^-)$ and $A^-A^-(B^-B^-)$, not nine as described previously. The covariance now takes the expectation $\tfrac{1}{2}a_A a_B + \tfrac{1}{4}d_A d_B$ which will give a correlation of unity between the two traits unless some proportion of their genetic variation is controlled by separate, independently segregating genes. Like linkage, the effect of pleiotropy will also be seen in the form of covariance only when there is a net imbalance between genes showing negative and positive pleiotropy. This means that pleiotropy for a set of genes could result in almost any genetic correlation. Some genes may affect the two traits in the same direction, while others may have opposite effects; some genes may have large effects on X but small effects on Y, while with other genes the effects may be reversed. Therefore, complete pleiotropy does not automatically imply that correlations between traits cannot be broken.

14.6 General conclusions about correlations

Correlations between traits have been extensively studied and have led to some general conclusions, of which we list some of the most important.

1. Correlations between characters are ubiquitous, particularly between fitness characters and their component traits.
2. The majority of commercially important correlations, e.g. those between the number and the size of eggs or between seed number, size and oil content, are generally undesirable and therefore impede selection.
3. Genetic correlations are rarely absolute and there is normally sufficient independent variation between traits to permit joint improvement in both.
4. Much of the covariation in small populations is due to correlated gene frequencies and loose linkages which are transient. They can normally be removed by random mating and keeping the population size large. Tight linkages and certain types of pleiotropy, on the other hand, are difficult to manipulate.

14.7 Threshold traits

We have assumed that quantitative traits are generally controlled by several, possibly many genes each with relatively small individual effect. Thus a population of individuals will include those with very few + alleles through to those that have + alleles at most genes and the relative frequencies of these individuals result in an approximately normal distribution for the trait. This is illustrated in Figure 14.1(i) for height of plants in an F_2 population scored in cm.

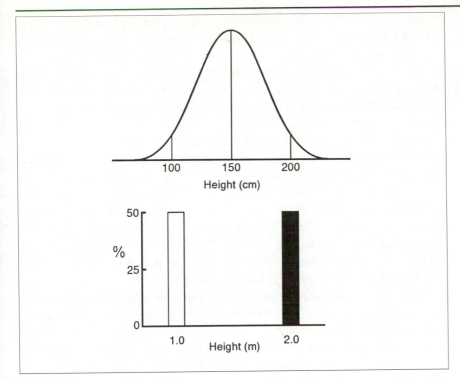

Figure 14.1 Distribution of height in a population of plants measured either to the nearest cm (i) or to the nearest m (ii) in order to show how a continuous, normally distributed variable would appear discontinuous on a different scale, so creating a threshold trait.

If we were to measure this population with a ruler that only measured to the nearest metre, then our previously normal distribution would now appear as just two classes; those up to 1.5 m would be in one class, i.e. recorded as 1 m high, and those above 1.5 m which would be recorded as 2 m high (Figure 14.1(ii)). A continuous distribution would have been converted into a discontinuous distribution simply by the method of scoring the trait. The true height of 1.5 m would be a threshold. Any plant with that height or less being ascribed to one height class and the rest to another.

The reader is probably asking why one would want to measure the trait in such an approximate fashion as to create this two-class effect; of course, no experimenter would deliberately do this in practice. However, in many cases one does not know the correct units in which to measure the trait or indeed one is only presented with the equivalent of a very coarse ruler. Take a disease such as schizophrenia in humans as an example. It is believed that this unpleasant condition is due to an underlying polygenic system in which the genes control the production of some critical product; the more + alleles an individual has, the more product is produced. Up to a certain product level or threshold, the personality is apparently normal, while above this level the clinical symptoms of schizophrenia appear. Another example would be whether or not an individual plant is fertile; it either is or it is not, although the underlying propensity to produce fertile shoots may be polygenic. In both cases, our measure of phenotype in terms of the presence of schizophrenic behaviour or a fertile shoot is equivalent to using a coarse ruler. Had we measured something closer to the level of the actual gene product, we would have observed a continuous, approximately normal distribution.

Clearly, with our original example of height measured with a coarse ruler, there could be more than two classes. Any plants that happened to be taller than 2.5 m would be recorded to have 3 m height. Thus there could be more than one threshold. For example, the number of offspring in a single gestation in cattle might be one, two or occasionally three. There are thus three classes which probably correspond to some underlying hormone level with three thresholds, below a certain threshold the cow will be infertile. Between this threshold and the second she has one offspring, etc.

There are many cases in plants and animals where one is dealing with threshold traits of this sort but it can be difficult to distinguish such traits in which there are just two or three phenotypes from traits controlled by a single gene. The latter should follow the normal rules for a single gene under Mendelian inheritance while a threshold trait with underlying polygenic control would be difficult to reconcile with such a model.

14.8 Handling threshold traits

We will assume that the trait we are measuring results from polygenic variation at a number of genes, the product of which constitutes the underlying, cryptic trait, while we measure the visible, threshold trait. The underlying trait is often called the liability. Let us assume that the liability is normally distributed in the population with mean μ and variance σ^2. Suppose that there is just one threshold and so we observe two classes of individual, which for generality we will call the affected and the unaffected, depending on whether they are above or below the threshold (see Figure 14.2).

If we observe 5% affected individuals in the population, then the threshold would be at 1.645σ above the mean liability, as obtained from tables of the normal deviate (Appendix E). We do not know what σ is, however, because we do not know the underlying units of liability. So, for convenience, we will say that σ is 1.0. Now consider

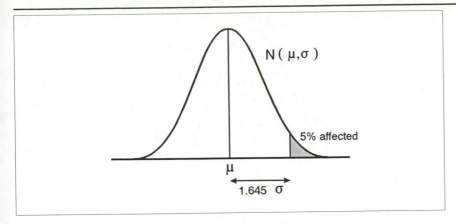

Figure 14.2 A threshold character in which 5% of the individuals are affected assuming an underlying normal distribution of liability.

another population in which the incidence of affected individuals is different, say 25%. By the same argument, the threshold in this case is 0.675σ above the mean liability of this population. Had there been 50% affected, then the threshold would have been exactly on the mean, i.e. 0σ away.

Let us further assume that the liability threshold is the same for all populations, while the genotypes in the different populations vary so that each population has a different mean liability but the variance of liability is constant, σ^2. It is more convenient to define the underlying liability threshold as the zero point around which our populations differ and hence we can illustrate the two populations as shown in Figure 14.3. Thus the mean of population 1, μ_1, is -1.645σ below the threshold and that of population 2, μ_2, is -0.675σ below. The difference between the two means, $\mu_2 - \mu_1$, being 0.970σ.

We do not know the value of σ and we have to assume that it is constant for all populations, but in situations in which this is likely to be approximately true, we can start to analyse the genetics of the trait. For example, consider a biparental family experiment in which each of the FS families is raised as a number of replicated plots, and that the proportion of plants infected with some disease in every plot is recorded. Assuming that this is a threshold trait, each plot can be assigned a mean deviation from the threshold as explained above for just two populations. Defining the underlying variance of the liability as unity, i.e. $\sigma = 1.0$, implies that we can ignore σ and continue to analyse the mean liabilities as the raw data as described in Chapter 4. This is illustrated in Table 14.6 for five FS families each with two replicate plots. The same approach can be used for any experimental design, and the genetics of liability analysed as for any other trait, yielding estimates of heritability, dominance, etc.

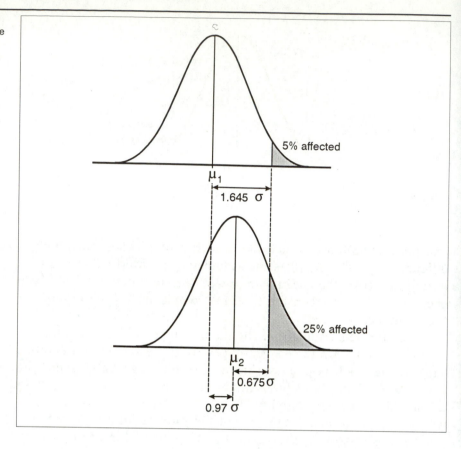

14.9 Two or more thresholds

With two or more thresholds it is possible to obtain more information about the genetics of the trait and to relax some of the assumptions. A typical example might be the number of calves born at one time to a

Table 14.6 A set of replicated FS families scored for % plants per plot infected with disease. Data transformed to give mean liability score

Family	Diseased plants (%)		Mean liability	
	Plot 1	Plot 2	Plot 1	Plot 2
1	5	8	−1.645	−1.405
2	15	20	−1.036	−0.842
3	80	65	0.842	0.385
4	25	20	−0.674	−0.842
5	6	15	−1.555	−1.036

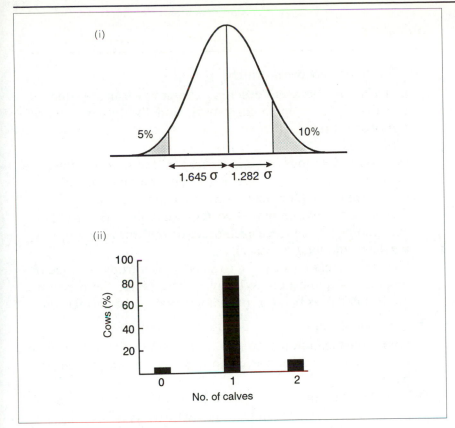

Figure 14.4 A threshold trait with two thresholds: (i) the underlying distribution; (ii) the three phenotypic classes.

cow after artificial insemination, none, one or two. Consider the illustration in Figure 14.4 in which there are 5% of cows which produce no young, 85% with one and 10% with two calves. The lower threshold, between zero and one calf is -1.645σ below the mean while the other threshold is 1.282σ above the mean. With two thresholds we have a choice of zero point and we will use the lower. Hence the mean of this population is 1.645σ and the upper threshold is at 2.927σ.

We still do not know the units of the underlying liability but we could define the difference in liability score between the two thresholds as being some arbitrary unit of liability, l. In our case the difference between the two thresholds is 2.927σ (i.e. $l = 2.927\sigma$), from which it follows that σ is $0.342l$ ($= l/2.927$) and $\sigma^2 = 0.117l^2$. It is necessary to assume that the thresholds are the same for all populations, where populations in this context could be different families. This particular population has a mean of 1.645σ ($= 0.56l$) and a variance of $0.117l^2$. With some family structure to the experimental material it would be possible to assign means and variances to every family in this way, and then to proceed to analyse the data as in Chapters 2–5.

Summary

(a) Correlations between traits

1. Correlations between different characters can be due to genetic and environmental causes, and the relative contributions of these factors to any observed correlation can be estimated.
2. Genetic correlations can impede response to selection for two or more traits selected together; useful combinations of economic traits are often negatively correlated genetically.
3. Genetic and environmental correlations may have the same sign and hence reinforce each other or they may have opposing signs and tend to cancel.
4. Genetic correlations can be caused by pleiotropy, linkage or linkage disequilibrium. Whatever their cause, there is normally some possibility of partially breaking the correlation.

(b) Threshold traits

5. Some traits exhibit just two or three distinct phenotypes in a segregating population even though the underlying genetic basis is polygenic.
6. This discontinuity is thought to be due to a threshold in the production of some underlying factor(s) produced by the polygenes. The number of + alleles determines an individual's liability; individuals above this threshold exhibiting one phenotype (the affected individuals) and those below, another.
7. With a single threshold, it is possible to use the proportion affected in a population to calculate the population mean in standard deviation units. This can be extended to calculate the means of all families in a given experiment, which allows standard FS and HS genetical analyses to be carried out.
8. With two or more thresholds, it is possible to calculate both the means and the variances of different families, so allowing more sophisticated analyses of the underlying liability.

References

1. Fisher, R.A. (1944) *Statistical Methods for Research Workers*, 3rd edn, Oliver and Boyd, Edinburgh.
2. Mather, K. (1965) *Statistical Analysis in Biology*, Methuen, London.

3. Mode, C.J. and Robinson, H.F. (1959) Pleiotropism and the genetic variance and covariance. *Biometrics*, **15**, 518–37.
4. Robertson, A. (1959) The sampling variance of the genetic correlation coefficient. *Biometrics*, **15**, 469–85.

Further reading

Falconer, D.S. (1989) *Introduction to Quantitative Genetics*, 3rd edn, Oliver and Boyd, Edinburgh.

15 Applications

We have devoted most of the preceding pages to the methods of analysing quantitative traits and interpreting the results of these analyses. The knowledge gained is of little value on its own and the purpose of this chapter is to describe some of the many ways in which it can be used. Largely, these applications are in the areas of animal and plant breeding, but we will also devote some space to the use of information on quantitative traits in the study of natural populations, including man.

The applications to breeding fall into the following categories:

(i) choosing the type of strain or cultivar to breed;
(ii) understanding the causes of heterosis (hybrid vigour);
(iii) predicting the breeding potential of different lines or populations;
(iv) identifying the most appropriate type of selection;
(v) predicting the rate of response and potential limits to selection; and
(vi) the use of genetic markers to aid selection.

The applications to natural populations concern:

(i) the potential response to natural selection;
(ii) the relation between the genetic architecture of the trait and past selection; and
(iii) the understanding of human behaviour or disease and their response to the environment.

We will now consider the applications of quantitative genetics to these various issues.

15.1 Choice of breeding objective

There are principally three types of varieties or strains that are used for commercial production: pure breeding lines, hybrids and populations.

The first two are largely confined to plants. Populations are the mainstream of animal breeding because herds of cattle are not easy to replace in a short span of time due to their high commercial value and long life-cycle cum productive period. Populations are also suitable as varieties in several outbreeding plant species such as many grasses, tree species and some crops which have natural self-compatibility barriers.

The choice of variety is governed by several factors of which, possibly, the most important is the breeding system. Intuitively, it makes sense to utilize the natural breeding mechanisms of the species in their improvement; working against the system may be both expensive and cumbersome. For instance, producing hybrids will be very costly and time consuming in cleistogamous species such as wheat, barley and rice, while it would be equally difficult to develop superior performing inbred lines of self-incompatible species like the forage grasses and some brassicas. Consequently, in the initial stages of crop improvement, the breeding system played an important role in determining the type of varieties produced. Hybrids of various kinds dominated the lists of varieties for outcrossing species while inbred lines were the main cultivars of self-pollinating crops. Table 15.1 lists the breeding system and the predominant types of variety that have been released for some of the most important crop species in the recent past. The breeding system was considered so important in breeding work that separate breeding and selection methodologies were developed for handling different types of crop. Self-pollinating species were improved by bulk or pedigree methods while breeders of outcrossing species employed mass or recurrent selection, sib-mating and top crossing [1].

Currently, however, commercial considerations and not the breeding system determine the type of variety in plants and scant regard is paid to the pollination system when producing inbred lines or hybrids. Numerous techniques, including bud pollination to avoid incompatibility, microspore or ovule culture to produce DH lines, micropropagation and embryo rescue/culture have been developed in order to bypass the breeding system to allow the production of pure breeding lines in those crops that are obligate outbreeders. Similarly, hybrid varieties have been released for cultivation in self-pollinating crops like rice and attempts have been made to do the same in other cereals; self-pollination can be inhibited by means of male sterility, chemical gametocides or temperature-sensitive fertility genes.

The main factors which determine the type of commercial variety now are the genetical control of variation in the species and commercial expediency. If the genetics of a species allows the exploitation of hybrid vigour through crossing, then hybrid varieties are preferred

Table 15.1 Examples of types of variety released for various crop species

Crop	Type of variety
(a) Self-pollinating species	
Wheat	Inbreds
Barley	Inbreds
Rice	Inbreds/hybrids
Oats	Inbreds
Sorghum	Inbreds/hybrids
Cotton	Inbreds/hybrids
Linseed	Inbreds
Tobacco	Inbreds
Tomato	Inbreds/hybrids
Peas	Inbreds
Chick pea	Inbreds
(b) Cross-pollinating species	
Maize	Hybrids/populations
Sunflower	Hybrids/populations
Pearl millet	Hybrids/populations
Rye	Inbreds/hybrids
Oilseed rape	Inbreds/populations
Brussels sprouts	Hybrids/populations
Swedes	Inbreds/populations
Carrot	Hybrids/populations
Onions	Hybrids/populations

because they are faster to produce, bring steady financial returns and allow the breeder to retain control of the valuable seed stocks, particularly pollen donors and male sterile lines. It is therefore the existence and extent of heterosis together with the cost of exploiting it which broadly determine the type of variety that the breeder produces.

15.2 The causes of heterosis

The most useful and widely accepted meaning of heterosis is the superiority of an F_1 hybrid over the mean performance of its better inbred parent. Using the familiar notation $\bar{P}_1$, $\bar{P}_2$ to refer to the means of the higher and lower scoring parents for some trait, positive heterosis implies that $\bar{F}_1 > \bar{P}_1$; for example the F_1 has a higher yield or superior quality than its better parent. There are situations where, for particular traits, the lower-scoring parent may be preferred, as for early maturity or low incidence of disease, and in such cases one may be interested in negative heterosis, $\bar{F}_1 < \bar{P}_2$. For simplicity, we will consider just positive heterosis as it is more common; the same factors, however, will apply to both types, only the signs of the parameters will change.

Whether positive or negative, this type of heterosis is referred to as **better-parent heterosis**. Unfortunately the term heterosis is also used occasionally by breeders either to describe the F_1 exceeding the mid-parental value of a particular cross or to the hybrid progeny of a top-cross between several inbred lines exceeding the average performance of the parental inbreds. Clearly, these are similar situations and such heterosis, **mid-parent heterosis**, requires nothing more than directional dominance. We will exclude these situations from the present discussion and thus will use the term heterosis specifically to describe better-parent heterosis.

Let us now consider the possible causes of heterosis and the evidence for their importance. We will start from the complete definitions of the means of F_1 and P_1 as given in Chapter 11.

$$\bar{F}_1 = m + [d] + [dd]$$

$$\bar{P}_1 = m + [a] + [aa].$$

Since heterosis is $\bar{F}_1 > \bar{P}_1$, it implies that $\bar{F}_1 - \bar{P}_1 > 0$. From the formulae above,

$$\bar{F}_1 - \bar{P}_1 = \{[d] + [dd]\} - \{[a] + [aa]\}. \qquad \text{[Eqn 15.1]}$$

What does this tell us about the possible causes of heterosis?

15.2.1 Heterosis without epistasis

When there is no evidence for epistasis (i.e. $[aa] = [dd] = 0$), and we have considered various ways in this book by which epistasis can be tested, heterosis simply requires that $[d] > [a]$, i.e. given our previous definitions, that $\sum d_i > r_a \sum a_i$. Obviously heterosis will be greatest when $\sum d_i$ is large relative to $\sum a_i$ and r_a is small, that is when the dominance ratio is large and the increasing alleles are dispersed in the parents. Dispersion is the most important factor. Assuming a constant dominance ratio of f $(= d/a)$, then $\sum d_i = f \sum a_i$ and there will be heterosis providing $f \sum a_i > r_a \sum a_i$, or $f > r_a$ and such a situation is likely to be common. For example, if the better parent has 25% of the undesirable alleles which distinguish the parents, r_a will be 0.5 and hence quite modest amounts of dominance will result in heterosis. Only if $r_a = 1.0$, when P_1 has all the increasing alleles, would it be necessary to invoke over-dominance to explain the heterosis.

Experimental evidence suggests that dominance ratios measured as $\sqrt{(4V_D^*/2V_A^*)}$ are invariably ≤ 1, despite the fact that most failures of the basic assumptions underlying the genetic models, such as linkage, epistasis and unequal allele frequencies, would tend to inflate the estimates [2]. Some examples for a variety of commercial traits in several crop species are given in Table 15.2. The higher dominance ratios are typical of traits related to reproductive or competitive ability and the possible reasons for this association of fitness traits with dominance will be discussed later. Other traits tend to have much lower dominance ratios, while what dominance there is tends to be ambi-directional which further reduces heterosis since it reduces $\sum d_i$. Thus, the size of the dominance ratio is, in general, a function of the trait rather than the breeding system of the species. When the heterosis is due to moderate dominance and allele dispersion, it may prove quite easy to produce one or many transgressive inbred lines from the cross which perform better than the F_1, and this has frequently been demonstrated in practice and will be illustrated in a later section.

15.2.2 Heterosis in the presence of epistasis

It has long been recognized that epistasis is commonly found to occur when the causes of heterosis in any specific case are examined [4]. Given the components in Equation 15.1, it is clear that epistatic situations in which $[dd] > [aa]$ will increase heterosis. Intuition might suggest that complementary epistasis would be the major determining factor, as it represents the situation where dominant increasing alleles at each of two genes have a proportionately greater effect when they occur together than would be expected from their

Table 15.2 Dominance ratios for some crop species

Crop	Character	Dominance ratio
(a) Self-pollinating species		
Tobacco	Leaf yield	0.55
Nicotiana rustica	Plant height	0.46
Barley	Grain yield	1.08
Swedes	Dry weight	0.50
Cotton	Boll number	1.14
(b) Cross-pollinating species		
Onions	Yield/plot	0.49
Brussels sprouts	Marketable yield	0.60
Maize	Seed yield	0.62*

* From [3]

individual effects. They complement each other's strengths, which is why we refer to it as complementary epistasis. It will be recalled from Chapter 11 that complementary epistasis is recognized when $[d]$ and $[dd]$ have the same sign; $[aa]$ and $[ad]$ are not very informative about the type of epistasis because they are affected by dispersion.

However, experimental data on epistasis and heterosis completely refute this intuitive expectation. In every single case presently known to us, where epistasis has been investigated in heterotic crosses, it has been found to be of a duplicate, not complementary type [2]. That is, whenever both $[d]$ and $[dd]$ are significant, they have opposite signs. There is no *a priori* reason why this should be so, despite the negative correlation between the coefficients of $[d]$ and $[dd]$, and it is easy to generate simulated situations where it does not hold. The interesting fact is that they do not occur in practice. Duplicate genes, gene dispersion and directional dominance appear to be the common factors underlying most cases of heterosis. The interaction actually opposes the dominance effect rather than reinforces it.

15.2.3 Other aspects of heterosis

We have considered one trait only but with several different traits, involving many genes altogether, the same principles apply. The possibilities of gene dispersion are, however, greater with more genes and it would become more difficult to combine all the better alleles in the same inbred line. We saw in Chapter 12 that heterozygotes were on average more stable over environments than inbreds and that heterosis tended to be more pronounced in poor environments. For these reasons hybrids can give short term gains at low cost, while ensuring that the breeder maintains control of the parental lines, so preventing the widespread use of farmers' own seed.

Even though it may be possible to produce inbred lines from an F_1 which out-yield the F_1, it is likely that any new hybrid produced by intercrossing these improved inbred lines will be better than the original F_1. This result is not, perhaps, intuitively obvious, but a brief example will illustrate the phenomenon. Suppose we start with two parents which differ for ten QTL each with equal effect $a = 2d = 1$, and that these alleles are in complete dispersion. The means of P_1 and F_1 will be m and $m + 5$ respectively, so there will be heterosis of 5 units. Now suppose this F_1 is allowed to produce a number of recombinant inbred lines (RILs) and that the best two happened each to be homozygous for eight increasing alleles, i.e. they would have a mean of $m + 6$, which is better than the F_1. Further suppose that the two remaining genes, homozygous for the decreasing alleles, are different in the two lines. The F_1 between these lines will be

Figure 15.1 The effect of several generations of selecting superior transgressive recombinant inbred lines for a heterotic trait illustrated for (i) height in *Nicotiana* and (ii) egg production in *Drosophila melanogaster*. In cycle 1, the score of the best parent (P_1) and the heterotic F_1 are given together with the best recombinant inbred line (RIL) extracted from the cross by inbreeding. The best line was then crossed to the second best to start cycle 2, where the score of the new F_1, and best transgressive line are shown, and so on. It can be seen that in each cycle, the F_1 exceeds the previous F_1 and the best RIL exceeds the previous best RIL, etc.

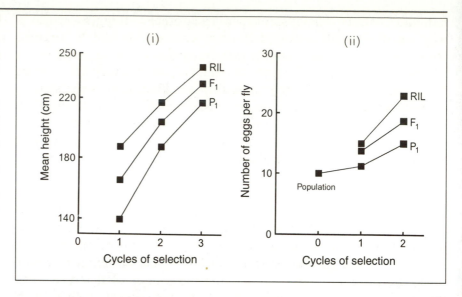

heterozygous for these four genes and homozygous for six increasing alleles and so its mean will be $m + 4d + 6a = m + 8$. This F_1 is better than its parents and the first F_1. However, it would still be possible to derive inbreds from this hybrid which would have nine or ten + alleles, i.e. they would be as good as or better than the second F_1, and so on. That this occurs in practice, even for highly heterotic traits, is illustrated in Figure 15.1 for both plant and animal data.

15.3 Predicting the breeding potential of crosses

Plant breeders are frequently interested in knowing the breeding potential of a cross between two inbred lines in terms of the potential performance of the best lines that can be derived from the cross by inbreeding or doubled haploid techniques. This information can be used on its own or to make choices between crosses and, as we have just seen, even when the cross exhibits heterosis it is possible to produce RILs which out-perform the F_1. The breeder may also be interested to know the potential of generating new F_1s from crosses among these superior derived RILs; such F_1s are referred to as **second cycle hybrids** (SCHs). We will now consider how a knowledge of the genetics of the trait in a given cross can be used for such predictions.

15.3.1 The performance of recombinant inbred lines, RILs

In the absence of epistasis, the best RIL that can be obtained from a cross, P_H, has a genetic value of $m + \sum a_i$ (Chapter 2). However, although an estimate of m is easily obtained from the basic generations, one normally only has an estimate of $[a]$ $(= r_a \sum a_i)$, not $\sum a_i$. One method that has been proposed for estimating $\sum a_i$ is to use the product of $[d]$ and the inverse of the dominance ratio, $(\sqrt{2V_A^*/4V_D^*})$ [5]. Assuming a constant dominance ratio at all loci, $f = d/a$, then

$$[d] \times \sqrt{(2V_A^*/4V_D^*)} = f \sum a_i \times 1/f = \sum a_i.$$

This procedure can provide an assessment of the range of inbred lines that could be produced from a highly heterotic cross, but the accuracy of such predictions is very dependent on the assumption of equal dominance at all loci and on the reliability of the dominance ratio, which is often low. Therefore, the procedure can be very unreliable.

An alternative approach is to determine the probability of extracting a recombinant inbred line that will perform better than some target genotype, such as one of the parents (P_1 or P_2), the F_1 or an existing commercial variety. This can be achieved by predicting the distribution of all the RILs that can be extracted from the cross by random selfing. In the absence of epistasis these RILs should be normally distributed with mean m and genetic variance, $2V_A^*$ (see Figure 15.2), both of which can be estimated from the basic generations or the F_3 generation of the cross (Chapters 2, 3 and 4) [6]. The probability of obtaining transgressive RILs which perform as well as or better than the target genotype is easily obtained from the standardized normal distribution using (target mean $- m)/\sqrt{(2V_A^*)}$ as a normal probability integral (Appendix E). We can illustrate the procedure using basic generations and F_3/TTC families of the cross V2 × V12 in *Nicotiana rustica* for the character final height, chosen because it was earlier identified as the most heterotic among a sample of 28 crosses [7]. The relevant parameters are given in Table 15.3.

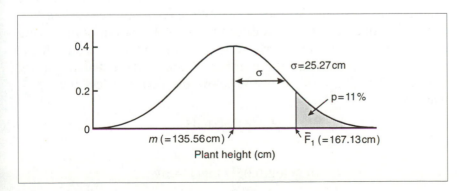

Figure 15.2 The proportion of SSD lines exceeding the performance of the F_1 for final height. The SSD lines are expected to have a mean (m) of 135.56 cm and an additive genetic variance ($2V_A^*$) of 638.57 (σ^2) based on information from early generations.

Table 15.3 Generation means and components of plant height in *Nicotiana rustica*

Generations		Parameters	
Family	Mean	Component	Estimate
P_1	146.46	m	135.56
P_2	107.42	V_A^*	319.29
F_1	167.13	V_D^*	124.82
F_2	155.50		

If we use the F_1 as the target genotype, then we calculate the normal deviate from these estimates as $(F_1 \text{ mean} - m)/\sqrt{(2V_A^*)} = (167.13 - 135.56)/25.27 = 1.25$, which gives the area in the tail of the distribution to be 0.11 or 11% (see Figure 15.2). Thus, 11% of the RILs extracted at random from the cross V2 × V12 are expected to have heights either equal to or larger than the F_1 mean. In other words, it should be relatively easy to produce an inbred line from the cross which was taller than the heterotic F_1. Consequently the most heterotic hybrid has only a transient advantage in this case. In a separate experiment conducted several years later, five F_7 families out of 59 were found to be taller than the F_1 hybrid, while we would predict 6.5, so confirming that the procedure works in practice. Many experiments with a wide range of crops have confirmed the robust reliability of this approach [8], which relies on the fact that both m and $\sqrt{(2V_A^*)}$ can be estimated with reasonable accuracy from the basic generations, or F_3 families, generations which would have to be raised in order to generate the RILs. Normally one is simply interested in a guide to the proportion of superior RILs that might be obtained, e.g. whether one should expect 5%, 0.5% or 0.05%, so that decisions can be made about the feasibility of the approach for a given cross and how many RILs should be raised to guarantee that some lines exceed the target score. For example, if 10% of the lines were predicted to exceed the target, how many lines should be produced to be fairly certain of observing at least one? Translating 'fairly certain' to mean 95% certain, this is equivalent to calculating the number of lines (n) needed to ensure that the probability of all being below the target is less than 5%. That is,

$$(1 - p)^n \leq 0.05$$

or

$$n \geq \log(0.05)/\log(1 - p).$$

Thus, if 10% are predicted to be superior, that is $p = 0.1$, then n should be ≥ 29 lines. With a 1% prediction, on the other hand, then at least 300 lines would be needed.

The breeder can apply this approach for other purposes too. Thus he or she may wish to identify from among a collection of hybrids those which have the greatest potential for producing superior RILs. He or she simply compares the predictions for different crosses and selects the best. Alternatively, they may wish to choose the most suitable generation (i.e. backcross or F_2) from which to start the inbreeding programme. This involves making predictions for the RILs derived from the F_2, $Bc_{1.1}$ and $Bc_{1.2}$ generations, or, indeed, any other generation, and comparing their potential for producing superior inbred lines. No additional information is needed for this except the parental scores or an estimate of the additive deviation [a]. These can be used to calculate the overall means and genetic variances of the RILs derived from, say, $Bc_{1.1}$ and $Bc_{1.2}$ generations ($Bc_{1.1}S_\infty$ and $Bc_{1.2}S_\infty$). To illustrate this we consider a single gene case and $Bc_{1.1}$ as an example. From Chapter 3 we know that $Bc_{1.1}$ will have A^+A^+ and A^+A^- genotypes in a 1:1 ratio. Selfing these genotypes will produce just two types of inbred line, i.e. A^+A^+ and A^-A^- with the expected frequencies of $\frac{3}{4} : \frac{1}{4}$. Thus the mean and variance of the RILs produced from $Bc_{1.1}$ (i.e. the $Bc_{1.1}S_\infty$ generation) will be:

$$\text{Mean}(Bc_{1.1}S_\infty) = \tfrac{3}{4}(m + a_A) + \tfrac{1}{4}(m - a_A)$$

$$= m + \tfrac{1}{2}a_A; \text{ and for many genes}$$

$$= m + \tfrac{1}{2}[a] = \tfrac{3}{4}\bar{P}_1 + \tfrac{1}{4}\bar{P}_2.$$

$$\text{Genetic variance}(Bc_{1.1}S_\infty) = \tfrac{3}{4}(m + a_A)^2 + \tfrac{1}{4}(m - a_A)^2 - (m + \tfrac{1}{2}a_A)^2$$

$$= \tfrac{3}{4}a_A^2$$

$$= 1\tfrac{1}{2}V_A^* \text{ for many genes.}$$

Similarly, the mean of $Bc_{1.2}S_\infty = m - \tfrac{1}{2}[a] = \tfrac{3}{4}\bar{P}_2 + \tfrac{1}{4}\bar{P}_1$, and the genetic variance of $Bc_{1.2}S_\infty = 1\tfrac{1}{2}V_A^*$.

The proportion of RILs predicted to exceed the target can be obtained for the backcrosses in the same way as for the F_∞ lines from the F_2. The method which predicts the highest number of RILs would normally be adopted.

15.3.2 Second cycle hybrids

At the end of the section on heterosis we showed that it was possible to select high-performing RILs from a particular heterotic cross and, by intercrossing them, to produce a second cycle hybrid (SCH) that performed even better than the original. In order to make general predictions about the SCHs that can be produced from crosses among the RILs derived from the first F_1, we need a method of defining their distribution in the same way as we defined the distribution of RILs above.

Starting from a single gene case, the RILs that we can extract from an F_2 are A^+A^+ and A^-A^-, in equal frequencies. The SCHs will be produced by crossing these lines in pair-wise combinations and their frequency and genetic values are as follows.

Cross	Frequency	Genetic value
$A^+A^+ \times A^+A^+$	$\frac{1}{4}$	$m + a_A$
$A^+A^+ \times A^-A^-$	$\frac{1}{4}$	$m + d_A$
$A^-A^- \times A^+A^+$	$\frac{1}{4}$	$m + d_A$
$A^-A^- \times A^-A^-$	$\frac{1}{4}$	$m - a_A$

These crosses will represent the three genotypes of the original F_2 with the same frequencies, i.e. $\frac{1}{4}$ of the families will be A^+A^+, $\frac{1}{2} A^+A^-$ and $\frac{1}{4}A^-A^-$, and the same will be true for all other genes, B, C, etc. Therefore if we were to randomly mate the RILs and take just one progeny from every family, we would effectively regenerate the original F_2. The population would thus have a mean $m + \frac{1}{2}[d]$ and variance $V_A^* + V_D^* + V_E$. However, each family is not just one individual but potentially contains many (r) genetically identical individuals, and this allows the mean of each family (i.e. individual genotype) to be obtained with considerable precision. The expected mean and variance of all these family means will be $m + \frac{1}{2}[d]$ and $V_A^* + V_D^*$ respectively [9]. Thus the expected distribution of SCHs will be approximately normally distributed with mean $m + \frac{1}{2}[d]$ and variance $V_A^* + V_D^*$, and so the approach used to determine the proportion of RILs exceeding some target can be used for SCHs also, only the mean and variance change. Using the data in Table 15.3, we obtain the prediction that 29% of the SCHs would exceed the F_1 mean compared with 11% of the RILs. Of course, in practice one would not randomly mate all the RILs, but simply intercross the best. The prediction has revealed that the superior SCHs will exist and crossing the best RILs will ensure that a maximum number of superior SCHs will actually be produced.

15.3.3 Predicting three- and four-way crosses
In addition to the identification of superior single crosses, the F_1 data can also be used to predict the performance of three-way and double (four-way) crosses. While numerous methods have been proposed to

Table 15.4 Prediction of three-way and double-crosses using single crosses

Single crosses		Multiparent crosses		
Cross	Mean	Cross	Predicted mean	Observed mean
V1 × V2	154.86 ± 1.41	(V1 × V12) × V5	158.85 ± 1.00	162.34 ± 2.05
V1 × V5	146.50 ± 1.18	(V2 × V12) × V1	158.80 ± 0.85	156.22 ± 1.69
V1 × V12	162.74 ± 0.94	(V1 × V5) × (V2 × V12)	162.26 ± 0.73	166.37 ± 2.14
V2 × V5	160.26 ± 1.74	(V1 × V2) × (V5 × V12)	162.36 ± 0.71	167.61 ± 1.91
V2 × V12	179.92 ± 1.67			
V5 × V12	171.19 ± 1.61			

make these predictions, the most reliable is based on the average of the non-parental crosses [10]. That is, the performance of a three-way cross of say $(1 \times 2) \times 3$ parentage is predicted from the mean of the 1×3 and 2×3 crosses and that of a double cross $(1 \times 2) \times (3 \times 4)$ is obtained from the average of 1×3, 2×3, 1×4 and 2×4. The data in Table 15.4 for final height of *Nicotiana rustica* crosses, taken from our unpublished work, illustrate the procedure. Clearly, the predicted values of the three-way and double-cross families fall within the error limits of the observed means in three out of four cases ($t < 1.96$ for df > 30).

15.3.4 General and specific combining ability

We have considered predicting the inbreeding and outcrossing potentials of generations derived from pair-wise crosses among inbred lines, but on many occasions, breeders have unrelated stocks that may have been obtained from germ-plasm collections or other sources that need evaluation. Also, they may have large samples of inbreds from various sources out of which they may wish to identify the best combination of lines for maximizing heterosis in a new hybrid variety. The pair-wise assessment of such a large number of lines in all possible combinations would be both time consuming and logistically inefficient. This practical problem has long been recognized and various attempts made to devise procedures for at least identifying and eliminating the poorer lines in the early stages.

One of the most favoured procedures for this purpose is the 'top cross'. It involves crossing each of a set of new lines with a widely based source population, which may be an open pollinated variety, a synthetic population or a double-cross population. The average performance of progeny produced by each line provides a measure of the **general combining ability** of each line (Chapter 5). This procedure will provide a measure of the general potential of different lines and, on the basis of such evaluation, more than half of the evaluated lines can be discarded without much risk of losing any highly valued material. Further, the evaluation process can be spread over sites and even across several environments, provided suitable check varieties are included in the trials.

Those lines with good combining ability are then crossed in all possible pair-wise combinations and the progenies evaluated to identify the best performing single crosses. At this stage, the crosses should be selected not only on the basis of their general superiority but also on the extent of improvement they display over and above that expected from the general combining abilities of the parents. This is selection for **specific combining ability**.

15.4 Effects of failed assumptions on predictions of RILs and SCHs

So far we have assumed that the additive genetic, dominance genetic and additive environmental model is sufficient to explain all the variability in the material. However, this is rarely true and epistasis, $G \times E$, etc. are frequently observed to affect genotypic expression. What effect do they have on the predictions of RILs and SCHs?

15.4.1 Epistasis

Epistasis does not affect the overall mean of RILs which is still expected to be m, but it will change the genetic variance from $2V_A^*$ to $2V_A^* + V_{AA}$ where V_{AA} is the additive $\times$ additive interaction variance, as defined in Chapter 13. Epistasis also makes the distribution marginally skewed, although it will still be a close approximation to a normal distribution.

An estimate of m can be obtained from model fitting to the basic generations as described in Chapter 2, while a slightly biased estimate can be obtained from the F_3 mean. The additive genetic variances obtained from the F_3 or the triple test-cross families are generally found to approximate closely to $2V_A^* + V_{AA}$ and, if anything, they are conservative estimates which lead to correspondingly conservative predictions [11].

15.4.2 Linkage disequilibrium

We showed in Chapter 10 that linkage disequilibrium (LD), the imbalance between repulsion and coupling linkages, has no effect on the estimates of $[a]$ and $[d]$ in the absence of epistasis. It does, however, bias the variances, coupling linkages inflate while repulsion linkages deflate the estimate of V_A^*. The dominance variance, V_D^*, on the other hand, will always be inflated unless dominance is ambi-directional at many loci.

What are the consequences for prediction? Generally we would expect repulsion linkages to predominate in situations where we are looking for transgressive segregation. We showed in Chapter 10 that although the estimate of V_A^* obtained from the basic generations was less than $\sum a^2$ in the presence of repulsion linkage, so also was the additive variance of the RILs. Estimates of V_A^* from the F_2 BIPs or F_3 families, on the other hand, will be much closer to that of the RILs, as shown in Chapter 10, and therefore will give robust reliable predictions, even in the presence of linkage disequilibrium. Predictions of the SCHs will be less affected by LD because the downward bias of V_A^* is at least partly cancelled by the inflation of V_D^*. In those cases, possibly the majority, where early generation variances underestimate the true variances of RILs or SCHs, the predicted gains will be conservative.

15.4.3 Genotype by environment interaction

Probably the major causes of disagreement between predictions and observations will be one or both of the following related types of $G \times E$ interaction. First, there may be differences between the design of the experiment at the prediction and assessment stages. Such differences arise from the necessity to evaluate limited samples of highly heterogeneous material at the early stages of a breeding programme in small plots or in individually randomized families because of the shortage of seed, while final assessment will be on a large plot or field basis. Second, there may be differential inter-genotypic competition in the prediction and assessment stages. Many genotypes tend to show markedly different performances in mixed compared to pure stands for agronomic traits. The early generation families are mixed stands while those at the end of a breeding pro-gramme are normally pure stands. Unfortunately, not much can be done to overcome these problems except that one should try to raise as large plots as possible in the early generation assessments.

15.5 Predicting the response to selection

If parents are selected from a population, how do we predict the likely performance of their progeny and if there is a response to selection, for how many generations will further response be achieved?

15.5.1 One pair of parents

Let us first consider the simple case of selecting two parents, male and female, from some population such as an F_2 or natural population, crossing them and raising their progeny in the same environment. We showed in Chapter 4 that, if we took a population and randomly mated the individuals in pairs to produce FS families, the regression of offspring family means onto the mid-parental values was an estimate of the narrow heritability (see Figure 4.3). We also showed that the mean of the parental and offspring populations should be the same, providing the environments are the same. These relationships are shown in Figure 15.3.

We can now use these relationships to predict the progeny mean of any pair of parents. Consider a pair of parents with mid-parental value X, which deviate from the population mean by S units. It can be seen from Figure 15.3 that their progeny should have a mean Y which deviates from the population mean by R units. S is termed the **selection differential** and R is the **response to selection**. Clearly the slope of the line, b $(= h_n^2)$, is R/S or, to re-arrange the equation,

$$R = h_n^2 S.$$

Therefore, knowing the population mean, the heritability and S, we can predict R and hence the mean of the progeny of the selected parents. This prediction will have a standard error which depends on the residual MS_R of the regression analysis ($s_Y = \sqrt{\{MS_R[1/r + 1/n + (X_i - \bar{X})^2/SS_X]\}}$ where n = number of X,Y values and r = family size [12]. Alternatively, if we know R and S we can estimate the regression coefficient b as R/S and this estimate

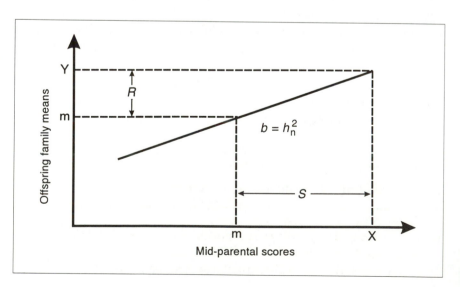

Figure 15.3 Predicting the response to selection. The relationship between the mean scores of the offspring and that of their mid-parents to illustrate the concept of selection differential (S) and response to selection (R).

of the heritability is known as the **realized heritability**, because it is the heritability realized by selection.

15.5.2 Selecting many parents

A more common situation is that in which the population is scored for some trait but only the highest scoring individuals are used for breeding. For example, the breeder may decide to select only the top 10% for breeding or breed from individuals whose scores for the trait exceed some target value. Either way, the breeder will know the proportion selected which, following convention, we will call p, as shown in Figure 15.4. As for the previous case of one pair of parents, we need to know the mean score of the selected individuals, X, and the deviation (S) of X from the population mean, μ. Providing the population from which the selected sample is taken is normally distributed for the trait, it is a simple matter to predict S knowing the phenotypic variance of the parents, as follows.

Consider the special case of a normally distributed population with mean $\mu = 0$ and phenotypic variance of $\sigma^2 = 1$. Knowing p, it is possible to calculate the distance from the mean to the point of truncation of the distribution (t), beyond which the selected sample

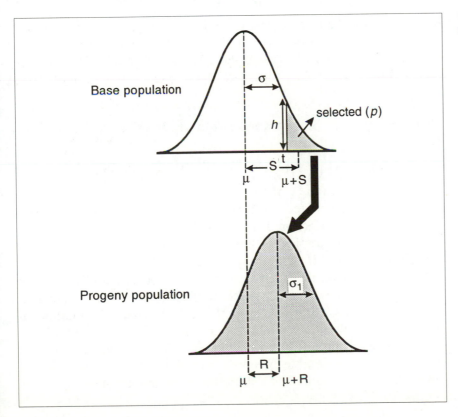

Figure 15.4 Predicting the response to mass selection.

lies, using tables of the normal probability integral (Appendix E). For example, if $p = 0.1$ then $t = 1.28$. In order to derive the mean of the selected sample we need to obtain the height of the ordinate (h) at t and divide h by p. Tables of the heights of the ordinate are available in many statistical texts [13]. In the present case, at $t = 1.28$, $h = 0.1758$ and the mean (h/p) of the selected sample is 1.758 above the population mean. This value of 1.758 is referred to as the **intensity of selection**, i, and is equivalent to S but for a population of unit variance. The actual value of S is $i.\sigma_p$, where σ_p is the phenotypic standard deviation of the population from which the selected sample is taken. Because i is simply a function of p, the values of i corresponding to various values of p are given in Appendix D in order to avoid the tedious use of different statistical tables.

To take an actual example, suppose we wished to select the tallest 10% from a population with a mean height (μ) of 65 cm, a phenotypic variance ($\sigma_p^2 = V_A^* + V_D^* + V_E$) of 144 (or $\sigma_p = 12$ cm) and a narrow heritability (h_n^2) of 0.3. From Appendix D or the paragraph above we find that $i = 1.755$, and therefore,

$$S = i.\sigma_p = 1.755 \times 12 \, \text{cm}$$

$$= 21.096 \, \text{cm}.$$

In other words, the mean of the selected sample is $\mu + S = 86.096$ cm. The response to selection, $R \, (= h_n^2 S)$, is 6.318 cm and so the expected mean of the progeny of the selected 10% is $\mu + R = 71.318$ cm. The genetical variance of the progeny of the selected parents is [14],

$$\{1 - \tfrac{1}{2}h^2 i(i - t)\}V_A^*,$$

where $t = $ the point of truncation in σ units. In this case the variance is,

$$\{1 - 0.15 \times 1.755(1.755 - 1.28)\}V_A^* = 0.87V_A^*.$$

This type of selection, where a proportion p of the population is selected and inter-mated to form the next generation, is called **individual** or **mass selection**. It is probably the oldest method of selection which has been applied by humans to improve their domesticated animals and plants. It is relatively easy and cheap to apply, and is reasonably effective, provided that h_n^2 is not too low and that there are no major common environmental effects. This type of selection is also typical of natural selection. It is a highly versatile method of improving populations. While the product of several rounds of selection can be marketed as a new variety, the improved population may also form a suitable breeding base for extracting samples of inbred lines for producing F_1 varieties or to initiate new 'synthetic populations'.

Mass selection can be applied to plant populations in two ways. First, both male and female parents may be selected before crossing. In other cases, it may not be possible to implement selection before pollination and thus selection can only be on maternal plants because the origin of pollen will be unknown. Therefore, selection before pollination will be twice as effective as selection after pollination.

When mass selection is repeated over a number of successive generations, it is called **recurrent selection**. In the latter case, the experimenter hopes to obtain a continuous shift in the population mean in the desired direction. In theory, the response $R = i.\sigma_P.h_n^2$ applies to only one cycle of selection because directional selection leads to changes in allele frequency at the various loci, and hence σ_P^2 and h_n^2 will change. However, it is generally observed that the same R can be used as a predictor for up to around five cycles of selection, i.e. the mean after five cycles of selection should be $\mu + 5R$. This does, however, depend on the selection intensity being kept low (e.g. $p > 0.2$), and the size of the selected sample being large in order to avoid the fixation of undesirable alleles by chance. Selecting a large proportion of the population will produce a small response (R) in each generation but the largest ultimate gain in performance. Intense selection will produce a rapid response that will plateau at a lower level than with weak selection (Figure 15.5). Selection will also favour those genes with large allelic effects and will tend to pick out sets of linked genes which have a large positive combined effect. In so doing it can easily fix undesirable alleles which may be embedded in or close to these linked blocks [15].

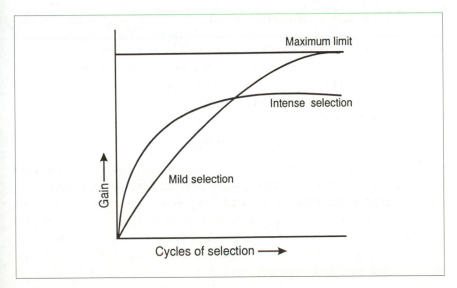

Figure 15.5 The likely long-term effects of intense versus mild selection.

15.5.3 Other types of selection

As well as individual selection, the other major types of selection that we will consider are:

 (i) family selection, i.e. selection on the family mean;
 (ii) within-family selection; and
(iii) combined or index selection.

To compare the efficiency of these methods, we have to define the phenotype of each individual (x) in terms of the overall mean (μ), the family effect (x_f) and the individual's own deviation from the family mean (x_w) as shown below.

$$x = \mu + x_f + x_w.$$

The overall mean is of little consequence for selection because it is a constant throughout. It is therefore the family and the within-family deviations that are manipulated to varying degrees by the different methods of selection. Clearly, family selection puts maximum emphasis on x_f, within family selection ignores the family performance and allocates maximum weight to x_w, mass selection gives equal importance to both x_f and x_w, and combined selection takes into account both x_f and x_w but allocates weights according to their comparative importance.

In terms of standard symbols, we can represent the response in each case using the following formulae.

Method	Response	Prediction
Mass selection	R	$i.\sigma_P.h_n^2$
Family selection	R_f	$i.\sigma_F.h_f^2$
Within-family selection	R_w	$i.\sigma_W.h_w^2$

Here, σ_f^2, h_f^2, σ_w^2 and h_w^2 are the variance and narrow heritability of the family means and the individuals within families respectively.

A similar equation for index selection, however, is complex but will be based on the expected genetic value, I_g, of each individual which is given by the formula,

$$I_g = h_f^2 x_f + h_w^2 x_w.$$

To compare the selection efficiencies of the various methods, we need to calculate their responses in a standardized form by redefining the various narrow heritabilities and phenotypic variances as proportions of h_n^2 and σ_P^2 and by taking into account the number of families and family size. While the details of the algebra can be found elsewhere [16–18], we summarize below the main conclusions that can be drawn from such comparisons.

(i) Index selection is the most efficient because it optimizes genetic information by allocating appropriate weights to the between- and within-family effects. However, its maximum superiority over any of the alternative methods under any specified conditions is rarely more than 20%.

(ii) Among the remaining three types, family selection gives a better response when h_n^2 and V_{EC} are low. In the case of HS families, this advantage is not realized unless the family size (r) is 20 or more.

(iii) Individual selection is more efficient than family or within family selection for intermediate to high heritabilities. It is unconditionally superior with HS families for smaller family sizes, i.e. $r < 10$.

(iv) Within-family selection is generally inferior and is always so for FS families where $r > 10$. Selection within HS families gives better response than FS families because the HS families display more additive genetic variance, $\frac{3}{4}V_A^*$ compared with $\frac{1}{2}V_A^*$. This difference is more marked when h_n^2 is high and V_{EC}, the common family environment, is very large.

Similar approaches can be applied to selection during inbreeding, such as predicting the means of the RILs produced from selected F_3 families [19].

15.6 Correlated response to selection

We have seen in Chapter 14 that characters of economic importance are often correlated with each other and therefore selection for one should affect the performance of the other. Intuitively, we can imagine that selection for increased score of character X will increment character Y when X and Y are positively correlated, and the score of Y will decrease if the correlation is negative. In general, change in Y of those individuals selected for X is given by the regression of Y on X; i.e. $b = SP_{(XY)}/SS_{(X)}$ and it follows from this that the correlated selection differential, $CS_{(Y)}$, is equal to $b.S_{(X)}$. The correlated response to selection $CR_{(Y)}$, like response to selection in general, however, will depend on the additive genetic effects only, and thus

$$CR_{(Y)} = R_{(X)}.b_{(A)} \qquad \text{[Eqn 15.2]}$$

where $b_{(A)}$ is the regression between the additive genetic values of individuals for X and Y. Now, representing h_n^2 of character X with

h_X^2 gives,

$$R_{(X)} = i.h_X^2.\sigma_{P(X)} = i(\sigma_{A(X)}^2/\sigma_{P(X)}^2)\sigma_{P(X)}$$

$$= i.\sigma_{A(X)}\{\sigma_{A(X)}/\sigma_{P(X)}\}$$

$$= i.\sigma_{A(X)}\sqrt{h_X^2}$$

$$= i.h_{(X)}.\sigma_{A(X)}. \qquad \text{[Eqn 15.3]}$$

Further, in general;

$$b = \sigma_{(XY)}/\sigma_{(X)}^2 = \sigma_{(XY)}.\sigma_{(Y)}/\sigma_{(X)}^2.\sigma_{(Y)}$$

$$= \{\sigma_{(XY)}/(\sigma_{(X)}.\sigma_{(Y)})\}(\sigma_{(Y)}/\sigma_{(X)})$$

$$= r_{(XY)}(\sigma_{(Y)}/\sigma_{(X)}). \qquad \text{[Eqn 15.4]}$$

Substituting the $b_{(A)}$ equivalent of Equation 15.4 in Equation 15.2 gives:

$$CR_{(Y)} = i.h_X.\sigma_{A(X)}\{r_{A(XY)}(\sigma_{A(Y)}/\sigma_{A(X)})\}$$

$$= i.h_X.\sigma_{A(Y)}.r_{A(XY)}. \qquad \text{[Eqn 15.5]}$$

Now, $h_Y^2 = \sigma_{A(Y)}^2/\sigma_{(Y)}^2$, or $h_Y = \sigma_{A(Y)}/\sigma_{(Y)}$.

Replacing $\sigma_{A(Y)}$ with $h_Y x \sigma_{(Y)}$ gives the response as:

$$CR_{(Y)} = i.h_X.h_Y.\sigma_{(Y)}.r_{A(XY)}. \qquad \text{[Eqn 15.6]}$$

This shows that the correlated response can be predicted when the narrow heritabilities and the additive genetic correlation between the characters are known. It is also apparent that the realized response in Y will not be great unless the correlation is strong and the heritability values are high, i.e. 0.5 or more. It is also possible to use Equation 15.3 to obtain estimates of the realized heritability or of the additive genetic correlation provided the remaining statistics are available.

15.7 Indirect selection

Breeders often seek improvement in complex characters such as yield by indirect selection, that is, they select say seed size or seed number instead of yield. The conditions under which such selection will be more effective than direct selection can be deduced easily from Equations 15.3 and 15.5, assuming that Y represents yield and X is the component which will be subjected to selection. Now, response of Y through indirect and direct selection will be;

$$CR_{(Y)} = i.h_X.\sigma_{A(Y)}.r_{A(XY)}$$

$$R_{(Y)} = i.h_Y.\sigma_{A(Y)}.$$

It follows from this that indirect selection will be better when

$$CR_{(Y)}/R_{(Y)} > 1$$

or

$$\{i.h_X.\sigma_{A(Y)}.r_{A(XY)}\}/\{i.h_Y.\sigma_{A(Y)}\} > 1.$$

Assuming that i is the same in both cases, indirect selection is expected to be more efficient when $r_{A(XY)}.h_X > h_Y$. Thus, indirect selection will be more effective when the correlation between the characters is very strong and the narrow heritability of the selected character is much larger than that of the unselected. As such situations are rarely observed in practice, indirect selection is unlikely to be better than direct selection and therefore should not be used unless the desired character is either difficult or costly to measure, or is sex-limited such as milk yield or egg production.

15.8 Multi-trait selection

Selection is often applied to more than one trait when improvement in the economic worth of an organism is sought. There are three main methods of multi-trait selection. The simplest is to select a different trait in each cycle of selection and so improve different traits one after the other. This type of selection, termed **tandem selection**, is considered highly effective but it is expensive in terms of resources and time. An alternative is **independent culling**, where selection is applied to several traits in the same generation, but the traits are selected successively. Those selected for the first trait are then further reduced in number by selecting for trait two, and so on. This procedure, however, can lead to a greater loss of good material because any individuals which fail to meet the set criteria for one trait will be rejected regardless of their good genetic values for other traits. The third method is the most complex to apply because it is based on an index equation similar to the one that we have given earlier for combined selection. This method is popularly called '**index selection**' as individuals are selected on the index scores and not their phenotypic values. The index takes the form,

$$I_g = \sum w_i P_i \qquad \text{[Eqn 15.7]}$$

where w_i and P_i represent the weight and the phenotypic score of the ith trait respectively.

These weights will take into account the economic value of each trait, the extent of the additive genetic variability displayed by the various traits and the genetic correlations between them. Index selection is the most efficient of all the multi-trait selection procedures and further details of how to develop and apply the index can be found elsewhere [20].

15.9 Marker-based selection

We have seen that many quantitative traits have low heritabilities and hence the high performance of a particular individual is no guarantee that it has desirable alleles at the relevant QTL. Many phenotypes are difficult to score on an individual, such as disease resistance, sex-limited traits, those which only reveal their effects late in life or development, which are recessive or have poor penetrance. If it was possible to associate marker loci, which were easily identifiable, with particular useful QTL alleles, then selection could be targeted at the marker loci instead and could be more efficient. Given the last sentence of section 15.7, we are assuming that the markers have a heritability of 1.0 and the correlations with the QTL are high. In Chapters 6, 7 and 8 we explained how genes, including QTL, could be mapped and associated with marker loci such as RFLPs, RAPDs, etc. and we shall now explain some of the uses of such linkages.

15.9.1 Genetic fingerprinting

Perhaps one of the most useful contributions that molecular markers have made to breeding, particularly plant breeding, is that of providing a check against contamination of material throughout a breeding programme. Until recently, one had to rely on careful techniques during crossing and harvesting to minimize the risk of uncontrolled pollination or mis-labelling. The availability of biochemical and molecular markers has provided a method of obtaining a **genetic fingerprint** of individual genotypes based on the particular pattern of alleles at a number of marker loci in different cultivars. This has often revealed errors in breeding material and should be used routinely to provide better quality control and avoid expensive mistakes [21].

15.9.2 Introgressing a single QTL

Another major potential use for marker-aided selection (MAS) is in introgressing alleles from an alien species or land-race into a commercial strain or cultivar. Before the availability of markers, this was achieved by backcrossing to the commercial strain as recurrent parent while selecting for the allele to be introgressed. Two problems arise in this process.

First, it may be difficult to recognize individuals which carry the favoured allele among the backcross progeny because of low heritability, poor penetrance or the allele is recessive. Providing close linkage can be established before backcrossing commences between the relevant QTL and a marker, or preferably two markers which are on either side of the QTL, then selection can be based on the markers alone during backcrossing. Clearly, the tighter the linkage the more reliable the approach, while selecting for two, close flanking markers will almost guarantee that the QTL is selected also because cross-over interference implies that no double cross-overs will occur within approximately 15 cM. Since selection is based on the marker genotype and not the trait, the backcross individuals can be scored at an early stage in development and there is no need for costly and inefficient trials.

The second problem is linkage to undesirable alleles. Apart from the QTL to be introgressed, the breeder wishes to introduce as little of the donor genome as possible into the recipient commercial strain because it may well carry undesirable alleles. At the same time, the breeder is trying to minimize the number of backcross generations involved in order to save time and resources. As backcrossing proceeds, the genome becomes closer to that of the recurrent parent but the region around the selected locus can retain segments of donor chromosome for many generations, because, as explained in Chapter 8, cross-over frequencies are low. Figure 15.6 shows the average length of donor chromosome, in cM, surrounding a selected marker after various generations of backcrossing [22]. These are only average figures and hide considerable variability, but clearly there could be a major part of the donor chromosome still associated with the selected region even after eight generations of backcrossing. This is known as **linkage drag**.

Using several well-spaced markers on the chromosome containing the critical QTL as well as the selective marker(s) for the QTL, together with three or four markers on each of the other chromosomes, the breeder can simultaneously select for the donor allele at the QTL

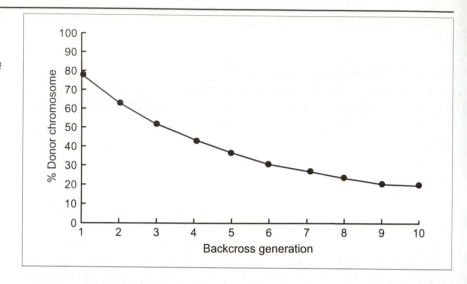

marker(s) and the recurrent parent alleles at all other markers. This not
only identifies backcross individuals which have short donor regions
around the QTL but also accelerates the return to homozygosity for
the recurrent parent genotype at all other chromosomes as well as
the chromosome receiving the donated gene. Using such an approach,
no more than two or three backcrosses may be needed to be sure that
the desired QTL has been introgressed with very little residual donor
chromosome. Although considerable numbers of markers may have
to be genotyped in the first backcross, they will be homozygous in
the selected individuals and hence will not have to be screened in subsequent backcrosses; effort can thus be concentrated on the region
around the introgressed QTL. It is always possible subsequently to
locate other polymorphic marker loci closer to the QTL in order to
reduce the donor segment even further.

15.9.3 Selecting for multiple QTL

The procedure above was predicated on the accurate location of the
QTL with respect to the selectable markers. As we saw in Chapter
7, such accuracy is difficult to obtain from segregating populations
unless the trait has a very high heritability and just one or two key
QTL are involved. Major genes for resistance are good candidates
for this. On the other hand, with traits such as grain yield, many
QTL may be segregating, and the QTL locations may have wide
confidence intervals. The most likely location of a useful QTL may
appear to be between a particular pair of selectable markers, based
on an F_2 or DH population, but it could actually be as far as 20 cM
away. Concentrating selection on these possible markers could be
quite ineffectual.

We saw in Chapter 7 that it was statistically difficult to resolve two or more linked QTL unless they were some distance apart on the chromosome and individually had high heritabilities. Frequently, linked QTL either appear as one large, ghost QTL if linked in coupling or they escape detection altogether if linked in repulsion, because they cancel each other's effect. This suggests that the most useful role for marker-aided selection in these situations is to use marker loci to identify whole chromosomes or chromosome arms that contain useful QTL alleles, and to select for these. Such an approach requires relatively few markers and will allow key chromosomes or chromosome arms to be accumulated quickly, providing a larger selection response than by mass selection alone [23,24].

When one is trying to select for several traits simultaneously, this approach can be used to construct an ideotype of the optimum genotype and determine the particular combination of chromosomes or chromosome arms which can best achieve this ideotype based on the marker information.

15.10 Genetic architecture of populations

15.10.1 Dominance and epistasis

Natural selection can, over long periods of time, have pronounced effects on the way genes express themselves in terms of their dominance and epistatic effects, and the particular type of non-additive control is referred to as the **genetic architecture** of the trait. As long ago as 1931, Fisher [25] suggested that the dominance of one allele over another was the result of selection for other genes which modify the dominance properties. The argument was roughly as follows. Normally, favoured wild-type alleles are dominant over their less-favoured alternatives. The present alleles must have replaced earlier dominant alleles against which they too would have been recessive initially, and hence their current dominance must have evolved. Experiments with Guinea fowl showed that the dominance of alternative alleles could, in fact, be modified.

This idea was subsequently developed by Mather [26] to account for the varied genetical architecture of quantitative traits. He argued that traits which were under directional selection such as high reproductive or competitive ability, so called **fitness traits** because they are positively correlated with fitness, would evolve dominance for the favoured alleles as Fisher had postulated. He further suggested that these genes would also show duplicate epistasis because, like dominance, duplicate epistasis would protect the individual from deleterious alleles arising from mutation or migration, which decrease fitness. On the other hand, traits under stabilizing selection (see Figure 1.5) in which intermediate phenotypes are favoured, would obtain no particular advantage from evolving dominance in any particular direction. Dominance at particular QTL would be left at the level which existed when the allele first arose and so dominance would be weak and ambi-directional. Similarly, epistasis would be weak. Many examples have been documented subsequently which confirm this prediction.

This view of the genetic architecture of traits can be used to infer the type of selection acting on novel traits in natural populations. Intelligence in humans shows directional dominance for high IQ suggesting, what one might have expected, namely that high intelligence has been under strong directional selection. Conversely, bristle number in *Drosophila* shows little dominance or epistasis, and what dominance there is, is ambi-directional. There is experimental evidence to suggest that this trait is under stabilizing selection [27].

15.10.2 Heritability

As with dominance, the heritability of a trait also indicates something of the natural selection that has acted on the trait in the past. Typically, traits closely related to fitness show lower narrow heritabilities and larger amounts of non-additive genetic and environmental variation than do traits under stabilizing selection. This is both because directional selection reduces genetic variation and also, by increasing the allele frequencies of the dominant alleles, it decreases the additive genetical variation as shown in Figure 9.1.

Heritability estimates, particularly for quantitative traits in plants, must always be treated with some caution. They measure the relative importance of the genetical variation to the total variation in a population and hence will depend on the absolute size of both the genetical and the environmental variation; the former can vary with the type of population while the latter is very dependent on the ambient conditions in which the trait was measured. One should not expect a close quantitative similarity between heritability estimates for the same traits in different populations, although in rough qualitative terms they may be similar. Measuring the heritabilities of traits under natural conditions in wild populations is very difficult while estimates obtained for the same population under controlled laboratory conditions could seriously mislead one about what is actually happening in the natural habitat.

Table 15.5 Some examples of heritability in various crop species

Species	Character	h_n^2	h_b^2
Onions	Yield/plot	0.36	0.40
	Bulb weight	0.37	0.43
Cotton	Seed yield	0.11	0.16
	Boll number	0.10	0.17
Tobacco	Plant weight	0.19	0.27
	Plant height	0.33	0.39
Nicotiana rustica	Plant height	0.42	0.51
	Flowering time	0.44	0.52
Sorghum	Plant weight	–	0.24
	Leaf length	–	0.49
Chick pea	Seed yield	0.40	0.52
	Seeds/pod	0.26	0.27
Brussels sprouts	Marketable yield	0.20	0.37
	Sprout number	0.22	0.45
Barley	Yield/plant	0.19	0.26
	Grain yield	0.54	0.65
Oilseed rape	Flowering time	0.63	–
	Nodes at flowering	0.53	–

Some examples of heritability estimates are given in Table 15.5 for plant species. Similar wide ranges of estimates can also be found in animals.

15.11 Human populations

In studying human populations we are not interested in performing selection experiments but a knowledge of the genetic control of traits such as intelligence, behaviour and susceptibility to disease could be a factor in guiding us in planning social or medical policies. In the remainder of this Chapter we will discuss a few important, though somewhat controversial issues simply to indicate the way in which a knowledge of quantitative genetics might influence our judgement.

Consider intelligence as an example. As we said in Chapter 9, the nature and extent of genetical variation in human intelligence has been a subject of considerable controversy, particularly over the past 30 years or so. A large number of different tests of intelligence have been devised, some of which measure general intelligence while others attempt to measure special skills such as spatial or linguistic ability. Considerable effort has been made to make the tests culture fair, that is, to prevent them from rating people on background or knowledge rather than innate ability, and there is general agreement that all these special skills correlate quite highly with general intelligence. The tests of intelligence produce a score or intelligence quotient (IQ), which correlates highly with what we generally understand by intelligence, such as the level of skill required for a particular job. That is, it ranks people well according to whether they are skilled professionals to unskilled manual workers. It also correlates with educational attainment and, of course, earnings because better-educated people end up with higher-paid jobs. There is little doubt that knowing a child's IQ is about the best predictor of adult work status and earnings that is yet available [28].

IQ also has a very high heritability. Estimates vary, but there are now many large-scale studies using different approaches as described in Chapter 9 which put the most likely value between 0.4 and 0.8 [29]. Moreover, the non-genetical variance involves only a very small component of common environment effects, such as home background, which are often postulated to be major determinants of intelligence in their own right [30]. This does not mean that IQ is independent of home background, but that it is the genes and hence the IQ which largely determine the home background, not the home background which determines IQ. There is genotype–environment correlation, in other words.

Such high genetical components and small common environment components do not suggest that there is any obvious panacea currently on the horizon that is going to improve people's IQ or remove the differences between people in any major way. On the other hand, it does not preclude the possibility. It does have social implications, however. Although genetic variation in IQ has presumably existed at its present level for many generations, only during this century have societies provided an environment where these differences in intelligence could result in individuals realizing their true potential. The highly intelligent son of a farm worker 100 years ago would have had little chance to receive an education and enter a career suited to his talents. He would probably have married the girl next door and worked on the farm like his father; the 'mute inglorious Milton' of Gray's *Elegy written in a country churchyard*. Now it would be much easier for his abilities to be recognized and for him to go to high school and university and probably enter a professional career. Society is allowing people to sort themselves according to ability, which, if the evidence on heritability is to be believed, means sorting themselves according to the number of plus QTLs they have for IQ.

IQ also correlates with career prospects, so to a greater extent than ever before, people are being ranked according to their genes both in educational achievement and in careers. This effect is coupled with strong assortative mating, because people tend to marry within the same education and job groups both through choice and opportunity. Now, a major effect of assortative mating is to increase the genetical variance in the population. We saw in Chapter 4 that the additive genetical variation between families with selfing is twice that with random mating. Assortative mating lies somewhere in between. Just as continued selfing causes the population to increase its genetical variation from $V_A^* + V_D^*$ to $2V_A^*$, so continual assortative mating increases the genetical variance, resulting in more people in the upper and lower tails of the IQ distribution. Given a knowledge of heritability, level of assortative mating and the correlation between IQ and job prospects, it should be possible for social scientists to explore the future consequences of these interacting effects.

Concern is frequently expressed that the proportion of University applicants from working class backgrounds is getting fewer rather than increasing with greater availability of education. It is suggested that this is evidence of financial pressures and lack of incentive which denies children the opportunities enjoyed by people from wealthier homes. But it is also probable that following two or three generations of selection for such a highly heritable trait as IQ, coupled with assortative mating and its association with people resorting themselves into different careers, that a significant proportion of the plus alleles for QTL associated with IQ may have been withdrawn from a major part of the population.

It has been argued that the differences in intelligence between different populations of man are difficult to reconcile with the evolutionary time since they separated between 20 000 to 50 000 years ago [31]. As conventionally measured, IQ has a mean of 100 and a standard deviation of 15 in white Caucasians, and there is evidence that mean IQ can range in other racial groups from 85 to 115. Using the formula given earlier for predicting selection response, R, the loss of just 1% against low IQ by natural selection (i.e. the top 99% survive to reproduce), would produce a very small selection differential, S, of 0.016. However, maintained over 1000 generations, or approximately 25 000 years, this would result in the mean IQ changing by 15 points, i.e. one standard deviation. Of course we have no idea how intense selection for intelligence has been over the last 20 000–50 000 years, but 'back of the envelope' calculations such as these would not suggest it needs to be very strong nor should it be very different in separated populations.

Quantitative genetics, with its hypothesis testing and predictive ability can clearly make a contribution to the debate on social issues as it can on evolutionary processes and breeding. Knowledge, whether palatable or not, is a better guide to coping with social problems than intuition or prejudice.

Summary

1. Better parent heterosis is normally associated with epistasis. This epistasis is invariably duplicate and never complementary in action and so opposes the effects of dominance.
2. In the absence of epistasis, heterosis appears to be mainly due to dispersed directionally dominant genes. This dominance may be partial to complete but little evidence can be found for over-dominance.
3. As a result of the findings in 1 and 2 above, it is often possible to produce inbred lines which have higher performance than their heterotic F_1 parent.
4. The performance of recombinant inbred lines (RILs) and their second cycle hybrids (SCHs) can be predicted from parameters obtained from early generations. Choices between crosses can be made using these predictions, allowing breeding resources to be concentrated on particular pedigrees.
5. The response to selection for individual traits can be predicted from a knowledge of the heritability and phenotypic variance. These predictions allow choices to be made between individual, family or combined selection strategies.
6. Theory suggests that indirect selection via a correlated trait is not worthwhile unless the correlated trait has both a very high heritability and a high genetic correlation with the trait to be improved.
7. Molecular markers may satisfy the requirements of 6 above. They can be used to assist the introgression of genes from distant relatives and facilitate selection of important QTL.
8. The genetical architecture of traits is probably determined by past natural selection and so the particular type of genetic control can be used to infer how the trait was selected in the past.
9. Knowledge of the genetics of human traits such as IQ provides a guide for exploring the consequences of changing social patterns in society.

References

1. Allard, R.W. (1960) *Principles of Plant Breeding*, John Wiley & Sons, New York.
2. Kearsey, M.J. and Pooni, H.S. (1992) The potential of inbred lines in the presence of heterosis, in *Reproductive Biology and Plant Breeding* (eds Y. Dattee, C. Dumas and A. Gallais), Springer-Verlag. London, pp. 371–85.
3. Gardner, C.O. (1963) Estimates of the genetic parameters in cross fertilised plants, and their implications in plant breeding, in *Statistical Genetics and Plant Breeding*. NAS NRC Publication No. 982, pp. 225–48.
4. Jinks, J.L. and Jones, R.M. (1958) Estimation of the components of heterosis. *Genetics*, **43** 223–34.
5. Jinks, J.L. and Perkins, J.M. (1972) Predicting the range of inbred lines. *Heredity*, **28**, 399–403.
6. Jinks, J.L. and Pooni, H.S. (1976) Predicting the properties of recombinant inbred lines derived by single seed descent. *Heredity*, **36**, 253–66.
7. Jinks, J.L. and Pooni, H.S. (1980) Comparing predictions of mean performance and environmental sensitivity of recombinant inbred lines based upon F_3 and triple test cross families. *Heredity*, **45**, 305–12.
8. Jinks, J.L. (1983) Biometrical genetics of heterosis, in *Heterosis* (ed. R. Frankel), Monographs on Theoretical and Applied Genetics, Springer-Verlag, Berlin, pp. 1–46.
9. Toledo, J.F.F. de, Pooni, H.S. and Jinks, J.L. (1984) Predicting the properties of second cycle hybrids produced by intercrossing random samples of recombinant inbred lines. *Heredity*, **53**, 283–92.
10. Eberhart, S.A. (1964) Theoretical relations among single, three-way, and double-cross hybrids. *Biometrics*, **20**, 522–39.
11. Pooni, H.S. and Jinks, J.L. (1979) Sources and biases of the predictors of the properties of recombinant inbreds produced by single seed descent. *Heredity*, **42**, 41–8.
12. Sokal, R.R. and Rohlf, F.J. (1980) *Biometry*, 2nd edn, W.H. Freeman & Co., New York.
13. Murdoch, J. and Barnes, J.A. (1974) *Statistical Tables*, Macmillan, New York.
14. Bulmer, M.G. (1980) *The Mathematical Theory of Quantitative Genetics*, Oxford University Press, Oxford.
15. Robertson, A. (1970) A theory of limits in artificial selection with many linked loci, in *Mathematical Topics in Population Genetics*, Vol. 1 (ed. K. Kojima), Springer-Verlag, Berlin, pp. 246–88.
16. Lerner, I.M. (1954) *Population Genetics and Animal Improvement*, Cambridge University Press, Cambridge.
17. Lush, J.L. (1947) Family merit and individual merit as bases for selection. *Am. Nat.*, **81**, 241–61, 362–79.
18. Robertson, A. (1955) Prediction equations in quantitative genetics. *Biometrics*, **11**, 95–8.

19. Toms, E.M., Cornish, M.A. and Kearsey, M.J. (1994) Comparison of family and sib selection during selfing. *Heredity*, **73**, 635–41.

20. Baker, R.J. (1986) *Selection Indices in Plant Breeding*, CRC Press, Boca Raton, Florida.

21. Welsh, J. and McClellan, M. (1990) Fingerprinting genomes using PCR with arbitrary primers. *Nucleic Acids Res.*, **18**, 7213–18.

22. Stam, P. and Zeven, A.C. (1981) The theoretical proportion of the donor genome in near isogenic lines of self-fertilisers bred by backcrossing. *Euphytica*, **30**, 227–38.

23. Lande, R. and Thompson, R. (1990) Efficiency of marker assisted selection in the improvement of quantitative traits. *Genetics*, **124**, 743–56.

24. Stuber, C.W. and Sisco, P.H. (1991) Marker facilitated transfer of QTL alleles between elite inbred lines and responses in hybrids. *Proceedings 46th Annual Corn and Sorghum Industry Research Conference*, American Seed Trade Assoc., **46**, 104–13.

25. Fisher, R.A. (1931) The evolution of dominance. *Biol. Rev.*, **6**, 345–68.

26. Mather, K. (1960) Evolution in polygenic systems. *Evoluzione e Genetica Colloquio Internazionale Roma*, 8–11 April, 1959, **47**, 131–52.

27. Kearsey, M.J. and Barnes, B.W. (1970) Variation for metrical characters in *Drosophila* populations. II Natural Selection. *Heredity*, **25**, 11–21.

28. McCall, R.B. (1977) Childhood IQ's as predictors of adult educational and occupational status. *Science*, **197**, 482–3.

29. Bouchard, T.J., Jr., Lykken, D.T., McGue, M., Segal, N.L. and Tellegen, A. (1990) Sources of human psychological differences: The Minnesota study of twins reared apart. *Science*, **250**, 223–8.

30. Plomin, R. and Bergeman, C.S. (1987) Why are children in the same family so different from one another? *Behavioural and Brain Sciences*, **10**, 1–60.

31. Gould, S.J. (1984) Human equality is a contingent fact of history. *Natural History* (November), pp. 26–83.

Further reading

Falconer, D.S. (1989) *Introduction to Quantitative Genetics*, 3rd edn, Longman, UK.

Frankel, R. (ed.) (1983) *Heterosis*. Monographs on Theoretical and Applied Genetics. Springer-Verlag, Berlin.

Hayward, M.D., Bosemark, N.O. and Romagosa, I. (eds) (1993) *Plant Breeding: Principles and Prospects*, Chapman & Hall, London.

Mather, K. and Jinks, J.L. (1982) *Biometrical Genetics*, Chapman & Hall, London.

Phillips, R.L. and Vasil, I.K. (eds) (1994) DNA-based markers in plants, in *Advances in Cellular and Molecular Biology of Plants*, Vol. 1, Kluwer Academic Publishers, Dordrecht, The Netherlands.

Experimental design 16

In this Chapter we will consider various aspects of the design and analysis of biometrical experiments so that trials can be planned in the most efficient manner and maximum information can be extracted from the data produced. Most researchers have a limited budget, finite space to carry out their trials or surveys and restricted time and resources. It is therefore essential that within these restraints the trials are designed to be cost effective and informative. Unfortunately, it is very easy to rush into an experiment with more zeal than objectivity, resulting in wasted resources and failure to detect important sources of variation. Failure to obtain significance in an experiment does not prove the effect is small or even absent, it just has not been detected to be significant.

16.1 Replication

A common question asked by anyone embarking on an experiment concerns the extent of replication required. Replication is, of course, essential to provide an unbiased estimate of the error variation, but the actual number of replicates will depend on a number of factors, as we shall see later. On many occasions one may be advised to use elaborate controlled environment equipment to reduce the error variation, but beware of such advice. While it is prudent to make reasonable attempts to reduce the error variation, i.e. one should not deliberately choose a heterogeneous trial area, further control can be very expensive while giving little reduction in error variance. This is because, as we saw in Chapter 3, much of the error variation is caused by factors which are not truly environmental, such as that due to statistical sampling, development and scoring. In practice it is invariably easier to reduce the variance of the treatment mean by a factor of two, for example, by doubling the replication than by increasing the control of the environment, even assuming that one knows what the critical environmental factors are which should be controlled. In commercial situations it is important to carry out trials under normal agricultural conditions with all the variability in management and sites that this may involve. Results obtained under controlled environmental conditions could be misleading. The standard error of a treatment mean, $\sqrt{(s^2(x)/r)}$, does not decrease linearly with the number of replicates, r. Thus, the standard error is reduced to a quarter by increasing the replication from say r to 16r while 64r replicates will be needed to reduce it to $\frac{1}{8}$th of its initial value.

16.1.1 The use of controls

A commonly used mechanism to reduce environmental error in plant trials is the inclusion of regular 'control' or 'check' varieties in the experiment. The idea is to use the difference between the score of a trial unit (individual or plot) and its nearest 'control' so that local environmental effects can be removed from the error. The key word here is **local**, as the method relies on the environmental variation being due to area effects which will affect both the trial unit and the 'control' to a similar extent. If x_i is the trial unit score and c_i is the nearest 'control' mean, then,

$$s^2(x_i - c_i) = s^2(x_i) + s^2(c_i) - 2s(x_ic_i),$$

where $s(x_ic_i)$ is the environmental covariance between x and c. If the environments of x and c are completely correlated, then this device will reduce $s^2(x_i - c_i)$ to zero, which is the aim of using controls. However, if there is no correlation, i.e. covariance is zero, the net result will be to increase the variance by a factor of 2; without the control the variance would be $s^2(x_i)$ while with the control it is $s^2(x_i) + s^2(c_i)$. In order for controls to reduce the error, $2s(x_ic_i)$ must be greater than $s^2(c_i)$, and this is probably not very common unless the trial site is a patchwork of very different environments. Before using such a device, it would be useful to compare errors before and after removing the scores of the 'controls'. It may prove better to replace the 'controls' with further replicates of the trial units.

16.1.2 The use of plots

Before attempting to answer the question of replicate size, we should consider what our basic unit of replication might be. If, for example, we have an experiment involving n families each containing r individuals we could, as we have in most of the illustrations used in this book, base the experiment on complete individual randomization. That is, all $n \times r$ individuals are allocated a position in the experiment at random. Other things being equal, this is in fact the most efficient design, as will be shown later. Although such a structure is easily achieved in plants it is much more difficult, if not impossible, in many animals, where a degree of maternal care is required until the young are weaned, and cross-fostering is the best that could be achieved in the past. *In vitro* fertilization may make complete individual randomization possible in mammals in the future, however. Even with plants, it may be necessary to keep individuals of the same family together in plots in order to simulate normal agronomic conditions. For example, a commercial crop of wheat would involve a virtually pure stand of one genotype where each plant is exposed to intense competition from near-identical neighbours. Thus if an F_3 was under study, for instance, it would be closer to a true farm situation if individuals from the same family were raised together in plots so that they would experience the type of competitive conditions encountered by the final cultivar. Complete randomization would require lower densities, because each plant would have to be separately identifiable, which will result in less competition because the density is less and the neighbours are different genotypes; the greatest competition occurs between like genotypes as they all have the same requirements. Such plots can also be managed and harvested in the same, or at least a very similar, manner to the normal crop so increasing the similarity with agronomic practice.

Plots are typically the unit of replication in most agronomic trials, although with early generation breeding material, the seed numbers available may be so low as to make plots of realistic size impossible. The breeder is forced either to use mini-plots, ear-rows, or even individual plant randomization, with all the problems these involve when it comes to extrapolating to field conditions.

Given that the experiment is made up of replicate plots, is it better to grow a few large plots of each family or several small plots? Suppose we have enough space to raise r individuals of every family, these can be spread over p plots of size k ($r = k \times p$). The ANOVA of such an experiment with n families is shown in Table 16.1. The variance ratio that tests for differences between families is,

$$F = MS_F/MS_P$$

with $n - 1$ and $n(p - 1)$ df.

The expected value of F is, therefore,

$$\epsilon F = (\sigma_W^2 + k\sigma_P^2 + r\sigma_F^2)/(\sigma_W^2 + k\sigma_P^2)$$
$$= 1 + (r\sigma_F^2)/(\sigma_W^2 + k\sigma_P^2).$$

Table 16.1 A generalized ANOVA of n families with p replicated plots per family and k individuals per plot. The total number of individuals per family is $k \times p = r$

Source	df	MS	ems
Between families	$n - 1$	MS_F	$\sigma_W^2 + k\sigma_P^2 + r\sigma_F^2$
Between plots within families	$n(p - 1)$	MS_P	$\sigma_W^2 + k\sigma_P^2$
Within plots	$np(k - 1)$	MS_W	σ_W^2

Clearly, F will increase as k decreases and will take its maximum value when $k = 1$, i.e. there are r plots of size 1. Also for a given value of r, as k decreases, p increases and with it the df of the denominator of F, so increasing the chance of F being significant given that there is a real difference between families. On both counts, therefore, the optimum design involves r plots of size 1.

This is the argument for complete individual randomization being the most efficient strategy on statistical considerations alone. Given that plots are necessary for agronomic reasons, then the same argument suggests we identify the **smallest** plot size that adequately mimics agricultural practice and have many replicate plots. This plot size will vary with the species and, to some extent, with the trait and can only be determined by experiment.

16.1.3 The use of blocks

Higher levels of replication, such as blocks, are used to enable known sources of heterogeneity to be removed from the error variance. Such sources might be known management or site differences or reflect units that can conveniently be handled at one time as with sowing or harvesting a crop. Equipment failure or sudden change in the weather may result in a discontinuity in handling a trial which could affect the evaluation of the trait. With blocks, these effects are easily removed while without blocks they can only be removed by regression techniques. It is always better to design a trial which has the capacity to cope with obvious contingencies than to rely on statistics to recover some order out of unreliable data. In practice, a very high proportion of cases where data require complex and unreasonable models can be traced to unreliable data arising from inadequate design control. If possible, it is always preferable to use balanced designs with equal sub-group sizes and constructed to permit orthogonal comparisons so that standard ANOVA can also be used to partition the component effects. The analysis of unbalanced designs requires additional and sometimes unwarranted assumptions to be made about the data as well as loss of degrees of freedom.

16.1.4 When to replicate

Randomization and replication should always be carried out as early as possible in the experiment because common environment effects can have a disproportionate effect at early stages. For example it was long ago shown with *Nicotiana* that 50% of the within-family environmental variation for flowering time had occurred in the first two weeks after sowing [1]. However, it is likely that common environmental effects among full sibs can arise from their sharing the same mother plant and being in the same packet of seed as with mammals given the same maternal care. If possible crosses should be replicated, using different mothers if that is feasible, while every attempt should be made to produce all the progeny for a trial under the same conditions. Seed of the same genotype but of different ages or from different sites can produce marked differences in traits at the end of the growing season. It follows from this, therefore, that seed quality differences between genotypes will be confused with genetic differences, which in turn can lead to spurious epistasis, $G \times E$ or other complications if not identified.

16.2 Power of experiments

Most experiments have two objectives; to test if a particular source of variation exists and, if it does, to estimate its magnitude. The power of an experiment relates to the first of these two objectives and tells us how likely a given experiment is to detect an effect of interest. Unfortunately the optimum structure of an experiment will not necessarily be the same for both detection and estimation.

In order to explain how the power of an experiment can be calculated we will consider a very simple example. Suppose we have two seed lots of a particular variety, one new and one old and we wish to see if they differ in germination rate. A trial is set up in which equal numbers of seed from the two lots are allowed to germinate and the number germinated (after a pre-determined length of time) are counted.

The null hypothesis initially is that the two lots of seed have the same expected germination and we could test the observed germination numbers for a 1 : 1 ratio by a χ^2 for 1 df. Given the normal significance level of $P = 0.05$ or 5%, should the χ^2 be less than 3.84 we would accept the null hypothesis and if it was 3.84 or more we would reject it.

By definition of the significance level, the null hypothesis would be rejected falsely on 5% of occasions and we would conclude that there was a difference when there was none. This is called a Type I error. Of course, we can always set more stringent standards and use a 2.5% or 1% significance level. If there really is a difference between the two seed lots, then the null hypothesis is not applicable and a significant result will be obtained on more than 5% of occasions. However, how much more often will clearly depend on the extent of the difference between the samples. A very small difference may result in the proportion of rejections increasing only, say, to 7%, while a very large difference might invariably make us correctly reject the null hypothesis. A Type II error arises when we accept the null hypothesis while in fact it is false. That is, the two seed lots do have different germination rates but our experiment does not detect the difference as being significant.

The **power** of an experiment is measured as the probability of correctly rejecting the null hypothesis when it is false. In our example, it will be the probability of getting a significant χ^2 when the two seed lots really do differ in germination rate. In order to calculate the power, we need to have some alternative hypothesis. It is not enough simply to ask if the two populations differ, we need to have some particular degree of difference in mind, as it must be easier to detect a large difference than a small one. It might be unreasonable to believe that the two seed lots were exactly the same, but the experimenter would be more concerned to show that the differences were not trivial and so a minimum difference that would be considered non-trivial has to be set. For example, he may only be concerned to detect a difference of 10% or more in germination rate, while smaller differences are not so crucial. How large should an experiment be in order to detect a difference of at least this magnitude if it exists? Therefore, the new hypothesis is that the difference is greater than or equal to 10%.

The general solution to this problem is simple and is illustrated in Table 16.2. Let the expected proportion of 'new' seed germinating be 1.0 while $(1 - Q)$ is the expected proportion of 'old' seed germinating; Q represents the proportionate difference in germination. We see from Table 16.2 that the expected number of 'new' n' and 'old' n'' seed

Table 16.2 An experiment to compare germination in old and new seed

Sample	New seed	Old seed	Total
Numbers at start	$\frac{1}{2}N$	$\frac{1}{2}N$	N
Proportion germinating	1	$1 - Q$	–
Expected number germinating	$\frac{1}{2}N$	$\frac{1}{2}N(1 - Q)$	$\frac{1}{2}N(2 - Q)$
Observed number germinating	x	y	x + y

germinating, are,

$$n' = N/(2 - Q) \quad \text{and}$$

$$n'' = N(1 - Q)/(2 - Q).$$

Given this, the expected value of χ^2 is

$$\chi^2 = (n' - n'')^2/N$$

$$= NQ^2/(2 - Q)^2.$$

We can calculate the value of N to give any expected value of χ^2 providing we know the value of Q, i.e.

$$N = \chi^2(2 - Q)^2/Q^2.$$

In our example Q was 0.1 (10% reduction) and hence to achieve significance at 5% we set χ^2 at 3.84, from which we obtain N = 1386.

However, if we were to score this number of seed and Q was 0.1, then we would only obtain an observed value of $\chi^2 > 3.84$ on 50% of occasions. We would have a 50% power, in other words, and would still fail to detect the difference on the remaining 50% of occasions. Generally, we only do the experiment once, so we would have only a 50:50 chance of detecting what is in fact quite a large decline in germination. Normally higher power is required and tables for calculating this are to be found in *Biometrika Tables* [2]. Thus, to be 95% (β) sure of detecting a significant χ^2 at the 5% level (α) we need to set χ^2 at 13.0. That is, if the expected value of χ^2 for 1 df is 13, then in 95% of experiments the observed χ^2 will be greater than 3.84, the 5% significance level. This will require scoring N = 4693 germinated seeds.

This experiment could also be analysed as a 2×2 contingency χ^2, where the classes will be 'old' versus 'new' seed lots and germinated versus ungerminated. Assuming all the 'new' seeds germinate, the reader may wish to calculate the size of experiment to give the same power as before. A major reduction will be found indicating that it is worth counting the 'corpses'.

We can use the same approach to illustrate an example involving a one-way ANOVA. Let us suppose we wish to compare the mean performance of lines produced from an F_1 by two different methods, e.g. SSD and anther culture, and we wish to construct an experiment with a given power to detect a difference of a particular magnitude. The appropriate ANOVA is as given in Table 4.1 from which we see that the variance ratio of the expected mean squares is,

$$F = MS_B/MS_W$$

$$= 1 + r\sigma_B^2/\sigma_W^2.$$

Providing the df of MS_W are large, F is approximately equal to χ^2 for 1 df, and hence we can essentially use the same approach as before.

For example, the overall mean, the additive genetic variance $(2V_A^*)$ and the environmental variance for height of a set of SSD lines extracted from a particular cross of *Nicotiana* are known to be 135, 676 and 324 respectively. We wish to design an experiment that will detect a reduction of at least 5% in the overall mean of the lines produced by anther culture, i.e. the mean height should be no more than 128.25.

The expected MS_B is,

$$(135.0 - 128.25)^2 r/2,$$

while the expected value of $MS_W = \sigma_W^2 = $ total variance of inbreds

$$= 676 + 324 = 1000.0.$$

Therefore,

$$\chi^2 = MS_B/MS_W = 6.75^2 r/2000$$

and $$r = 2000/45.5625\chi^2.$$

If we take the value of χ^2 that we used before (13.0) to give us 95% power, then this translates to 571 replicates of each treatment.

16.3 Power of biometrical experiments based on ANOVA

We have introduced the idea of power calculations for non-biometrical examples, and providing it is possible to construct a model, there is no reason why it cannot be extended to biometrical situations also. For example, if we consider the use of an NCII to detect dominance we could ask how powerful a given-sized experiment would be to detect dominance of a particular magnitude. As before, we have to set the minimum amount of dominance we would wish to detect and this will have to be set in the context of the additive genetic and environmental variances.

It will be recalled from Table 5.5 that the F test for dominance in an NCII is,

$$F = MS_{MF}/MS_W \text{ for } (n_1 - 1)(n_2 - 1) \text{ and } n_1 n_2 (r - 1) \text{ df,}$$

so that the expected value of

$$F = 1 + r\sigma_{MF}^2/\sigma_W^2$$

where,

$$\sigma_{MF}^2 = \tfrac{1}{4} V_D^*$$

$$\sigma_W^2 = \tfrac{1}{2} V_A^* + \tfrac{3}{4} V_D^* + V_E.$$

Call this expected value of F, λ. From the values of r, V_A^*, V_D^* and V_E, the expected value of F, that is λ, can be calculated and the power of any particular experiment to detect significant dominance can be obtained. For the purposes of illustration, consider an absurdly small NCII involving five males crossed to three females with 11 replicate individuals per family. Thus,

$$\epsilon F = 1 + 11\sigma_{MF}^2/\sigma_W^2 \text{ for 8 and 150 df.}$$

With these df the minimum value of the variance ratio that we require for it to be significant is $F > 2.0$.

We now need to define the genetical model and this would normally be based on some prior knowledge of the material, such as the phenotypic variance of the trait and its broad heritability. In this case we will assume that the trait has a phenotypic variance of 10.0, a broad heritability of 75% and complete dominance. This implies that $V_A^* = 5.0$ and $V_D^* = V_E = 2.5$, which results in ems of,

$$\sigma_{MF}^2 = 10/16$$

$$\sigma_W^2 = 110/16$$

and hence,

$$\epsilon F = 1 + (11 \times 10/16)/(110/16)$$

$$= 2.0.$$

In this contrived illustration, therefore, the expected value of F just happens to be the same as the table value of F for 5% significance. In order to calculate the power of the test we need to know how often the F value should exceed 2.0 given that its expected value (λ) is 2.0.

This is fairly straightforward because the distribution of F given $\lambda > 1.0$, is the same as the usual distribution of F based on the Null hypothesis where $\lambda = 1.0$, except that all the F values along the abscissa are λ times larger as shown in Figure 16.1. Thus, with 8 and 150 df the value of F for 5% significance, assuming the NH ($\lambda = 1.0$), is 2.0, while the 5% value of F given the alternative hypothesis based on the model ($\lambda = 2.0$) will be $\lambda \times 2.0 = 4.0$. Conversely, therefore, to calculate the power of the experiment, we simply divide the F value for 5% significance by λ and obtain the area in the tail of the F distribution in the usual way. In this example, $\lambda = 2.0$, and

Figure 16.1 Distribution of
variance ratio for (i) the null
hypothesis $\lambda = 0$ and (ii) on the
alternative hypothesis of $\lambda \neq 0$.
(iii) The two distributions
combined.

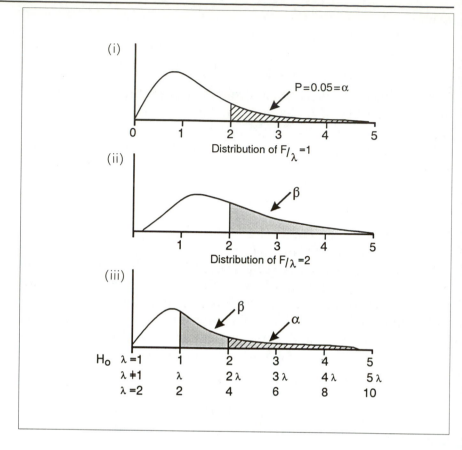

F at 5% is 2.0, therefore the appropriate value of F to look up is $F/\lambda = 1$. We need to know the probability of $F > 1.0$ given the NH ($\lambda = 1.0$). This is not easily found from tables but can be calculated from the F distribution using standard computer software. With 8 and 150 df, $F > 1.0$ on 44% of occasions, i.e. β is 44%.

So, given the genetical model and an NCII of the size described we would expect to detect significant dominance on 44% of occasions. Conversely, we would fail to detect it on 56% of occasions, despite the large amount of dominance present. We could explore other ways of carrying out the experiment, however, to see how they affect our likely success rate. Let us suppose that we have limited field space and that we can only raise a total of about 165 progeny as in the first experiment. What would happen if we used five males, 11 females and three sibs per family? The appropriate F value for significance at 5% is now 1.5, while λ becomes 1.2727. The power in this case can be obtained from the probability of $F > 1.5/1.2727$ with 40 and 110 df. This proves to be 25%, much worse than before.

This illustrates how important the structure of the experiment is in determining the successful outcome of the work. By simply using a small family size we have managed to halve the power of the experiment, despite the total size and, presumably, the effort and resources staying the same.

So far we have considered the power of significance tests. A different approach is to consider how to design an experiment so as to minimize the variance of the parameters to be estimated. The approach can again be illustrated with the NCII example used earlier.

An estimate of V_D^* can be obtained in the usual way (see Chapter 5) from the observed values of MS_{MF} and MS_W obtained from the ANOVA as,

$$V_D^* = (MS_{MF} - MS_W) \times 4/r.$$

The expected variance of V_D^* (s_{VD}^2) is therefore,

$$s_{VD}^2 = (4/r)^2 \{s^2(MS_{MF}) + s^2(MS_W)\}.$$

There is no covariance term because the MS in the ANOVA are orthogonal. Now the expected variance of a MS is twice the expected value of the MS squared divided by its df,

$$s_{MS}^2 = 2(\epsilon MS)^2/df.$$

Thus,

$$s_{VD}^2 = (4/r)^2 2\{(\epsilon MS_{MF})^2/df_1 + (\epsilon MS_W)^2/df_2\}, \qquad \text{[Eqn 16.1]}$$

where df_1 and df_2 are the degrees of freedom of MS_{MF} and MS_W respectively.

If we take the first example, with five males, three females and $r = 11$ sibs per family, and use the same genetical model with $\sigma_W^2 = 110/16$ and $\sigma_{MF}^2 = 10/16$, then,

$$\epsilon MS_{MF} = \sigma_W^2 + r\sigma_{MF}^2$$

$$= 220/16$$

$$\epsilon MS_W = 110/16$$

$$df_1 = 8\,df$$

$$df_2 = 150\,df.$$

Substituting in Equation 14.1 gives $s_{VD}^2 = 6.33$, i.e. a standard error of 2.52. Since V_D^* was set at 2.50, a standard error of 2.52 does not inspire confidence. Had we used the alternative design with five males, 11 females and three sibs per family, the corresponding standard error proves to be 2.89, marginally worse. We have used an unrealistically small experiment for the purposes of illustration, so we should not be surprised that its power and reliability are low.

16.4 Reliability of the additive genetic variance

The estimation of the additive genetic variance or the narrow heritability is so central to quantitative genetics that considerable effort has been devoted to optimizing experiments to estimate them. Two of the most frequently used analyses for this purpose are FS, i.e. BIPs, and HS, such as the NCI. The variance of the estimate of the additive genetic variance is easily computed from these two designs using the same approach as in Equation 16.1. Thus for FS,

$$s_{VA}^2 = 8(\epsilon MS_B^2/df_1 + \epsilon MS_W^2/df_2)/r^2 \qquad \text{[Eqn 16.2]}$$

where $\epsilon MS_B = \frac{1}{2}V_A^*$ and $\epsilon MS_W = $ Total variance $- \frac{1}{2}V_A^*$, while for HS,

$$s_{VA}^2 = 32(\epsilon MS_M^2/df_1 + \epsilon MS_W^2/df_2)/r^2 \qquad \text{[Eqn 16.3]}$$

where $\epsilon MS_B = \frac{1}{4}V_A^*$ and $\epsilon MS_W = $ Total variance $- \frac{1}{4}V_A^*$.

In the latter we are assuming that all FS families are of size 1.

Using these formulae, the standard deviation of the additive genetic variance has been computed as a proportion of the expected additive variation for various experimental sizes in Appendices A and B for FS and HS families respectively. For any particular family structure, the expected standard error of the additive genetic variation can be ascertained. As one would expect, the precision of the estimates increase, i.e. the standard deviations decrease, as the number of families increase for a given family size or as the family size increases for a given number of families. There are however optimum strategies. For example, if one could raise 1000 individuals in a FS experiment it is clearly better, if the heritability is 20%, to use 100 families of size 10 (sd = 0.27) rather than 50 families of size 20 (sd = 0.29), 200 families of size 5 (sd = 0.29) or 500 families of size 2 (sd = 0.45). Irrespective of the total size the optimum strategy is to have the family size, $r = 2/h^2$ for FS and $4/h^2$ for HS families. Thus if h^2 is 0.2, the optimum r for FS will be 10 which is what we obtained above for an experiment of 1000 individuals. Had there been enough resources to raise 2000, for instance, the optimum strategy would still be to have r = 10 as 200 families of size 10 give greater precision than any other combination as can be seen from Appendix A. HS are considerably less precise than FS and at the optimum family sizes they are half as precise. If several progeny are raised from each mother, the precision for the same total size of experiment is even less.

Calculations such as these take little time and can easily be programmed on a PC [3]. They enable the efficiency of the trials to be maximized before they are carried out and can avoid a costly waste of time and resources on an inappropriate design. It may even convince the experimenter that the objective is unattainable and the trial can be aborted before any resources are committed. Often one hopes to obtain answers to several questions at the same time. Unfortunately it is generally found that the best design for answering one question will be sub-optimal for others. Thus in optimizing conditions for studying dominance with an NCII we may be creating difficulties with respect to additive variation. Another approach to optimizing the power of an experiment is that of computer simulation. This has the advantage of allowing more complex experimental situations to be explored, particularly those in which it is difficult to obtain simple statistical solutions.

16.5 Data analysis

We now wish to turn to another aspect of the experiment, namely the analysis of the data, and to make a few points which may prove useful in practice.

An obvious point, but one which experience shows is worth emphasizing nonetheless, is the importance of checking that the data are correct. Most statistical software will detect outliers. So, really aberrant values can be corrected, but it is important to check the data for more minor transcription errors either by having the data entered twice and getting the computer to check for discrepancies or by the old-fashioned, but reliable method of calling out the data to someone else to check against a listing. Inaccurate data can create real difficulties in the final interpretation of the results which are quite unnecessary.

Although normality of the underlying data is a formal requirement for all ANOVA, they are very robust against quite serious non-normality. Table 16.3 illustrates the two-way ANOVA of a set of NCII data using the raw scores (i) and the same individual scores but transformed into natural logarithms to the base e. Although the absolute values of the MS have changed, the F and P values have altered very little.

Table 16.3 Two-way ANOVA with replicates carried out on the raw data (normally distributed) and on the $\log_e$ of the raw data to show the robustness of the analysis to non-normality

Source	df	MS (raw)	F	P	MS ($\log_e$)	F	P
Rows	4	301.01	5.69	*	0.0524	5.29	*
Columns	4	312.93	5.92	*	0.0548	5.54	*
R × C	16	52.89	0.62	n.s.	0.0099	0.63	n.s.
Error	100	84.75	–		0.0158	–	

One important situation where a transformation of the scale will be necessary, however, is when the basic data are proportions, i.e. measures of disease attack or lodging. This is because distributions of proportions are very skewed when their mean value is near 0 or 1. The appropriate transformation, arc-sine or angular, not only eliminates non-normality but can also provide a theoretical error against which to assess the empirical error. Scale transformation and its uses are well explained in most texts of statistics.

Summary

1. The efficiency of any experiment, and biometrical experiments are no exception, can be increased greatly by the choice of suitable design and by additional replication.
2. The standard error of the treatment mean can be considerably reduced by using up to 10 replicates while further decrease requires a geometric increase in replication.
3. The use of frequent 'control' varieties in a trial may reduce the experimental error but will most probably increase it unless there are marked 'area effects'.
4. Blocking should be used to remove known sources of heterogeneity in an experiment.
5. The power of an experiment to detect an effect of interest can be explored before the experiment is started and the most powerful design should be used to avoid wasting manpower and valuable resources.
6. It is always preferable to have reliable data from a well-designed, balanced and powerful experiment than to rely on statistical procedures to bring order out of experimental chaos.

References

1. Mather, K. and Jinks, J.L. (1982) *Biometrical Genetics*, 3rd edn, Chapman & Hall, London
2. Pearson, E.S. and Hartley, H.O. (1972) *Biometrika Tables for Statisticians*, Cambridge University Press, London.
3. Kearsey, M.J. (1970) Experimental sizes for detecting dominance variation. *Heredity*, **25**, 529–42.

Further reading

Mead, R., Curnow, R.N. and Hasted, A.M. (1993) *Statistical Methods in Agriculture and Experimental Biology*, Chapman & Hall, London.

Appendix A

Precision of h_n^2 with FS families

Expected standard deviation of the estimate of the additive genetical variation, as a proportion of the true additive variation, when estimated from various numbers of FS families of different sizes

h^2	reps (r)	Number of FS families (n)						
		50	100	150	200	500	750	1000
60%	2	0.50	0.35	0.28	0.25	0.16	0.13	0.11
	5	0.30	0.21	0.17	0.15	0.09	0.08	0.07
	10	0.25	0.18	0.14	0.12	0.08	0.06	0.06
	20	0.22	0.16	0.13	0.11	0.07	0.06	0.05
40%	2	0.73	0.51	0.42	0.36	0.23	0.19	0.16
	5	0.37	0.26	0.21	0.18	0.12	0.10	0.08
	10	0.28	0.20	0.16	0.14	0.09	0.07	0.06
	20	0.24	0.17	0.14	0.12	0.08	0.06	0.05
20%	2	1.43	1.00	0.82	0.71	0.45	0.37	0.32
	5	0.59	0.42	0.34	0.29	0.19	0.15	0.13
	10	0.39	0.27	0.22	0.19	0.12	0.10	0.09
	20	0.29	0.21	0.17	0.15	0.09	0.08	0.07
10%	2	2.85	2.01	1.64	1.42	0.90	0.73	0.63
	5	1.04	0.73	0.60	0.52	0.33	0.27	0.23
	10	0.60	0.42	0.34	0.30	0.19	0.15	0.13
	20	0.40	0.28	0.23	0.20	0.12	0.10	0.09
5%	2	5.67	3.99	3.26	2.82	1.78	1.45	1.26
	5	1.94	1.37	1.11	0.96	0.61	0.50	0.43
	10	1.02	0.72	0.58	0.51	0.32	0.26	0.23
	20	0.60	0.42	0.34	0.30	0.19	0.15	0.13
	40	0.40	0.28	0.23	0.20	0.13	0.10	0.09

Appendix B

Expected standard deviation of the estimate of the additive genetical variation, as a proportion of the true additive variation, when estimated from various numbers of HS families of different sizes. All FS families are of size one

h^2	reps (r)	Number of HS families (n)						
		50	100	150	200	500	750	1000
60%	2	0.96	0.68	0.55	0.48	0.30	0.25	0.21
	5	0.45	0.31	0.26	0.22	0.14	0.11	0.10
	10	0.32	0.22	0.18	0.16	0.10	0.08	0.07
	20	0.26	0.18	0.15	0.13	0.08	0.07	0.06
40%	2	1.43	1.01	0.82	0.71	0.45	0.37	0.32
	5	0.59	0.42	0.34	0.29	0.19	0.15	0.13
	10	0.39	0.27	0.22	0.19	0.13	0.10	0.09
	20	0.30	0.21	0.17	0.15	0.09	0.08	0.07
20%	2	2.84	2.01	1.64	1.42	0.90	0.73	0.63
	5	1.04	0.73	0.60	0.52	0.33	0.27	0.23
	10	0.60	0.42	0.34	0.30	0.19	0.15	0.13
	20	0.40	0.28	0.23	0.20	0.12	0.10	0.09
10%	2	5.66	3.99	3.26	2.82	1.78	1.45	1.26
	5	1.94	1.37	1.11	0.96	0.61	0.50	0.43
	10	1.02	0.72	0.58	0.51	0.32	0.26	0.23
	20	0.60	0.43	0.34	0.30	0.19	0.15	0.13
5%	2	11.32	7.98	6.51	5.64	3.56	2.91	2.52
	5	3.74	2.63	2.15	1.86	1.17	0.96	0.83
	10	1.86	1.32	1.07	0.93	0.59	0.48	0.41
	20	1.01	0.71	0.58	0.50	0.32	0.26	0.22
	40	0.60	0.42	0.34	0.30	0.19	0.15	0.13

Appendix C

Statistical tables: F, χ^2, t

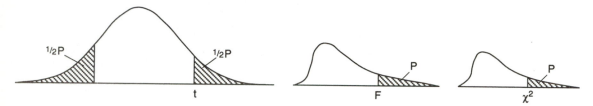

Critical values of F, t and χ^2 distributions. Top line 5%, bottom line 1% values. F is a universal test and t, c and χ^2 are its special cases. In general, $t^2_{(n2)} \equiv F_{(1,n2)}$, $\chi^2_{(n1)} \equiv n_1 \times F_{(n1,\infty)}$ and $c \equiv t_{(\infty)}$

| df (n_2) | t | \multicolumn{14}{c}{F values {df (n_1) for MS}} |
		1	2	3	4	5	6	7	8	9	10	20	50	100	∞
1	12.71	161	200	216	225	230	234	237	239	241	242	248	252	253	254
	63.66	4052	5000	5403	5625	5764	5859	5928	5981	6023	6056	6209	6302	6334	6366
2	4.30	18.51	19.00	19.16	19.25	19.30	19.33	19.35	19.37	19.38	19.39	19.44	19.47	19.49	19.50
	9.93	98.49	99.00	99.17	99.25	99.30	99.33	99.36	99.37	99.39	99.40	99.45	99.48	99.49	99.50
3	3.18	10.13	9.55	9.28	9.12	9.01	8.94	8.88	8.84	8.81	8.79	8.66	8.58	8.56	8.53
	5.84	34.12	30.82	29.46	28.71	28.24	27.91	27.67	27.49	27.34	27.23	26.69	26.35	26.23	26.12
4	2.78	7.71	6.94	6.59	6.39	6.26	6.16	6.09	6.04	6.00	5.96	5.80	5.70	5.66	5.63
	4.60	21.20	18.00	16.69	15.98	15.52	15.21	14.98	14.80	14.66	14.54	14.02	13.69	13.57	13.46
5	2.57	6.61	5.79	5.41	5.19	5.05	4.95	4.88	4.82	4.77	4.74	4.56	4.44	4.40	4.36
	4.03	16.26	13.27	12.06	11.39	10.97	10.67	10.45	10.29	10.15	10.05	9.55	9.24	9.13	9.02
6	2.45	5.99	5.14	4.76	4.53	4.39	4.28	4.21	4.15	4.10	4.06	3.87	3.75	3.71	3.67
	3.71	13.74	10.92	9.78	9.15	8.75	8.47	8.26	8.10	7.98	7.87	7.39	7.09	6.99	6.88
7	2.37	5.59	4.74	4.35	4.12	3.97	3.87	3.79	3.73	3.68	3.64	3.44	3.32	3.28	3.23
	3.50	12.25	9.55	8.45	7.85	7.46	7.19	7.00	6.84	6.71	6.62	6.15	5.85	5.75	5.65
8	2.31	5.32	4.46	4.07	3.84	3.69	3.58	3.50	3.44	3.39	3.35	3.15	3.03	2.98	2.93
	3.36	11.26	8.65	7.59	7.01	6.63	6.37	6.19	6.03	5.91	5.82	5.36	5.06	4.96	4.86
9	2.26	5.12	4.26	3.86	3.63	3.48	3.37	3.29	3.23	3.18	3.14	2.93	2.80	2.76	2.71
	3.25	10.56	8.02	6.99	6.42	6.06	5.80	5.62	5.47	5.35	5.26	4.80	4.51	4.41	4.31
10	2.23	4.96	4.10	3.71	3.48	3.33	3.22	3.14	3.07	3.02	2.98	2.77	2.64	2.59	2.54
	3.17	10.04	7.56	6.55	5.99	5.64	5.39	5.21	5.06	4.95	4.85	4.41	4.12	4.01	3.91
11	2.20	4.84	3.98	3.59	3.36	3.20	3.09	3.01	2.95	2.90	2.85	2.65	2.50	2.45	2.40
	3.11	9.65	7.20	6.22	5.67	5.32	5.07	4.88	4.94	4.63	4.54	4.10	3.80	3.70	3.60
12	2.18	4.75	3.89	3.49	3.26	3.11	3.00	2.91	2.85	2.80	2.75	2.54	2.40	2.35	2.30
	3.06	9.33	6.93	5.95	5.41	5.06	4.82	4.65	4.50	4.39	4.30	3.86	3.56	3.46	3.36
13	2.16	4.67	3.81	3.41	3.18	3.03	2.92	2.83	2.77	2.71	2.67	2.46	2.32	2.26	2.21
	3.01	9.07	6.70	5.74	5.20	4.86	4.62	4.44	4.30	4.19	4.10	3.67	3.37	3.27	3.16
14	2.15	4.60	3.74	3.34	3.11	2.96	2.85	2.76	2.70	2.65	2.60	2.39	2.24	2.19	2.13
	2.98	8.86	6.51	5.56	5.03	4.69	4.46	4.28	4.14	4.03	3.94	3.51	3.21	3.11	3.00
15	2.13	4.54	3.68	3.29	3.06	2.90	2.79	2.71	2.64	2.59	2.54	2.33	2.18	2.12	2.07
	2.95	8.68	6.36	5.42	4.89	4.56	4.32	4.14	4.00	3.89	3.80	3.36	3.07	2.97	2.87

df (n_2)	t	1	2	3	4	5	6	7	8	9	10	20	50	100	∞
							F values {df (n_1) for MS}								
16	2.12	4.49	3.63	3.24	3.01	2.85	2.74	2.66	2.59	2.54	2.49	2.28	2.13	2.07	2.01
	2.92	8.53	6.23	5.29	4.77	4.44	4.20	4.03	3.89	3.78	3.69	3.25	2.96	2.86	2.75
17	2.11	4.45	3.59	3.20	2.96	2.81	2.70	2.61	2.55	2.49	2.45	2.23	2.08	2.02	1.96
	2.90	8.40	6.11	5.18	4.67	4.34	4.10	3.93	3.79	3.68	3.59	3.16	2.86	2.76	2.65
18	2.10	4.41	3.55	3.16	2.93	2.77	2.66	2.58	2.51	2.46	2.41	2.19	2.04	1.98	1.92
	2.88	8.28	6.01	5.09	4.58	4.25	4.01	3.85	3.71	3.60	3.51	3.07	2.78	2.68	2.57
19	2.09	4.38	3.52	3.13	2.90	2.74	2.63	2.54	2.48	2.43	2.38	2.15	2.00	1.94	1.88
	2.86	8.18	5.93	5.01	4.50	4.17	3.94	3.77	3.63	3.52	3.43	3.00	2.70	2.60	2.49
20	2.09	4.35	3.49	3.10	2.87	2.71	2.60	2.51	2.45	2.40	2.35	2.12	1.96	1.90	1.84
	2.85	8.10	5.85	4.94	4.43	4.10	3.87	3.71	3.56	3.45	3.37	2.94	2.63	2.53	2.42
21	2.08	4.32	3.47	3.07	2.84	2.68	2.57	2.49	2.42	2.37	2.32	2.09	1.93	1.87	1.81
	2.83	8.02	5.78	4.87	4.37	4.04	3.81	3.65	3.51	3.40	3.31	2.88	2.58	2.47	2.36
22	2.07	4.30	3.44	3.05	2.82	2.66	2.55	2.46	2.40	2.35	2.30	2.07	1.91	1.84	1.78
	2.82	7.94	5.72	4.82	4.31	3.99	3.76	3.59	3.45	3.35	3.26	2.83	2.53	2.42	2.31
23	2.07	4.28	3.42	3.03	2.80	2.64	2.53	2.44	2.38	2.32	2.28	2.04	1.88	1.82	1.76
	2.81	7.88	5.66	4.76	4.26	3.94	3.71	3.54	3.41	3.30	3.21	2.78	2.48	2.37	2.26
24	2.06	4.26	3.40	3.01	2.78	2.62	2.51	2.42	2.36	2.30	2.26	2.02	1.86	1.80	1.73
	2.80	7.82	5.61	4.72	4.22	3.90	3.67	3.50	3.36	3.25	3.17	2.74	2.44	2.33	2.21
25	2.06	4.24	3.39	2.99	2.76	2.60	2.49	2.40	2.34	2.28	2.24	2.00	1.84	1.77	1.71
	2.79	7.77	5.57	4.68	4.18	3.86	3.63	3.46	3.32	3.21	3.13	2.70	2.40	2.29	2.17
26	2.06	4.23	3.37	2.98	2.74	2.59	2.47	2.39	2.32	2.27	2.22	1.99	1.82	1.76	1.69
	2.78	7.72	5.53	4.64	4.14	3.82	3.59	3.42	3.29	3.17	3.09	2.66	2.36	2.25	2.13
27	2.05	4.21	3.35	2.96	2.73	2.57	2.46	2.37	2.30	2.25	2.20	1.97	1.80	1.74	1.67
	2.77	7.68	5.49	4.60	4.11	3.79	3.56	3.39	3.26	3.14	3.06	2.63	2.33	2.21	2.10
28	2.05	4.20	3.34	2.95	2.71	2.56	2.44	2.36	2.29	2.24	2.19	1.96	1.78	1.72	1.65
	2.76	7.64	5.45	4.57	4.07	3.76	3.53	3.36	3.23	3.11	3.03	2.60	2.30	2.18	2.06
29	2.05	4.18	3.33	2.93	2.70	2.54	2.43	2.35	2.28	2.22	2.18	1.94	1.77	1.71	1.64
	2.76	7.60	5.42	4.54	4.04	3.73	3.50	3.33	3.20	3.08	3.00	2.57	2.27	2.15	2.03
30	2.04	4.17	3.32	2.92	2.69	2.53	2.42	2.34	2.27	2.21	2.16	1.93	1.76	1.69	1.62
	2.75	7.56	5.39	4.51	4.02	3.70	3.47	3.30	3.17	3.06	2.98	2.55	2.24	2.13	2.01
40	2.02	4.08	3.23	2.84	2.61	2.45	2.34	2.25	2.18	2.12	2.07	1.84	1.66	1.59	1.51
	2.70	7.31	5.18	4.31	3.83	3.51	3.29	3.12	2.99	2.88	2.80	2.37	2.05	1.94	1.81
50	2.01	4.03	3.18	2.79	2.56	2.40	2.29	2.20	2.13	2.07	2.02	1.78	1.60	1.52	1.44
	2.68	7.17	5.06	4.20	3.72	3.41	3.18	3.02	2.88	2.78	2.70	2.26	1.94	1.82	1.68
60	2.00	4.00	3.15	2.76	2.52	2.37	2.25	2.17	2.10	2.04	1.99	1.75	1.56	1.48	1.39
	2.66	7.08	4.98	4.13	3.65	3.34	3.12	2.95	2.82	2.72	2.63	2.20	1.87	1.74	1.60
120	1.98	3.92	3.07	2.68	2.45	2.29	2.18	2.09	2.02	1.96	1.91	1.66	1.46	1.37	1.25
	2.62	6.85	4.79	3.95	3.48	3.17	2.96	2.79	2.66	2.56	2.47	2.03	1.70	1.56	1.38
∞	1.96	3.84	2.99	2.60	2.37	2.21	2.09	2.01	1.94	1.88	1.83	1.57	1.35	1.24	1.00
	2.58	6.63	4.60	3.78	3.32	3.02	2.80	2.64	2.51	2.41	2.32	1.87	1.52	1.36	1.00
$\chi^2_{(n1)}$		3.84	5.99	7.82	9.49	11.07	12.59	14.07	15.51	16.92	18.31	31.41	67.51	124.3	∞
		6.64	9.21	11.35	13.28	15.09	16.81	18.48	20.09	21.67	23.21	37.57	76.15	135.8	∞

Appendix D

Normal deviate and intesity of selection, (*i*)

Values of normal deviate (x) and intensity of selection (*i*) for truncated selection where top P% of individuals are selected from a random mating population

P%	x	*i*	P%	x	*i*
0.5	2.576	2.892	6	1.555	1.985
1.0	2.326	2.665	7	1.476	1.918
1.5	2.17	2.527	8	1.405	1.858
2.0	2.054	2.421	9	1.341	1.804
2.5	1.96	2.336	10	1.282	1.755
3.0	1.881	2.268	15	1.036	1.554
3.5	1.818	2.174	20	0.842	1.400
4.0	1.751	2.154	25	0.674	1.271
4.5	1.694	2.127	30	0.524	1.159
5.0	1.645	2.063	50	0.000	0.798

Appendix E

Area under the normal curve

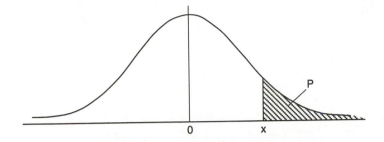

Area (P%) in the RHS tail of the normal curve for the given values of x

x	P%	x	P%	x	P%	x	P%	x	P%
0.00	50.0	0.55	29.1	1.10	13.6	1.65	4.9	2.30	1.1
0.05	48.0	0.60	27.4	1.15	12.5	1.70	4.5	2.35	0.9
0.10	46.0	0.65	25.8	1.20	11.5	1.75	4.0	2.40	0.8
0.15	44.0	0.70	24.2	1.25	10.6	1.80	3.6	2.45	0.7
0.20	42.0	0.75	22.7	1.30	9.7	1.90	2.9	2.50	0.6
0.25	40.0	0.80	21.2	1.35	8.9	2.00	2.3	2.55	0.5
0.30	38.2	0.85	19.8	1.40	8.1	2.05	2.0	2.60	0.5
0.35	36.3	0.90	18.4	1.45	7.4	2.10	1.8	2.70	0.3
0.40	34.5	0.95	17.1	1.50	6.7	2.15	1.6	2.80	0.3
0.45	32.6	1.00	15.9	1.55	6.1	2.20	1.4	2.90	0.2
0.50	30.9	1.05	14.7	1.60	5.5	2.25	1.2	3.00	0.1

Appendix F

The weighted least squares procedure

For a three-parameter model, we have the regression equation:

$$y_i = c + b_1 \cdot x_{1i} + b_2 \cdot x_{2i}$$

where

y_i = generation mean

$c = m$

$b_1 = [a]$

$b_2 = [d]$

x_{1i} = coefficients of $[a]$

x_{2i} = coefficients of $[d]$.

In addition, weights have to be taken into account when solving for m, $[a]$ and $[d]$. For the six basic generations, this information is summarized in matrix form as:

$Y = 6 \times 1$ vector of generation means
$M = 3 \times 1$ vector of parameters
$C = 6 \times 3$ matrix of coefficients
$W = 6 \times 6$ diagonal matrix of weights.

The matrix form of the data considered in Chapter 2 (Table 2.5) is:

$$
\begin{array}{cccc}
Y & C & W & M \\
\begin{bmatrix} 69.44 \\ 59.04 \\ 83.44 \\ 74.36 \\ 76.02 \\ 71.28 \end{bmatrix}
&
\begin{bmatrix} 1 & 1 & 0 \\ 1 & -1 & 0 \\ 1 & 0 & 1 \\ 1 & 0 & \frac{1}{2} \\ 1 & \frac{1}{2} & \frac{1}{2} \\ 1 & -\frac{1}{2} & \frac{1}{2} \end{bmatrix}
&
\begin{bmatrix} 1.6742 & 0 & 0 & 0 & 0 & 0 \\ 0 & 1.5218 & 0 & 0 & 0 & 0 \\ 0 & 0 & 1.9301 & 0 & 0 & 0 \\ 0 & 0 & 0 & 0.9925 & 0 & 0 \\ 0 & 0 & 0 & 0 & 1.2338 & 0 \\ 0 & 0 & 0 & 0 & 0 & 1.1010 \end{bmatrix}
&
\begin{bmatrix} m \\ [a] \\ [d] \end{bmatrix}
\end{array}
$$

The solution is obtained in the form of matrix X which gives unbiased estimates of m, $[a]$ and $[d]$ parameters. Now, $X = (C'WC)^{-1} \cdot (C'WY)$

where C′ is the transpose of C ($= 3 \times 6$ matrix) and $^{-1}$ represents the inverse of a matrix. C′WC and C′WY are obtained as follows.

C′WC is an **information matrix** of dimension 3×3 whose nine values are calculated as:

Entry $1,1 = \{\sum (\text{Coefficient of } m)^2 \times \text{wt}_i\}$
$$= 1 \times 1.6742 + 1 \times 1.5248 + \ldots + 1 \times 1.1010 = 8.4534$$

Entry $1,2 = \{\sum (\text{Coefficient of } m) \times (\text{coefficient of } [a]) \times \text{wt}_i\}$
$$= (1) \times (1) \times 1.6742 + (1) \times (-1) \times 1.5248$$
$$+ (1) \times (0) \times 1.9301 + \ldots + (1)(-\tfrac{1}{2}) \times 1.1010$$
$$= 0.21880$$

Similarly,

Entry $1,3 = \{\sum (\text{Coefficient of } m) \times (\text{coefficient of } [d]) \times \text{wt}_i\}$
$$= 3.59375$$

Entry $2,2 = \{\sum (\text{Coefficient of } [a])^2 \times \text{wt}_i\} = 3.7797$

Entry $2,3 = \{\sum (\text{Coefficient of } [a]) \times (\text{coefficient of } [d]) \times \text{wt}_i\}$
$$= 0.03320$$

Entry $3,3 = \{\sum (\text{Coefficient of } [d])^2 \times \text{wt}_i\} = 2.76192$

Entry $2,1 = $ Entry $1,2$

Entry $3,1 = $ Entry $1,3$

Entry $3,2 = $ Entry $2,3$

So,

$$\text{C′WC} = \begin{bmatrix} 8.45340 & 0.21880 & 3.59375 \\ 0.21880 & 3.77970 & 0.03320 \\ 3.59375 & 0.03320 & 2.76192 \end{bmatrix}$$

C′WY is a 3×1 vector whose values are calculated as:

Entry $1,1 = \{\sum (\text{Coefficient of } m) \times \text{wt}_i \times y_i\}$
$$= 1 \times 1.6742 \times 69.44 + \ldots + 1 \times 1.1010 \times 71.28 = 613.226$$

Entry $2,1 = \{\sum (\text{Coefficient of } [a]) \times \text{wt}_i \times y_i\} = 34.066$

Entry $3,1 = \{\sum (\text{Coefficient of } [d]) \times \text{wt}_i \times y_i\} = 284.085$

The inversion of $C'WC$ above can be obtained using the standard statistical packages, and

$$(C'WC)^{-1} = \begin{bmatrix} 0.265313 & -0.012327 & -0.345070 \\ -0.012327 & 0.265172 & 0.012853 \\ -0.345070 & 0.012853 & 0.810909 \end{bmatrix}$$

$(C'WC)^{-1}$ is generally referred to as the **variance–covariance matrix** and its multiplication with $C'WY$ gives the following estimates:

$$m = 0.265313 \times 613.226 - 0.012327 \times 34.066 - 0.345070 \times 284.085$$
$$= 64.25$$
$$[a] = -0.012327 \times 613.226 + 0.265172 \times 34.066 + 0.012853$$
$$\times 284.085 = 5.13$$
$$[d] = -0.345070 \times 613.226 + 0.012853 \times 34.066 + 0.810909$$
$$\times 284.085 = 19.20$$

Using these parameter estimates, the expected means of the various generations become:

$$\bar{P}_1 \ (\text{e.g.} = m + [a]) = 69.38, \quad \bar{P}_2 = 59.12, \quad \bar{F}_1 = 83.45, \quad \bar{F}_2 = 73.85,$$
$$\bar{B}c_{1.1} = 76.42 \text{ and } \bar{B}c_{1.2} = 71.29$$

Now,

$$\chi^2 = \sum (\text{Observed mean} - \text{Expected mean})^2 \times \text{wt}_i$$
$$= (69.44 - 69.38)^2 \times 1.6742 + \ldots + (71.28 - 71.29)^2 \times 1.1010$$
$$= 0.47$$

Degrees of freedom
$$= \text{number of generations} - \text{number of parameters} = 6 - 3 = 3$$

The $\chi^2_{(3)}$ value of 0.47 is non-significant indicating that the m, $[a]$ and $[d]$ model fits adequately to the given data.

The diagonal items in the variance–covariance matrix are the variances of the parameter estimates provided that the χ^2 is not significant, i.e. the model is adequate. So, in the present case,

$$\text{se}(m) = \sqrt{0.265313} = \pm 0.52$$
$$t = m/\text{se}(m) \ (\text{for df} > 100) = 123^{***}$$

$$\text{se}([a]) = \sqrt{0.265172} = \pm 0.51$$
$$t = [a]/\text{se}([a]) \ (\text{for df} > 100) = 10^{***}$$

$$\text{se}([d]) = \sqrt{0.810909} = \pm 0.90$$
$$t = [d]/\text{se}([d]) \ (\text{for df} > 100) = 21^{***}$$

Student's t tests reveal that each estimate is highly significant.

On the other hand, if the χ^2 was significant, it would be necessary to multiply each standard error above by $\sqrt{(\chi^2/\text{df})}$. This is equivalent to using the residual MS as error in a standard regresssion ANOVA when the residual MS is significant. However, a model with a significant χ^2 is seldom worth considering as there are clearly other parameters that are contributing to the variation. [Rowe, K.E. and Alexander, W.L. (1980) *Crop Science*, **20**, 109–10; Pooni, H.S. (1991) Proceedings of the Eighth Meeting of Eucarpia Section: Biometrics in Plant Breeding, Eucarpia, Brno, Czechoslovakia, pp. 307–17]

Symbols

N.B. The symbols in brackets ( ) are the equivalent terms in Mather and Jinks notation (Mather, K. and Jinks, J.L. (1982) *Biometrical Genetics*, 3rd edn, Chapman & Hall, London).

Symbol	Description
A, B ... K	Genes
A^+, A^-, B^+ and B^-	Alleles of genes A and B
$[aa]$ $(= [i])$	Additive × additive interaction
aa_{AB} $(= i_{AB})$	Interaction between a_A and a_B
ad_{AB} $(= j_{AB})$	Interaction between a_A and d_B
$a_A, a_B, . a_i ... a_K$	Additive genetic effects of genes A, B, etc.
$[a]$ $(= [d])$	Additive genetic effect of means
$[ad]$ $(= [j])$	Additive × dominance interaction
ANOVA	Analysis of Variance
A, **B**, etc	Scaling tests
$Bc_{1.1}$, $Bc_{n.1}$	First and subsequent backcrosses of P_1 with F_1
$Bc_{1.2}$, $Bc_{n.2}$	First and subsequent backcrosses of P_2 with F_1
b	Regression coefficient
β_a	Regression of ge_{aj} on e_j
β_d	Regression of ge_{dj} on e_j
β_{a_i}	Regression of ge_{aij} on e_j
BIPs	Biparental families
$[c]$	Cytoplasmic effect
$\chi^2_{(df)}$	Chi-squared
$[d]$ $(= [h])$	Dominance effect of means
$d_A, d_B, ... d_K$	Dominance effects of various genes
dd_{AB} $(= l_{AB})$	Interaction between d_A and d_B

$[dd]\ (= [l])$	Dominance × dominance interaction
$d_i \div a_i = f_i$	Dominance ratio at gene i
df	Degrees of freedom
ems	Expected mean squares
e_j	Effect of the jth environment
$\epsilon\bar{F}_2$	Expected mean of F_2 generation
$F_1(1 \times 2),\ RF_1(2 \times 1)$	Reciprocal crosses between P_1 and P_2
F	Variance ratio
FS	Full-sib families
g_i	Genotypic score
ge_{aj}	Additive × environmental interaction component
ge_{dj}	Dominance × environmental interaction component
ge_{aij}	Interaction of the ith line with the jth environment
GCA	General combining ability
HS	Half-sib families
h_n^2, h_b^2	Narrow and broad heritability
i	Intensity of selection
k, k'	Total number of segregating genes
m	Mid-parent or mean of random F_∞ families of a single cross
MS	Mean square
μ	Origin for populations where allele frequencies are unknown
n, n1, n', etc	Sample sizes
N	Number of generations or parameters
N	Population size
NCI, NCII	North Carolina-I, II, etc. designs
n.s.	$P > 0.05$
$\bar{P}_1, \bar{P}_2$	Means of parental lines
p, q	Frequencies of A^+ and A^- alleles
P	Probability
$\bar{P}_{1j}$, etc	P_1 mean in jth environment
QTL	Quantitative Trait Locus
r	Family size
r	Correlation coefficient
r_a, r_{aa}, r_{ad}	Coefficients of association/dispersion for the additive, additive × additive and additive × dominance effects of genes

R	Recombination frequency
R	Selection response
σ_P^2	Population phenotypic variance
σ_T^2	Total phenotypic variance of population or generation
σ_B^2	Between-families component of variance
σ_W^2	Within-families component of variance
σ_{gF2}^2	Genetic variance of F_2 generation
$\sigma_{1F4}^2, \sigma_{2F4}^2$	Rank variances
σ_M^2	Between males component of variance
σ_F^2	Between females component of variance
σ_{MF}^2	Males × females interaction component of variance
$\sigma_{F/M}^2$	Females within males component of variance
σ_S^2	Variance of test-cross sums
σ_D^2	Variance of test-cross differences
σ_{TS}^2	Testers × Sums interaction variance
s^2	Sample variance
sd	Standard deviation $(= s)$
$s_{\bar{F}_1}^2$, etc.	Variance of F_1 mean, etc.
SS	Sum of squares of deviations
SP	Sum of products of deviations
$S_{(XY)}$	Covariance of X, Y
SCA	Specific combining ability
S	Selection differential
$\sum\limits^{k} a_i^2$	$2V_A^*$
$\sum\limits^{k} d_i^2$	$4V_D^*$
$t_{(df)}$	Student's t value
t_{HS}, etc.	Intraclass correlation
VR	Variance ratio
V_G	Genetic variance
$V_A^* \ (= \tfrac{1}{2}D)$	Additive genetic variance of F_2 generation
$V_A \ (= \tfrac{1}{2}D_R)$	Additive variance of a random mating population
$V_D^* \ (= \tfrac{1}{4}H)$	Non-additive (dominance) variance of F_2 generation

$V_D \ (= \frac{1}{4}H_R)$	Non-additive (dominance) variance of a random mating population
$V_A^\dagger, V_A^\ddagger, V_D^\dagger, V_D^\ddagger$	Additive and dominance variance with linkage
V_{AA}	Additive $\times$ additive interaction variance
$V_{AD} \ (= \frac{1}{2}F)$	Covariance of the additive and dominance effects of various genes
$V_E \ (= E_W)$	Environmental component of within family variance
$V_{EC} \ (= E_B)$	Variance component due to common environmental effects of the families
$V_{EM} \ (= E_M)$	Variance component due to maternal effects
$\sqrt{(4V_D^*/2V_A^*)} \left(= \sqrt{\dfrac{H}{D}} \right)$	Dominance ratio
wt_i	Weight of the ith generation
X_1, X_2, X_i, etc.	Phenotypic scores of individuals
$\bar{x}$	Sample mean
X, Y	Sex chromosomes
x_i, x_{1i}, x_{2i}, etc.	Independent variables
y_i	Dependent variable
Y_1, Y_2, Y_i, etc.	Phenotypic scores of offspring
z	Fisher's transformation of the correlation coefficient
*	$0.05 \geq P > 0.01$
**	$0.01 \geq P > 0.001$
***	$P \leq 0.001$
n.s.	$P > 0.05$

Problems

The problems below are designed to check the reader's understanding of the statistical and genetical principles involved in the analysis and interpretation of data. They are linked to particular chapters of the book as indicated by the numbers in brackets, [3] etc., at the end of each question. Answers are provided in the following section.

1. The following are the flowering time (days) of individuals from two inbred lines of sunflower (P_1 and P_2) and their F_1 generation. Calculate the mean, variance, variance of the mean and standard error of the mean for each generation. [2]

Inbred P_1: sample size = 20
99 103 108 96 86 90 96 95 100 87 102 106 100 86 99 91 93 93 91 92

Inbred P_2: sample size = 20
50 38 31 42 36 29 46 32 38 41 48 34 42 35 41 51 38 46 38 31

Hybrid F_1: sample size = 40
93 93 83 94 94 102 94 89 83 89 84 77 88 83 93 83 90 83 84 94
85 90 87 73 75 95 78 89 94 96 99 85 94 87 90 88 83 95 93 82

2. The means and their standard errors of the six basic generations from a cross between two inbred lines are as follows:

Family	Mean	se	Sample size
P_1	95.65	1.4314	20
P_2	40.98	1.4497	20
F_1	88.28	1.0315	40
F_2	79.39	1.1213	80
$Bc_{1.1}$	92.69	0.7420	80
$Bc_{1.2}$	64.88	1.2006	80

(i) Carry out **A**, **B** and **C** scaling tests to check the adequacy of the additive–dominance model.

(ii) Estimate, m, $[a]$ and $[d]$ either by solving simultaneous equations or by WLS.
(iii) Is there any evidence for micro $G \times E$ interaction?
(iv) Using the standard errors and the numbers of individuals in each family above, estimate V_A^*, V_D^*, V_{AD} and V_E.
(v) Estimate the heritabilities and dominance ratio.
(vi) What can you conclude about the genetical control of this trait? [2,3]

3. In a recent trial of winter wheat, two pure breeding varieties and their F_1, F_2 and F_3 generations had the following means and variances for the character seedling weight (g):

	Mean weight	Within variance (s^2)	Sibs/generation(n)
P_1	7.12	9.82	80
P_2	3.12	3.92	80
F_1 (1 × 2)	6.16	6.63	80
F_2	8.02	1.97	250
F_3	5.35	see below	500

Item	df	MS	ems
Between families	99	40.70	$\sigma_W^2 + 5\sigma_B^2$
Within families	400	2.01	σ_W^2

Analysis of these data revealed three serious departures from a simple additive–dominance–environmental model. List these departures and explain how you arrived at your conclusions. [2,3]

4. A biparental crossing programme on an F_2 population of poppies produced 50 FS families each of size 20. Analysis of the flowering times (days) of these 1000 individuals and their parents yielded the following statistics:

SS of offspring family means   = 1412.87
SS within FS families   = 72690.00
SS of mid-parental scores   = 2030.93
SP of offspring means onto mid-parental scores = 648.45
Mean of all offspring   = 78.70
Mean of all parents   = 78.34

Analyse and interpret these data as far as you can [4].

5. A randomly mating population was investigated by carrying out a NCI in which 20 males were chosen at random and each crossed to a different pair of females. Five progeny were raised from each family.

Analysis of variance of the data yielded the following SS:

Source	SS
Between male HS groups	1122.9
Between female within male HS groups	544.0
Within FS families	2300.0

(i) Complete the ANOVA together with the *ems* and estimate significant components.
(ii) Interpret the genetical control of this trait [5].

6. Twenty DH lines were derived by anther culture from an F_1 and ten were crossed as male parents to the remaining ten as female in an NCII design. Five plants were raised from every family in a fully randomized design and analysis of their heights (cm) yielded the following ANOVA.

Source	SS
Between male HS families (M)	5913
Between female HS families (F)	5751
M × F	3888
Within FS families	13200

Analyse and interpret as far as you are able [5].

7. In a triple-test-cross experiment, 75 F_2 individuals were each crossed to the two parents and their F_1. Ten progeny were raised from each FS family. The average variances within FS families from crosses to P_1, P_2 and F_1 were 14.425, 14.575 and 20.5 respectively. ANOVA of the data produced the following (corrected) MS:

Source	MS	*ems*
Additive	86.5	$\sigma_W^2 + 20\sigma_S^2$
Dominance	82.5	$\sigma_W^2 + 20\sigma_D^2$
Epistasis	7.3	$\sigma_{W1}^2 + 60\sigma_{EPI}^2$

Analyse and interpret as far as you can [5].

8. In problem 7, the P_1, P_2 and F_1 generations took the following means.

$$\bar{P}_1 = 24.83, \qquad \bar{P}_2 = 15.17, \qquad \bar{F}_1 = 29.67$$

Using estimates of V_A^* and V_D^* from problem 7, calculate the proportions of RILs and SCHs with scores $\geq P_1$ mean, $\geq F_1$ mean or $\leq P_2$ mean. [15].

9. The WLS estimates of six parameters of a digenic interaction model for whole plant weight (g) in a cross between two pure breeding lines of oats are given below:

$$m = 165.0 \pm 1.8 \qquad [aa] = -6.0 \pm 2.4$$
$$[a] = 6.0 \pm 2.0 \qquad [ad] = 3.0 \pm 3.4$$
$$[d] = 43.0 \pm 10.8 \qquad [dd] = -23.0 \pm 8.4$$

Each estimate has at least 100 degrees of freedom.

(i) What type of epistasis is present in this cross?
(ii) What is the magnitude and direction of expected heterosis?
(iii) What are the possible causes of heterosis in this cross?
(iv) What is the expected numerical difference between the mid-parent and the mean of the F_∞ generation?
(v) What are the expected means of the F_3, $Bc_{2.1}$ and $Bc_{2.2}$ generations? [11,15]

10. The following highly significant estimates were obtained for the genetical and environmental parameters of yield (g) in rye grass:

$$m = 41.0 \qquad V_A^* = 50$$
$$[a] = 10.0 \qquad V_D^* = 10$$
$$[d] = 14.0 \qquad V_E = 40$$

Both models are adequate.

(i) What are the expected means and variances of the F_1 and F_2 generations?
(ii) What is the expected mean of the best recombinant inbred line that can be obtained from this material?
(iii) What proportion of the inbred lines produced by single seed descent would be expected to outyield the original F_1?
(iv) Estimate the broad and narrow heritability.
(v) If the top 10% of the F_2 individuals were selected and randomly mated, what would be the expected mean of their progeny? [4, 15]

11. The following estimates of the genetical components of mean performance (MP) and environmental sensitivity (ES) were obtained from the height measurements (cm) of two pure breeding lines of wheat and their F_1 hybrid in 15 contrasting environments.

	MP	ES
	$m = 80.00$	
	$[a] = 6.00$	$\beta_a = 0.40$
	$[d] = 8.00$	$\beta_d = 0.20$

Each component is highly significant. Use these components to estimate the following:

(i) The value of e_j in which P_1 and P_2 have the same height.
(ii) The value of e_j beyond which F_1 ceases to be heterotic.
(iii) The mean heights of the F_3 and F_∞ generations in an average environment.
(iv) A large sample of RILs derived from the F_1 were also raised in the above experiment and the highest and lowest β_a among their ES values were 0.80 and -0.76 respectively. A highly significant correlation $r = 0.35$ was also detected between the $m + [a]$ and β_a values of the RILs.

What can you deduce about the inheritance of MP and ES from this information? [12]

12. In an experiment designed to test the presence of linkage, 100 F_2 individuals of rice were each crossed to its ancestral genotypes, P_1, P_2 and F_1. Single individuals from these 300 families and 100 plants each of the F_2, $Bc_{1.1}$ and $Bc_{1.2}$ generations were raised in a randomized block and scored for flowering time. These data provided the following variances (each with 99 df):

Generation	F_2	$F_2 \times F_1$	$Bc_{1.1}$	$F_2 \times P_1$	$Bc_{1.2}$	$F_2 \times P_2$
Variance	50.25	80.52	25.00	40.57	62.70	82.44

Interpret these results as far as you can. [6]

13. In an experiment to locate QTLs controlling yield in wheat, 50 doubled haploid lines were produced by anther culture from an F_1 between two true breeding varieties.

Four RFLP loci on chromosome 5D were used with the RFLP alleles from the higher scoring parent being designated $+$ and those from the other parent, $-$. The genotypes of the 50 doubled haploid (homozygous) lines and their yields (obtained from large-scale plot trials) are shown in the table below.

(i) Calculate the recombination frequencies between all pairs of loci.
(ii) Convert these RF to map distances (in cM Haldane) using the appropriate formula and construct the map of these loci drawn to scale.

(iii) Calculate the mean yield effect associated with each allele at the RFLP loci, locate the QTL involved and estimate its likely effect on yield. [7]

RFLP – Yield data (g) of the 50 doubled haploid lines mentioned above

DH line	RFLP loci				Mean yield	DH line	RFLP loci				Mean yield
	1	2	3	4			1	2	3	4	
1	+	−	+	+	12	26	+	+	+	+	11
2	+	+	+	+	13	27	+	+	+	−	13
3	−	−	−	+	7	28	+	+	+	+	12
4	+	−	−	+	11	29	−	−	−	−	8
5	+	+	+	+	13	30	−	−	−	−	9
6	−	−	−	−	7	31	−	+	+	−	7
7	−	−	−	−	9	32	+	+	+	+	13
8	+	+	+	+	12	33	−	+	+	−	13
9	−	+	+	−	12	34	−	−	−	−	9
10	+	+	+	−	11	35	−	+	+	+	9
11	+	+	+	+	11	36	+	+	+	+	12
12	−	−	−	−	8	37	−	−	−	−	9
13	−	+	−	−	9	38	+	−	−	+	12
14	+	+	+	+	13	39	+	−	−	+	9
15	−	−	−	−	11	40	−	−	−	−	7
16	−	−	+	−	9	41	−	−	−	+	9
17	−	−	−	−	7	42	+	+	+	+	11
18	+	+	+	+	13	43	−	−	−	−	7
19	+	−	+	−	11	44	+	+	+	+	12
20	+	−	+	+	11	45	−	−	−	−	9
21	+	+	+	+	11	46	−	+	−	−	8
22	+	+	+	+	12	47	+	+	+	+	11
23	+	−	−	+	7	48	−	−	−	−	7
24	−	−	−	−	7	49	−	+	−	−	7
25	−	−	−	−	9	50	+	−	+	+	13

Answers to problems

Problem 1

Family	n	Mean	s^2	s^2 (mean)	se $\{= \sqrt{s^2 \text{ (mean)}}\}$
P_1	20	95.65	40.98	2.0488	1.4314
P_2	20	39.35	42.03	2.1014	1.4496
F_1	40	88.28	42.56	1.0641	1.0315

Problem 2

(i) Scaling tests

$$\mathbf{A} = 2\bar{B}c_{1.1} - \bar{F}_1 - \bar{P}_1 \quad = 1.45 \pm 2.31 \text{ n.s.}$$
$$\mathbf{B} = 2\bar{B}c_{1.2} - \bar{F}_1 - \bar{P}_2 \quad = 0.50 \pm 2.99 \text{ n.s.}$$
$$\mathbf{C} = 4\bar{F}_2 - 2\bar{F}_1 - \bar{P}_1 - \bar{P}_2 = 4.37 \pm 5.34 \text{ n.s.}$$

(ii) (a) From P_1, P_2 and F_1 means by simultaneous equations:

$$m = \text{mid parent} \qquad\qquad = 68.32 \pm 1.02^{***}$$

$$[a] = \text{average parental difference} = 27.34 \pm 1.02^{***}$$

$$[d] = \bar{F}_1 - \text{mid parent} \qquad\quad = 19.96 \pm 1.45^{***}$$

(b) By WLS method:

$$m = 68.74 \pm 0.85^{***}$$
$$[a] = 27.54 \pm 0.81^{***}$$
$$[d] = 19.98 \pm 1.45^{***}$$

Goodness of fit $\chi^2_{(3)} = 0.82 \text{ n.s.}$

(iii) By multiplying $(se)^2$ by sample size, we get:

$$s^2_{P_1} = 40.98, \quad s^2_{P_2} = 42.03, \quad s^2_{F_1} = 42.56$$
$$s^2_{F_2} = 100.59, \quad s^2_{Bc_{1.1}} = 44.05, \quad s^2_{Bc_{1.2}} = 115.32$$

Using the largest and the smallest variance among the $s^2_{P_1}$, $s^2_{P_2}$ and $s^2_{F_1}$;

$F_{(39,19)} = 1.04$ which has $P \gg 0.05$. As the F is artificially inflated by the use of the extreme variances, its correct probability will be equal to observed $P \times$ total number of pairs of variances [$= 3$ factorial] $= \gg 0.05 \times 6 = \gg 0.30$ n.s.

So, micro $G \times E$ is not significant.

(iv)

$$V_E = \tfrac{1}{4}(s_{P_1}^2 + s_{P_2}^2 + 2s_{F_1}^2) = 42.03$$

$$V_A^* = 2s_{F_2}^2 - s_{Bc_{1.1}}^2 - s_{Bc_{1.2}}^2 = 41.8$$

$$V_D^* = (s_{Bc_{1.1}}^2 + s_{Bc_{1.2}}^2 - s_{F_2}^2 - V_E) = 16.75$$

$$V_{AD} = \tfrac{1}{2}(s_{Bc_{1.2}}^2 - s_{Bc_{1.1}}^2) = 35.64$$

(v) $h_n^2 = 0.42$, $h_b^2 = 0.58$
Dom. ratio $= \sqrt{(4V_D^*/2V_A^*)} = 0.90$
Potence ratio $= [d]/[a] = 0.73$

(vi) a. Additive–dominance gene action
b. Predominantly unidirectional dominance for high score
c. No heterosis
d. High gene association in the parental lines
e. Moderate to high heritability.

Problem 3
a. Means of the filial generations are out of sequence. Usually, we expect

either $\quad \bar{F}_1 > \bar{F}_2 > \bar{F}_3 \quad$ or $\quad \bar{F}_1 < \bar{F}_2 < \bar{F}_3$

when there are no complications. In the present case, the F_2 mean is too high compared with those of the F_1 and F_3. The F_3 mean is exactly where it should be in relation to the F_1 mean and mid-parent.
b. F_2 variance is significantly smaller than the environmental variance, when it should be larger due to genetic segregation.
c. Within F_3 variance is again too small.

Problem 4
ANOVA to test the significance of differences between FS families

Source	df	SS (uncorrected)	SS (corrected)	MS	F	ems
Between FS families	49	1412.87	28257.4	576.68	7.54***	$\sigma_W^2 + 20\sigma_B^2$
Within FS families	950		72690.0	76.52		σ_W^2

The F test shows that FS families differ genetically. These genetic differences can be due to the additive and non-additive effects of genes. The former can be tested and separated using the parent–progeny regression.

$$
\begin{aligned}
&\text{SSx } (= \text{SS of mid-parent scores}) && = 2030.93 \\
&\text{SSy } (= \text{SS of offspring family means}) && = 1412.87 \\
&\text{SPxy} && = 648.45 \\
&b = \text{SPxy/SSx} && = 0.3193 \\
&\text{Regression SS} = b \times \text{SPxy} && = 207.04 \\
&\text{Remainder SS} = 1412.87 - 207.04 && = 1205.83 \\
&\text{se}(b) = \sqrt{(\text{Remainder MS/SSx})} && \\
&\qquad = \sqrt{\{1205.83/(48 \times 2030.93)\}} && = 0.1112 \\
&\text{Corrected Regression SS} = 207.04 \times 20 && = 4140.8 \\
&\text{Corrected Remainder SS} = 1205.83 \times 20 && = 24116.6
\end{aligned}
$$

ANOVA to test regression and remainder

Source	df	SS	MS	F
Between FS families	49	28257.4	576.68	7.54***
Regression	1	4140.8	4140.80	8.24***
Remainder	48	24116.6	502.43	6.57***
Within FS families	950	72690.0	76.52	

Both regression and remainder are highly significant indicating that variation among FS families is caused by the additive as well as the non-additive effects of the segregating genes. Assuming the adequacy of the additive–dominance–environment model:

$$
\begin{aligned}
&\sigma_W^2 = \tfrac{1}{2}V_A^* + \tfrac{3}{4}V_D^* + V_E && = 76.52 \\
&\sigma_B^2 = \tfrac{1}{2}V_A^* + \tfrac{1}{4}V_D^* + (V_{EC}) && = 25.01 \\
&\text{P/O Covariance} = \tfrac{1}{2}V_A^* = 648.45/49 && = 13.23 \\
&\therefore V_A^* && = 26.46 \\
&V_D^* = 4(\sigma_B^2 - \tfrac{1}{2}V_A^*) \text{ assuming } V_{EC} = 0 && = 47.12 \\
&V_E = (\sigma_W^2 - \tfrac{1}{2}V_A^* - \tfrac{3}{4}V_D^*) && = 27.95 \\
&h_b^2 = (V_A^* + V_D^*)/(\sigma_W^2 + \sigma_B^2) && = 0.7247 \\
&h_n^2 = \text{regression coefficient} && = 0.3193 \pm 0.1112
\end{aligned}
$$

$$
\begin{aligned}
&\text{Next, Overall means of the parents} && = 78.34 \\
&s^2 \text{ (overall mean)} = \{(2030.93)/(49 \times 50)\} && = 0.8290 \\
&\text{Offspring mean} && = 78.70 \\
&s^2 \text{ (Offspring mean)} = (576.68/1000) && = 0.5767 \\
&t_{(\infty df)} = (78.70 - 78.34)/\sqrt{(0.829 + 0.5767)} && = 0.30 \text{ n.s.}
\end{aligned}
$$

The t test confirms that these means do not differ significantly.

Conclusions:

(i) The character is highly heritable and both the additive and non-additive effects are significant.

(ii) The trait shows moderate narrow heritability and therefore will respond well to selection.

(iii) The design does not provide tests of epistasis, $G \times E$ and maternal effects, etc. and they have been assumed absent.

(iv) Comparison of the overall means of the parents and progeny indicates that:

   (a) sampling and pairing of parents was truly at random;

   (b) the population was randomly mating;

   (c) either there was no epistasis or there was no linkage disequilibrium between genes showing epistasis.

Problem 5

(i) ANOVA for the NCI design

Source	df	SS	MS	F	ems
Between male HS groups	19	1122.9	59.10	2.17*	$\sigma_W^2 + 5\sigma_{F/M}^2 + 10\sigma_M^2$
Between females within male HS groups	20	544.0	27.20	1.89*	$\sigma_W^2 + 5\sigma_{F/M}^2$
Within FS families	160	2300.0	14.375		σ_W^2

Both between males and between females within males MS are significant at 5% level indicating that σ_M^2 and $\sigma_{F/M}^2$ are not equal to zero. So,

$$\sigma_W^2 = \tfrac{1}{2}V_A^* + \tfrac{3}{4}V_D^* + V_E \qquad\qquad = 14.375$$
$$\sigma_{F/M}^2 = \tfrac{1}{4}V_A^* + \tfrac{1}{4}V_D^* + (V_{EC}) \qquad = 2.565$$
$$\sigma_M^2 = \tfrac{1}{4}V_A^* \qquad\qquad\qquad\qquad = 3.190$$
$$\therefore V_A^* \qquad\qquad\qquad\qquad\qquad = 12.76$$
$$V_D^* = 4(\sigma_{F/M}^2 - \sigma_M^2) \text{ assuming } V_{EC} = 0 \qquad = -2.50$$

As variance cannot take a negative value, we assume $V_D^* = 0$

$$V_E = (\sigma_W^2 - 2\sigma_M^2) \qquad\qquad\qquad = 7.995$$
$$h_n^2 = V_A^*/(\sigma_M^2 + \sigma_{F/M}^2 + \sigma_W^2) \qquad = 0.6339$$

(ii) (a) The trait under study is predominantly controlled by the additive effects of genes.

   (b) The heritability is high and consequently selection can be highly effective.

   (c) The design does not allow tests of epistasis, linkage or maternal effects.

Problem 6
ANOVA for the NCII design

Source	df	SS	MS	F	ems
Between male HS groups	9	5913	657	13.69***	$\sigma_W^2 + 10\sigma_{MF}^2 + 100\sigma_M^2$
Between female HS groups	9	5751	639	13.31***	$\sigma_W^2 + 10\sigma_{MF}^2 + 100\sigma_F^2$
M × F interaction	81	3888	48	1.45**	$\sigma_W^2 + 10\sigma_{MF}^2$
Within FS families	400	13200	33		σ_W^2

Both main effects and their interaction are highly significant. The main effects test additive gene action in the present case and interaction indicates the presence of non-additive effects.

$$\sigma_W^2 = V_E \qquad\qquad\qquad = 33.00$$
$$\sigma_{MF}^2 = V_D^* + (V_{EC}) = (48 - 33)/10 \quad = 1.50$$
$$\sigma_M^2 = \tfrac{1}{2} V_A^* = (657 - 48)/100 \quad = 6.09$$
$$\sigma_F^2 = \tfrac{1}{2} V_A^* + (V_{EM}) = (639 - 48)/100 = 5.91$$

Test of V_{EM}: when V_{EM} is significant, it will make $\sigma_F^2 > \sigma_M^2$ and consequently, F = Between females MS/Between males MS will be larger than one and significant. In the present case, the value of F is 0.97, i.e. n.s. and we conclude that V_{EM} is non-existent.

There is no comparable test of V_{EC} but its magnitude is likely to be minimized by the full randomization practised in the experiment. Therefore, V_{EC} can be assumed negligible.

The design does not allow a test of epistasis and we assume that non-allelic interaction is absent.

$$h_b^2 = (\sigma_F^2 + \sigma_M^2 + \sigma_{MF}^2)/(\sigma_W^2 + \sigma_F^2 + \sigma_M^2 + \sigma_{MF}^2) \quad = 0.2903$$
$$h_n^2 = (\sigma_F^2 + \sigma_M^2)/(\sigma_W^2 + \sigma_F^2 + \sigma_M^2 + \sigma_{MF}^2) \quad\qquad = 0.2581$$
$$\text{Dominance ratio} = \sqrt{(4V_D^*/2V_A^*)} = \sqrt{(6.0/24.00)} = 0.5000$$

Conclusions:
(a) The trait is moderately heritable.
(b) The genes controlling the trait show, on average, partial dominance.
(c) There are no maternal effects.
(d) There are no tests of epistasis and linkage.

Problem 7
Tests of heterogeneity of the within-family variances to determine if we need to use a different error MS to test the epistatic effects.

Within variance of $F_2 \times P_1$ versus within variance of $F_2 \times P_2$

$$F \text{ (for 675 and 675 df)} = 1.01 \text{ n.s. (one-tailed test)}$$
$$\text{Pooled variance} = 14.50$$

Within variance of $F_2 \times F_1$ versus within variance of $F_2 \times$ Parents

$$F \text{ (for 675 and 1350 df)} = 1.41^{***} \text{ (one-tailed test)}$$

Significance of this test indicates that error for the epistasis MS will be

$$\text{Error MS} = (1/6)(14.425 + 14.575 + 4 \times 20.5) = 18.50$$

Tests of the additive, dominance and epistatic effects

Source	df	MS	F
Additive	74	86.5	5.97***
Dominance	74	82.5	5.69***
Error$_1$	1350	14.5	
Epistasis	75	7.3	<1 n.s.
Error$_2$	2025	18.5	

Both additive and dominance effects are present but there is no interaction.

$$4\sigma_S^2 = V_A^* \qquad\qquad = 14.40$$
$$2\sigma_D^2 = V_D^* \qquad\qquad = 6.80$$
$$V_E = \text{Error}_1 - \tfrac{1}{4}V_A^* - \tfrac{1}{2}V_D^* \quad = 7.50$$
$$h_b^2 = (V_A^* + V_D^*)/(V_A^* + V_D^* + V_E) = 0.7387$$
$$h_n^2 = (V_A^*)/(V_A^* + V_D^* + V_E) \quad = 0.5017$$
$$\text{Dominance ratio} = \sqrt{(4V_D^*/2V_A^*)} = \sqrt{(27.2/28.8)} = 0.9718$$

Conclusions:

(a) The character is highly heritable.
(b) Significantly larger variance of crosses involving F_1 also confirms genetic segregation within the test-cross families.
(c) The genes show almost complete dominance.
(d) The direction of dominance cannot be determined from the available data.
(e) Similarity of within FS family variances of crosses involving P_1 and P_2 suggests a high degree of allele dispersion between the parental lines.

Problem 8

Assuming no epistasis at means level, $m =$ mid parent	$= 20$

So, the genotypic distribution of RILs has mean $= m$ $= 20$
    and standard deviation $= \sqrt{(2V_A^*)} = \sqrt{28.80}$ $= 5.37$

$\therefore$ abscissa$(P_1) = (24.83 - 20)/5.37 = 0.90$
    and proportion of RILs scoring $\geq \bar{P}_1$ $= 18.4\%$

Similarly, abscissa$(F_1) = (29.67 - 20)/5.37 = 1.80$
    and proportion of RILs scoring $\geq \bar{F}_1$ $= 3.6\%$

and abscissa$(P_2) = (15.17 - 20)/5.37 = -0.90$
    and proportion of RILs scoring $\leq \bar{P}_2$ $= 18.4\%$

In the case of SCHs, the overall mean $= m + \frac{1}{2}[d]$
    $= \frac{1}{2}(F_1$ mean $+$ mid parent$)$ $= 24.84$
and standard deviation $= \sqrt{(V_A^* + V_D^*)} = \sqrt{21.20}$ $= 4.60$

$\therefore$ abscissa$(P_1) = (24.83 - 24.84)/4.60 = 0.00$
    and proportion of SCHs scoring $\geq \bar{P}_1$ $= 50\%$

Similarly, abscissa$(F_1) = (29.67 - 24.84)/4.60 = 1.05$
    and proportion of SCHs scoring $\geq \bar{F}_1$ $= 14.7\%$

and abscissa$(P_2) = (15.17 - 24.84)/4.60 = -2.10$
    and proportion of SCHs scoring $\leq \bar{P}_2$ $= 1.8\%$

Problem 9

Student's t tests show that all parameters are significant except $[ad]$.
(i) Opposite signs of $[d]$ and $[dd]$ (both are significant) indicate the presence of duplicate epistasis.
(ii) As $[d] + [dd]$ is positive, there can only be positive heterosis whose magnitude is given by $[d] + [dd] - [a] - [aa] = 20$ units.
(iii) Unidirectional dominance (large $[d]$), high gene dispersion ($[a]$, $[aa]$ and $[ad]$ much smaller than $[d]$ and $[dd]$) and duplicate epistasis (makes $[aa]$ negative which increases positive heterosis) are the main causes of heterosis in this cross.
(iv) $= [aa] = -6.0$
(v) $\bar{F}_3 = m + \frac{1}{4}[d] + (\frac{1}{4})^2[dd] = 174.3125$

    $Bc_{2.1} = (Bc_{1.1} \times P_1)$,
      $\therefore \bar{Bc}_{2.1} = \frac{1}{2}\{(m + \frac{1}{2}[a] + \frac{1}{2}[d]) + (m + [a] + [d])\}$
   $= m + \frac{3}{4}[a] + \frac{1}{4}[d]$ on the additive–dominance model.

    Extension to an epistatic model gives

    $\bar{Bc}_{2.1} = m + \frac{3}{4}[a] + \frac{1}{4}[d] + (9/16)[aa] + (3/16)[ad] + (1/16)[dd]$
      $= 176.00$

Similarly,

$$\bar{B}c_{2.2} = m - \tfrac{3}{4}[a] + \tfrac{1}{4}[d] + (9/16)[aa] - (3/16)[ad] + (1/16)[dd]$$
$$= 165.875$$

Problem 10

(i) $\bar{F}_1 = 41.0 + 14.0 = 55.0 \quad \bar{F}_2 = 41.0 + 7.0 = 48.0$
$s_{F_1}^2 = 40 \qquad\qquad\qquad s_{F_2}^2 = 50 + 10 + 40 = 100$

(ii) $\sum a_i = [d] \times \sqrt{(2V_A^*/4V_D^*)} = 14 \times 1.5811 = 22.14$
$\bar{P}_{BEST} = m + \sum a = 41 + 22.14 \qquad\qquad = 63.14$

(iii) Abscissa $= [d]/\sqrt{(2V_A^*)} = 1.4$; P in the upper tail of the normal distribution $= 0.081 = 8.1\%$

(iv) $h_n^2 = 0.50, h_b^2 = 0.60$

(v) $\sigma_P^2 = 100, \sigma_P = 10, i = 1.755$
$\therefore S = i \times \sigma_P = 17.55$ and $R = h_n^2 \times S = 8.775$
$\therefore \bar{F}_2 + R \qquad\qquad\qquad = 48.775$

Problem 11

(i) $\bar{P}_1 = \bar{P}_2$ when $[a] = -\beta a.e_j$, or $e_j = -6/0.4 = -15.00$
(ii) F_1 will be no longer heterotic when $\bar{F}_1 = \bar{P}_1$
or $([d] - [a]) = (\beta a - \beta d)e_j$ or $e_j = 2/(0.2) = 10.00$ or higher
(iii) In an average environment $e_j = 0$, so,

$$\bar{F}_3 = m + \tfrac{1}{4}[d] = 82.00$$
$$\bar{F}_\infty = m \qquad\quad = 80.00$$

Problem 12

When there is no linkage disequilibrium, variation in the base generation and its randomly mated progeny is not expected to differ. In the present case,

for F_2 versus $F_2 \times F_1$: $F = 1.60^*$ (for one-tailed test)
for $Bc_{1.1}$ versus $F_2 \times P_1$: $F = 1.62^*$ (for one-tailed test)
for $Bc_{1.2}$ versus $F_2 \times P_2$: $F = 1.31$ n.s. (for one-tailed test)

Two of the three tests are significant, indicating that genes are linked predominantly in one phase in the original cross. The third test is not significant but the variances show the same trend as for the other generations. As the variances of $F_2 \times F_1$, etc. have gone through additional cycles of recombination compared with $F_2(F_1 \times F_1)$, etc. and the former variances are consistently larger, we conclude that there were more repulsion than coupling linkages in the original cross.

Problem 13

The marker genes are in the sequence: 2, 3, 1, 4

cM $\longleftarrow$ 19.3 $\longrightarrow$ $\longleftarrow$ 22.3 $\longrightarrow$ $\longleftarrow$ 13.7 $\longrightarrow$

2	3	1	4

Marker regression locates the QTL at 32 cM, with an additive effect of 1.89. At this position the Regression ANOVA is

Source	df	MS	F
Add. Reg	1	347.9208	325.4584
Residual	2	10.7194	10.0273
Error	46	1.0690	

The trait means associated with each marker are:

Position	Mean(1)	n(1)	Mean(3)	n(3)
1 0.00	11.6000	(25)	8.5200	(25)
Add(i) = 1.5400				
2 19.30	11.2083	(24)	9.0000	(26)
Add(i) = 1.1042				
3 41.60	11.5769	(26)	8.4167	(24)
Add(i) = 1.5801				
4 55.30	11.2000	(25)	8.9200	(25)
Add(i) = 1.1400				

Index